360
Topics in Current Chemistry

Aims and Scope

The series Topics in Current Chemistry presents critical reviews of the present and future trends in modern chemical research. The scope of coverage includes all areas of chemical science including the interfaces with related disciplines such as biology, medicine and materials science.

The goal of each thematic volume is to give the non-specialist reader, whether at the university or in industry, a comprehensive overview of an area where new insights are emerging that are of interest to larger scientific audience.

Thus each review within the volume critically surveys one aspect of that topic and places it within the context of the volume as a whole. The most significant developments of the last 5 to 10 years should be presented. A description of the laboratory procedures involved is often useful to the reader. The coverage should not be exhaustive in data, but should rather be conceptual, concentrating on the methodological thinking that will allow the non-specialist reader to understand the information presented.

Discussion of possible future research directions in the area is welcome.

Review articles for the individual volumes are invited by the volume editors.

Readership: research chemists at universities or in industry, graduate students.

More information about this series at
http://www.springer.com/series/128

Jean-Luc Montchamp
Editor

Phosphorus Chemistry I

Asymmetric Synthesis and Bioactive Compounds

With contributions by

N. Bakalara · V. Dive · D. Georgiadis · O.I. Kolodiazhnyi · J.-L. Pirat · D. Virieux · J.-N. Volle · A.J. Wiemer · D.F. Wiemer

Springer

Editor
Jean-Luc Montchamp
Department of Chemistry
Texas Christian University
Fort Worth, Texas
USA

ISSN 0340-1022 ISSN 1436-5049 (electronic)
Topics in Current Chemistry
ISBN 978-3-319-15472-5 ISBN 978-3-319-15473-2 (eBook)
DOI 10.1007/978-3-319-15473-2

Library of Congress Control Number: 2015934476

Springer Cham Heidelberg New York Dordrecht London

Printed on acid-free paper

Springer International Publishing AG Switzerland is part of Springer Science+Business Media (www.springer.com)

Preface

The element phosphorus was discovered early in the seventeenth century, as chronicled in John Emsley's excellent and fascinating book "The 13th Element". Phosphorus chemistry is often perceived as an "old Dame" in the chemical arts. This is because many important reactions were discovered at the turn of the twentieth century, largely from the contributions of Russian chemists. Since then, phosphorus chemistry has unfortunately been considered a mature and specialized field. However, phosphorus being key to all life processes, tremendous opportunities for phosphorus research have remained ever since. So the "old Dame" is now experiencing a second (or third?) youth.

It has been my privilege to guest-edit these two volumes for Topics in Current Chemistry on phosphorus chemistry, and I am grateful to Mike Krische who gave me this opportunity. *Topics in Current Chemistry* has already produced several excellent volumes on various aspects of phosphorus chemistry. The present volumes are dedicated to various topics in organophosphorus chemistry. The first volume concerns biologically-related topics (phosphinopeptides, phosphinic acids, prodrugs) as well as P-asymmetric compounds (also called P-stereogenic, P-chiral, etc. – there is still some intense discussion about how to name this field!). The second volume deals with various synthetic methods and phosphorus functionalities (P-BH_3, phospho-aldol, H- and C-phosphonates, phosphorus tethers in synthesis, and C–H to C–P transformations). In some cases, prior reviews were available on some of the topics. However, the present chapters constitute the best, most up-to-date and in-depth resource in the field, which has been growing rapidly in the past 5–10 years. I believe these volumes will be important additions to library shelves, both institutional and personal. The writings are appropriate for experts and interested students alike.

These chapters have been written by internationally recognized leading experts in the field, both European and American researchers having contributed to the volumes. I wish to thank personally all these authors for spending countless hours to produce these outstanding chapters, which are important contributions to the chemical literature. Clearly, phosphorus chemistry is not only alive and well, but has a promising future and offers great potential for scientific discoveries. Too often people assume that a well-researched topic no longer has anything to offer. I think these chapters prove that nothing could be further from the truth. Because phosphorus is such an important element, chemical research in organophosphorus chemistry has a very bright future indeed. The "old Dame" will remain young!

Fort Worth, TX, USA
9 November 2014

Jean-Luc Montchamp

Contents

Top Curr Chem (2015) 360: 1–38
DOI: 10.1007/128_2014_571

Published online: 5 November 2014

Phosphinic Peptides as Potent Inhibitors of Zinc-Metalloproteases

Dimitris Georgiadis and Vincent Dive

Abstract The development of transition-state analogs is a major objective in enzymology, not only for developing potent inhibitors of enzymes but also for dissecting enzyme catalytic mechanisms. Phosphinic peptides, which share closed structural similarities with the transition-state of peptide substrate upon hydrolysis, have thus been considered for identifying potent inhibitors of proteases. Focusing on the zinc-proteases family, this review presents the most important synthetic efforts performed to obtain the desired compounds. Crystal structures of the phosphinic peptides in interaction with their zinc-protease targets are reported to illustrate the structural features which may explain the potency of these compounds and how they contribute to uncover key enzyme catalytic residues. Based on a remarkable metabolic stability, phosphinic peptides can be used to probe the in vivo function of zinc-proteases. Progress on chemistry and better understanding on the functional roles of zinc-proteases should allow transferring these compounds from shelf to clinic.

Keywords Inhibitors · Phosphinic peptides · Zn-proteases

Contents

D. Georgiadis (✉)
Department of Chemistry, Laboratory of Organic Chemistry, University of Athens,
Panepistimiopolis Zografou 15771, Athens, Greece
e-mail: dgeorgia@chem.uoa.gr

V. Dive
CEA-Saclay, Service d'Ingénierie Moléculaire des Protéines Labex LERMIT,
CEA-DSV-iBiTecS 91191 Gif/Yvette, France

1 Introduction

Proteases have fundamental roles in multiple biological processes and are associated with a wide variety of pathological conditions, justifying impressive efforts over the past 50 years to develop synthetic inhibitors able to block potently the uncontrolled activity of these enzymes in pathological conditions [1]. Our view of the proteolytic world has expanded considerably after the recognition that, beyond their nonspecific functions in protein catabolism, proteases act as processing enzymes carrying out highly selective cleavage of specific substrates and influence cell behavior, survival, and death. The sequencing of the human genome and certain model organisms has revealed that the impressive diversity in protease functions derives from the evolutionary invention of a large number of structurally and catalytically diverse enzymes with the common ability to hydrolyze peptide bonds [2]. Thus, the human degradome, the complete set of proteases produced by human cells, consists of at least 569 proteases and homologues distributed into five classes: 194 metalloproteases, and 176 serine, 150 cysteine, 28 threonine, and 21 aspartic proteases [3]. Among the 194 metalloproteases, the largest family in humans, 121 are secreted and present in the extracellular or pericellular compartments, thus making them rather accessible targets for small synthetic inhibitors. Despite this favorable context, only one zinc-metallopeptidase "angiotensin-converting enzyme (ACE)" has been successfully targeted by synthetic inhibitors, leading to drugs currently approved for clinical use (Fig. 1) [4]. These ACE drug inhibitors belong to different classes of compounds characterized by the nature of

Fig. 1 The chemical structure of fosinoprilat

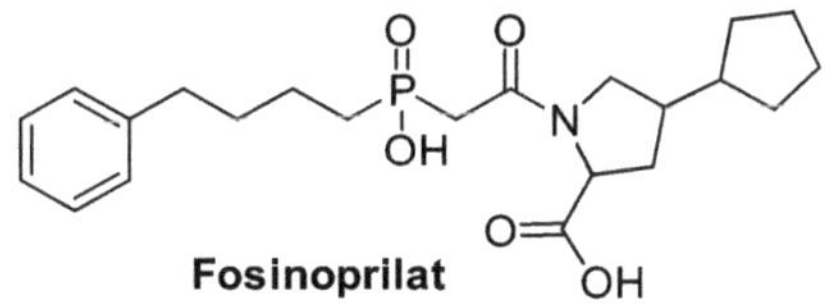

the zinc-chelating group present in their chemical structure. Interestingly, the ACE inhibitor "fosinoprilat" represents the first, and still unique, phosphinic pseudopeptide used as a protease-targeted drug.

Since fosinoprilat development, phosphinic peptide chemistry on solid-support has been exploited to allow easier identification of potent inhibitors for various zinc-metalloproteases. In this review, the synthetic routes to obtain the desired phosphinic peptides are reported, as well as several examples showing that this class of compounds behaves as extremely potent inhibitors toward different sub-families of zinc-metalloproteases. Moreover, the development of these compounds has stimulated efforts to obtain crystal structures of these inhibitors in interaction with their targets, leading currently to about 111 crystal structures in the protein data bank. Analysis of these structures illustrates how phosphinic peptides, as transition-state analogues, have provided important clues on the active site residues involved in catalytic processes, leading to a global view on the catalytic mechanisms of this class of enzymes. Finally, recent use of phosphinic peptide inhibitors for in vivo studies is discussed.

Phosphinic pseudo-peptides have also been used to derive potent inhibitors of the human immunodeficiency virus (HIV) protease [5], and more recently as inhibitors of hepatitis C virus (HCV), but these inhibitors are outside the scope of this review [6].

2 General Properties of Phosphinic Peptides

After the ACE inhibitor success story in the (1980s), the favored approach to develop inhibitors of zinc-metalloproteases was based on the use of a peptidic moiety grafted in the right position with a zinc-chelating group. Thus, thiol, carboxylate, and hydroxamic groups have been used, leading in most cases to potent inhibitors (Fig. 2a) [7–9]. However, as illustrated in Fig. 2a, the occupancy of a limited part of the enzyme active site could be a potential drawback of this approach. For chemistry reasons, such zinc-chelating groups are either on the N- or C-terminal part of the inhibitor, but cannot be placed in the middle of a peptide sequence. In contrast, phosphinic chemistry offers the possibility to probe the whole active site (primed and unprimed subsites), with no limitation on the inhibitor size (Fig. 2b). As compared to other zinc-chelating groups, the phosphoryl group acts as a weak zinc-chelating group; as a consequence, if other inhibitor substituents are not providing sufficient optimal interactions with enzyme active site moieties, the

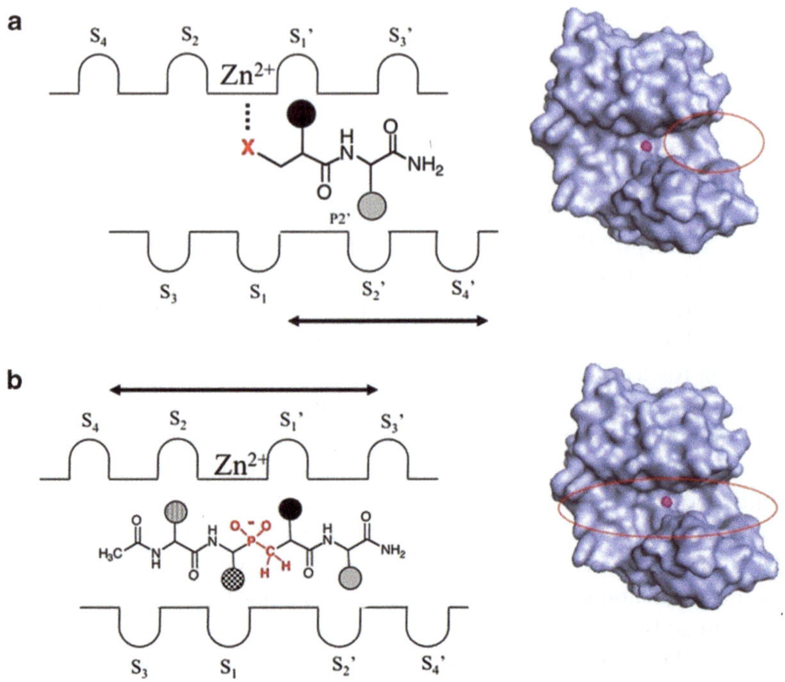

Fig. 2 (**a**) Schematic representation of the zinc-protease active site with subsite (S) nomenclature. X features a zinc-chelating group such as thiolate, carboxylate or hydroxamate. (**b**) Generic structure of a phosphinic peptide covering the S_2 to S_2' subsites of the active site

inhibitor potency will remain low. In contrast, the use of the hydroxamic group, a "strong" zinc-chelating ligand, has provided highly potent inhibitors of zinc-metalloproteases, such as matrix metalloproteases (MMPs) [10]. However, in most cases, this family of inhibitors exhibited extremely low selectivity. In those cases, it can be suspected that the inhibitor potency mostly relies on the presence of the hydroxamic group and its potent interaction with the zinc ion itself, favoring inhibitor interaction with various zinc-metalloproteases [10].

Another motivation to use phosphorus-containing peptides is related to their good structural analogy with what is called "the substrate structure in the transition-state." As shown in Fig. 3, three possibilities can be envisaged and have been explored to develop potent inhibitors. Phosphonamide peptides seem the most obvious choice, as the analogy is almost perfect [11, 12]. If such compounds have been proved to yield potent inhibitors towards some zinc-metalloproteases, the limited stability of the P–N bond through hydrolysis has prevented further use of this chemistry [13–15]. Phosphonates have been reported to yield one of the most potent inhibitor ever reported for carboxypeptidase A (Ki in the fM range) [13]. However, when tested with endoproteases, phosphonates turned out to be less potent [15].

These remarks have historically justified the development of the chemistry of phosphinic peptides as stable compounds and the ability to inhibit zinc-proteases potently [16]. Indeed, it has been found that the replacement of an NH by a CH_2

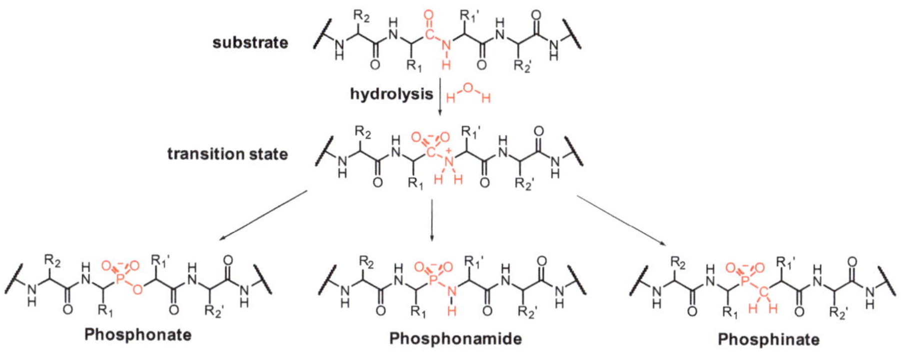

Fig. 3 Structural similarities between the transition-state and different phosphorus-containing peptides

group has only a marginal effect on inhibitor potency. These results were supported by the prediction on the relative affinity of phosphonamide vs phosphinic peptide inhibitors based on free-energy-perturbation simulations. From this study, it has been argued that the loss of NH active-site interaction, ongoing from a phosphonamide to a phosphinic group, should be balanced by the difference in solvation energy between the two groups [17]. This prediction was then confirmed experimentally [18]. One principle to guide these efforts has been to develop the solid-phase peptide chemistry of this class of pseudopeptides and thus appropriate building blocks for use in this chemistry.

3 Synthetic Chemistry of Phosphinic Peptides

3.1 General Considerations

From a synthetic point of view, phosphinic peptides are densely functionalized structures with increasing degree of complexity, depending on the presence of additional functional groups on their side chains. Therefore, the synthesis of such molecules has been a research area of intense activity in both industry and academia for almost 30 years now. Among the several excellent reviews in the literature concerning phosphinic peptides, the prototype work of Yiotakis [19] and the more recent article by Mucha [20] have attempted a systematic approach to the synthetic aspects of these structures. The discussion that follows aims at a concise description of the recent developments in the field and focuses on all current findings and future perspectives.

Figure 4 summarizes the main challenges imposed by a phosphinic peptide structure. According to a categorization proposed by Yiotakis [19], assembly of the pseudopeptidic backbone can be realized by two different approaches: (1) the NP+C strategy, which is the construction of the basic pseudodipeptidic unit by applying addition reactions of aminophosphorus species to suitable carbon

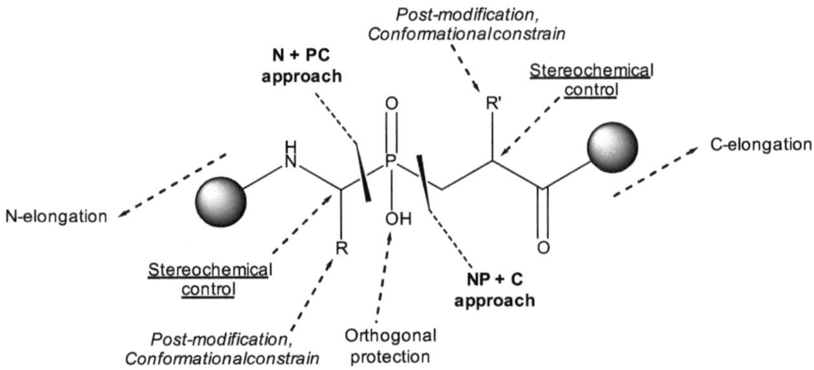

Fig. 4 Schematic representation of synthetic challenges in phosphinic peptides

reactants and (2) the less common N+PC strategy which involves incorporation of amino terminal fragments through P–C bond formation on preformed phosphopropionates. These two complementary strategies offer flexibility in terms of diversification requirements in medicinal projects since the NP+C strategy facilitates P_1' diversification whereas the N+PC strategy is more suitable for rapid P_1 diversification. Besides this consideration, the choice of strategy often depends on special requirements concerning the nature of R and R′ (Fig. 4). Beyond the strategic choice of backbone construction, several recent reports have dealt with the late diversification of side chain groups by post-modification of suitably functionalized precursors, a fruitful approach for rapid production and screening of pseudopeptidic libraries. In addition, another issue directly connected with advances in orthogonal protection of hydroxyphosphinyl moiety is the facile integration of phosphinodipeptidic building blocks into longer peptide structures via N- and/or C-elongation. Furthermore, efforts toward stereochemical control of P_1 and P_1' position stereogenic centers of phosphinic dipeptides have recently been intensified, since most classical NP+C and N+PC synthetic routes offer limited possibilities. Finally, the recent syntheses of phosphinic peptides containing conformational constraints or other non-classical structural features (i.e., thiophosphinates, extended transition-state analogues (TSAs)) have enriched the structural inventory which is now available to medicinal synthetic chemists aiming at efficient inhibitor optimization.

3.2 The NP+C Approach

The NP+C approach (Fig. 4) is undoubtedly the predominant strategy for the preparation of the main pseudodipeptidic unit. This approach usually involves the mild reaction of phosphorus nucleophiles (**2** or **3**) derived in situ from amino-protected aminophosphinic acids with carbon electrophiles such as acrylic

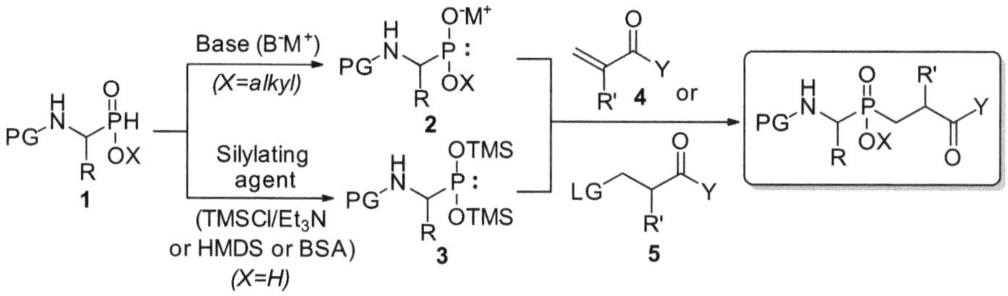

Scheme 1 Main synthetic routes toward phosphinic pseudodipeptides by the NP+C approach

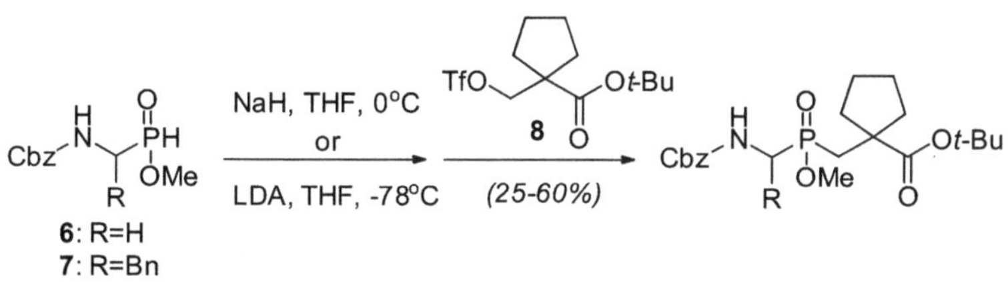

Scheme 2 Main synthetic routes toward phosphinic pseudodipeptides by the NP+C approach (*Cbz*: benzyloxycarbonyl, *Bn*: benzyl, *Tf*: trifluoromethanesulfonyl, *THF*: tetrahydrofuran, *LDA*: lithium diisopropylamide)

derivatives **4** or propionates **5** bearing a leaving group (LG) in their β-position (Scheme 1). As shown in Scheme 1, aminophosphinic acid conversion to nucleophilic tervalent species can be mediated either by deprotonation of aminophosphinic esters by mild bases (i.e., RONa, NaH) or by silylating agents (i.e., hexamethyldisilazane (HMDS), trimethylsilyl chloride (TMS-Cl)/Et_3N, *N,O*-bis(trimethylsilyl)acetamide (BSA)) which lead to reactive bis(TMS)phosphonites **3**. The latter method was introduced by industry researchers 30 years ago [21] and still constitutes the most reliable method for the formation of the key P–C bond in the synthesis of phosphinic pseudopeptides.

The use of easily accessible acrylates of type **4** is far from advantageous as compared to propionic electrophiles of type **5**, since it is compatible with a wide variety of R′ groups. In the case of propionic electrophiles, steric bulk at R′ seems to inhibit the course of the reaction. Indeed, with the only exception of the reaction of triflate **8** with deprotonated aminophosphorus **6** or **7** as reported by McKittrick and coworkers [22] (Scheme 2), it has been stated that the reaction is not productive when $R' \neq H$ [23].

Another significant advantage of the NP+C strategy is the availability of α-aminophosphinic acids with a wide variety of R groups and the possibility to obtain them in stereochemically pure form in bulk quantities. The addition of hypophosphorous acid to diphenylmethyl (DPM) imines proposed by Baylis and co-workers [24] offers this possibility, since it involves inexpensive starting materials, good overall yields, and high optical purity after resolution by recrystallization of the diastereoisomeric salts derived from Cbz-protected aminophosphinic

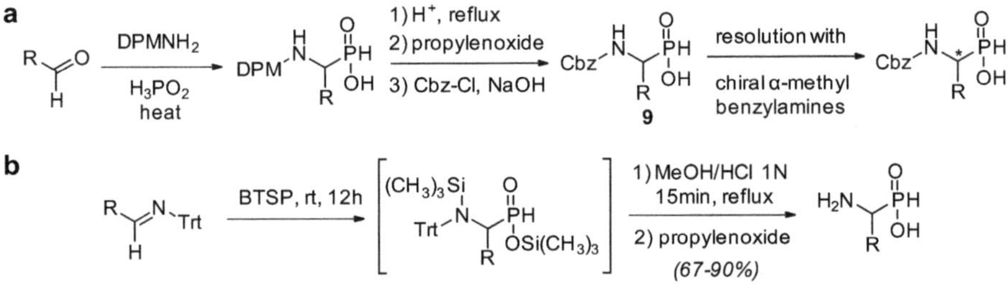

Scheme 3 Synthesis of aminophosphinic acids with the protocols proposed by (**a**) Baylis et al. [24] and (**b**) Jiao et al. [32]

acids (**9**) with (+)- or (−)-α-methylbenzylamine (Scheme 3a). Aminophosphinic acids with side chains corresponding to most natural aminoacids have been obtained by this method [24–26]. Alternatively, DPM imines can be replaced by oximes, leading directly to aminophosphinic acids without the need for deprotection [27–31]. A complementary milder methodology toward aminophosphinic acids is based on the reaction of bis(trimethylsilyl)phosphonite (BTSP) with trityl (Trt) [32, 33], DPM [34], or other imines [35–38] (Scheme 3b). A noteworthy advantage of this approach is that the use of Trt group facilitates the deprotection step toward the generation of the final compounds.

Recently, many research groups have focused their efforts on the development of stereoselective routes leading to optically pure aminophosphinic acids. With this aim, Yamagishi and co-workers recently devised a practical methodology for the preparation of optically pure *N*-protected 1,1-diethoxyethyl(aminomethyl) phosphinates (**12**) [39] and their participation in diastereoselective alkylation reactions [40] which were first studied several years ago by McCleery and Tuck [41] (Scheme 4). In particular, they managed to obtain on a gram-scale and 99 % enantiomeric excess (ee) compound **11**, after addition of paraformaldehyde to 1,1-diethoxyethyl-*H*-phosphinate (**10**) and subsequent lipase-catalyzed resolution of the resulting racemic alcohol. Conversion of **11** to substrate **12** in four steps afforded a valuable substrate suitable for lithium bis(trimethylsilyl)amide (LHMDS)-promoted alkylation performed in a diastereoselective fashion (dr = 10:1) (Scheme 4).

In another recent example, Yao and Yuan reported a high-yielding method for the asymmetric addition of phosphinate **10** to *N*-*tert*-butylsulfinyl imines (**13**) catalyzed by Rb_2CO_3 (Scheme 5) [42]. This protocol tolerates a wide range of functionalized imines and affords the final products in high to excellent optical purities.

The synthesis of aminophosphinic acids with subsequent formation of the second P–C bond via an NP+C strategy has also been investigated as a shorter and faster alternative toward phosphinic peptides. The research group of Haemers coupled the addition of BTSP to tritylimines with a subsequent Michael addition of acrylates by activating in situ intermediate *N*-Trt-*N*-TMS-protected silyl aminophosphinates with BSA and adding acrylates to the resulting phosphonite **14**

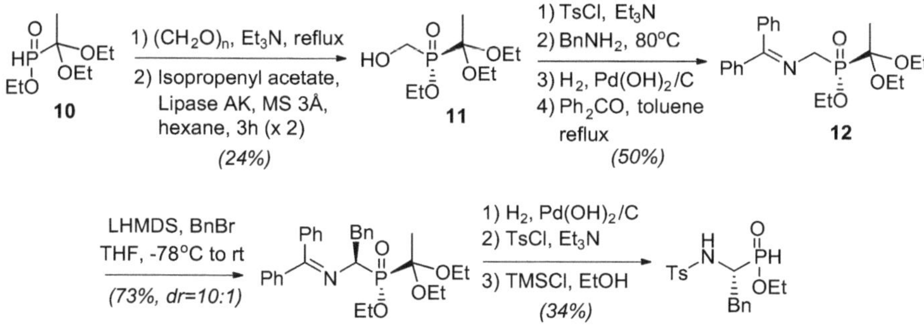

Scheme 4 Diastereoselective synthesis of α-aminophosphinic acids by Yamagishi et al. [39, 40] (Ts: 4-toluenesulfonyl)

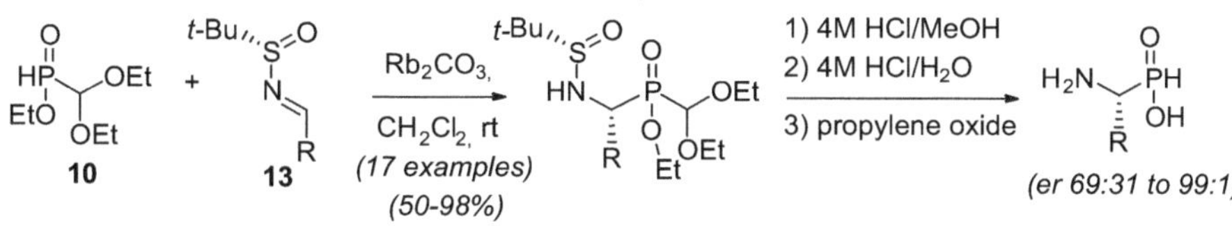

Scheme 5 Diastereoselective synthesis of α-aminophosphinic acids by Yao and Yuan [42]

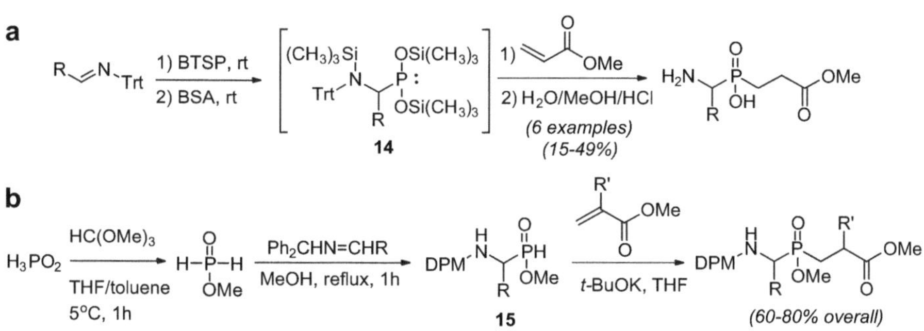

Scheme 6 NP+C strategy associated with α-aminophosphinic acid preparation by (**a**) Borloo et al. [35] and (**b**) Cristau et al. [43, 44]

(Scheme 6a) [35]. In contrast, Cristau and coworkers followed a different approach for developing a "one-pot," three step synthesis of protected phosphinopeptides (Scheme 6b) [43, 44]. This protocol involves sequential *O*-alkylation of H_3PO_2, addition of alkyl hypophosphites to imines and finally Michael addition of the resulting DPM-protected aminophosphinates **15** to acrylates.

Aiming to find a synthetic protocol for the complete preparation of phosphinic peptides on solid phase, including P–C bond formation, Meldal's research group has investigated the possibility of applying the NP+C strategy on solid phase [45]. The researchers extended preliminary results of Dorff and coworkers [46] who managed to perform conjugate addition of $FmocGlyPO_2H_2$ to resin-bound

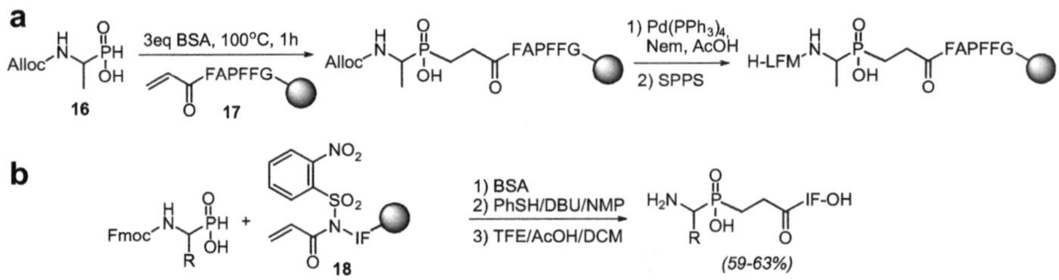

Scheme 7 Application of NP+C strategy to the solid phase synthesis of phosphinic peptides by (**a**) Buchardt et al. [45] and (**b**) Manzenrieder et al. [47] (Nem: *N*-ethyl morpholine, SPPS: solid phase peptide synthesis, *DBU*: 1,8-diazabicycloundec-7-ene, *NMP*: *N*-methyl-2-pyrrolidone, *TFE*: trifluoroethanol, *DCM*: dichloromethane)

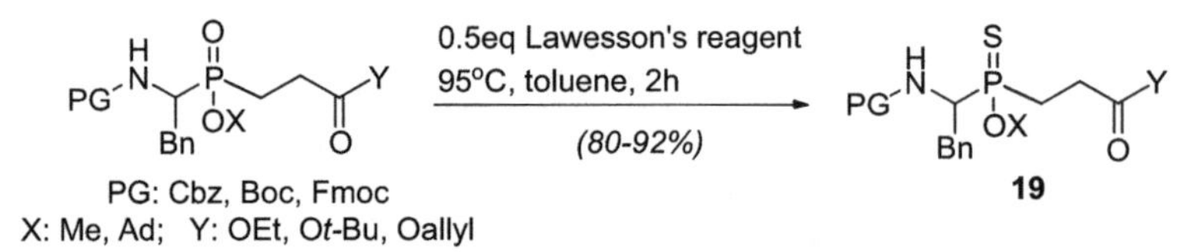

Scheme 8 Synthesis of thiophosphinic pseudodipeptides by Vassiliou et al. [48] (Boc: *tert*-butyloxycarbonyl, Ad: 1-adamantyl)

acrylates using BSA as a silylating agent and presented a versatile protocol for the synthesis of longer peptides, as shown in Scheme 7a. In particular, they observed that the use of excess BSA at elevated temperatures allowed addition of allyloxycarbonyl (Alloc)-protected aminophosphinic acids (i.e., **16**) to resin-bound acryloylated peptides (**17**). *N*-Elongation after Pd-catalyzed Alloc removal was achieved by using pentafluorophenyl (Pfp) esters of 9-fluorenylmethoxycarbonyl (Fmoc) aminoacids and 3,4-dihydro-3-hydroxy-4-oxo-1,2,3-benzotriazine (Dhbt-OH) as a catalyst. It was recently reported that the slow Michael addition step can be accelerated by adding the electron-withdrawing group *o*-nitrobenzenesulfonyl (*o*NBS) onto the nitrogen that bears the acryloyl group (**18**), but exact conditions are not given (Scheme 7b) [47].

Another recent interesting application of the NP+C strategy is the transformation of phosphinic pseudodipeptides prepared by such an approach to thiophosphinates (**19**), a new class of pseudopeptides with promising perspectives as enzyme inhibitors [48]. To this end, Vassiliou et al. observed that careful thionation by using Lawesson's reagent can afford the target molecules without affecting ester or urethane carbonyl groups (Scheme 8). The authors conducted a complete study of the behavior of these molecules under standard deprotection and coupling conditions, which may serve as a valuable guide for future use of these molecules in medicinal projects.

Variations of the electrophiles employed in the NP+C strategy can furnish various dehydro pseudopeptides, a class of molecules that have been tested for their inhibitory properties but, in addition, as substrates for further functionalization

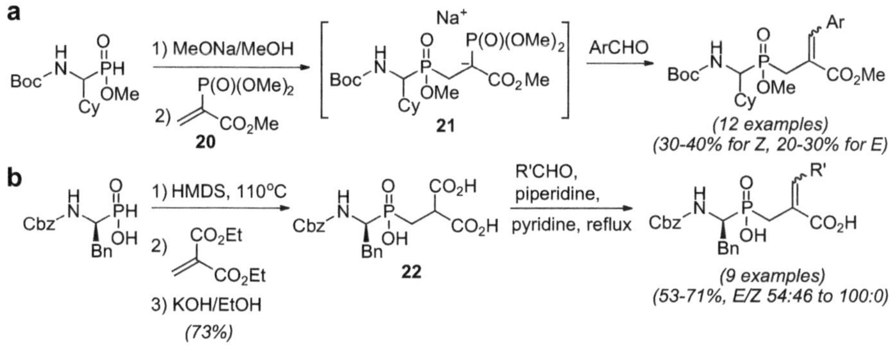

Scheme 9 Synthesis of dehydro pseudodipeptides using variations of the NP+C strategy by (**a**) Gurulingappa et al. [50, 51] and (**b**) Matziari et al. [52] (Cy: cyclohexyl)

(see below). Khan and co-workers, looking for renal dipeptidase inhibitors, utilized an older protocol of Schoen and Parsons [49] to synthesize a series of phosphinic dehydropeptide analogues (Scheme 9a) [50, 51]. The protocol involves a Michael addition of phosphorus nucleophiles to 2-trimethylphosphonoacrylate **20**, followed by a Horner–Wadsworth–Emmons olefination of the intermediate phosphonate carbanions **21** with various aldehydes. Another protocol leading to similar products involves the Knoevenagel-type condensation of malonic acid derivatives **22** with aldehydes, as described by Matziari and co-workers (Scheme 9b) [52].

3.3 *The N+PC Approach*

The synthesis of phosphinic peptides by a reverse sequence of P–C bond formation events (N+PC approach) is a less frequently applied strategy which may offer important diversification possibilities. In particular, an amidoalkylation condensation reaction between amides, aldehydes, and alkylphosphinic acids (the three-component Kabachnik–Fields reaction) affords in a single step the main pseudopeptidic backbone, thus facilitating fast screening of the nature of P_1 position. In 1996, Chen and Coward observed that a mixture of benzyl carbamates, aldehydes, and alkylphosphinic acid **23** in AcCl can lead to Cbz-protected phosphinic pseudodipeptides **24** (Scheme 10a) [53]. This method was adjusted by Matziari et al. to the synthesis of Fmoc-protected phosphinic building blocks **25** and peptides thereof (Scheme 10b) [54].

Ragulin and co-workers observed that acetic anhydride can be employed when acetamide participates as an amide partner for an amidoalkylation condensation and proceeded in the establishment of optimized and milder conditions for carbamates as well [55]. To this end, the authors observed that the reaction can be accelerated by *p*-TsOH in neat acetic anhydride (Scheme 11, method **A**) [56]. Based on the observation that the reaction proceeds through *N*,*N*′-benzylidene bis(carbamate)

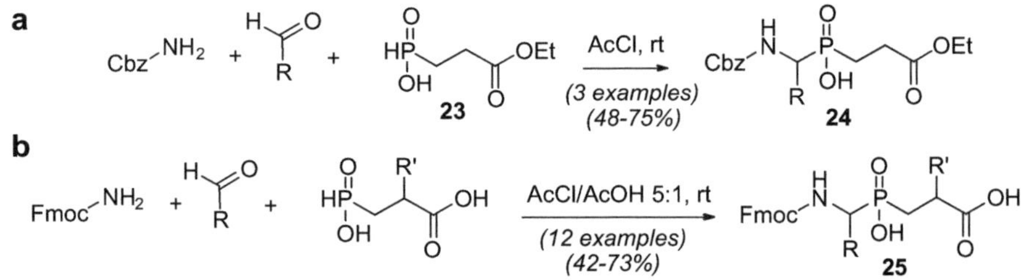

Scheme 10 Application of the N+PC strategy to the synthesis of phosphinic pseudodipeptides by (**a**) Chen et al. [53] and (**b**) Matziari et al. [54]

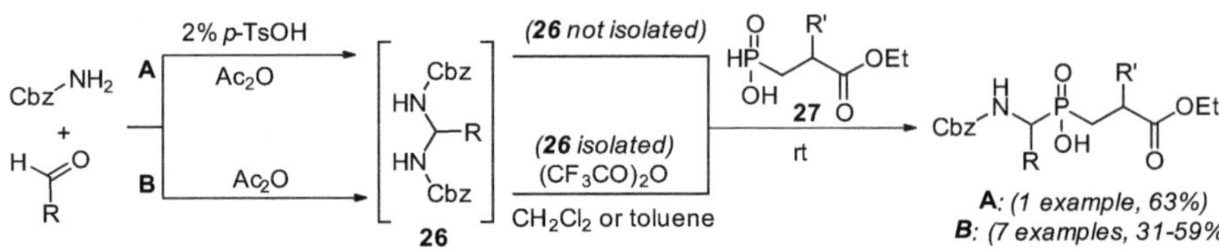

Scheme 11 Optimization of the amidoalkylation reaction by the research group of Ragulin [56, 57]

intermediate **26**, a milder version of the reaction was presented where biscarbamates **26** may react with alkylphosphinic acids **27** in toluene or dichloromethane with stoichiometric quantities of trifluoroacetic anhydride (Scheme 11 method **B**) [57].

3.4 *Orthogonal Protection and Implications to Peptide Elongation*

The presence of a hydroxyphosphinyl group within the backbone of a phosphinic peptide causes an additional problem during their synthesis, which is the behavior of phosphinic group during coupling reactions and consequently the need of its protection in the solid phase or solution peptide synthetic schemes. Several studies have been reported on this subject, dealing with the problem of orthogonality from three different points of view: (1) adjusting synthetic routes to alkyl protecting groups of phosphinic acids used during P–C bond-forming reactions, (2) introducing a suitable protecting group after P–C bond forming reactions, and (3) developing peptide-coupling conditions tolerant to free hydroxyphosphinyl moieties.

Concerning the first approach, methyl or alkyl protection of the phosphinic group requires suitable masking of the carboxylic terminus of the main pseudodipeptidic unit which could be selectively removed prior to *C*-elongation. Selective deprotection of carboxylic esters in the presence of alkyl phosphinates include enzymatic hydrolysis of methyl carboxylates [18, 58], controlled alkaline hydrolysis of ethyl carboxylates [59, 60], acidic cleavage of 3,4-dimethoxybenzyl

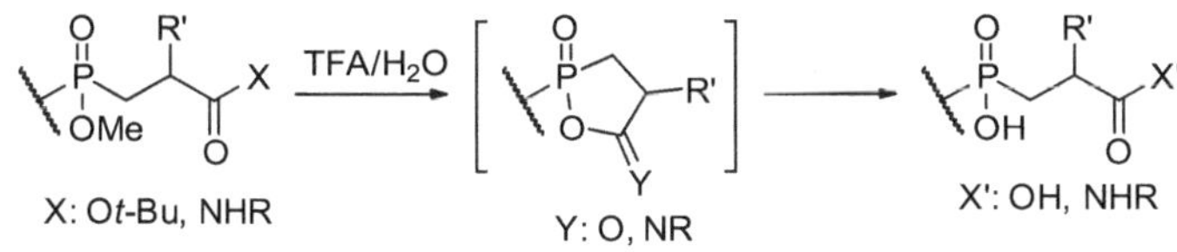

Scheme 12 Intramolecular acid-catalyzed deprotection of methyl phosphinates [64, 65] (TFA: trifluoroacetic acid)

carboxylates [61], and hydrogenolysis of benzyl carboxylates [62]. Acidic cleavage (TFA) of *tert*-butyl carboxylates has also been employed [63], but in several cases simultaneous phosphinic ester cleavage is observed. It has been demonstrated that this side-reaction is intramolecularly accelerated by the released carboxylic acid through the formation of five-membered mixed anhydrides (Scheme 12) [64]. Final deprotection of alkylphosphinates after peptide development is usually performed under strong alkaline or nucleophilic conditions (i.e., TMSBr, *n*-PrSLi) [60, 63, 64]. Intramolecular amide-assisted TFA hydrolysis through cyclic imidate structures has also been reported as a clean alternative for demethylation of methyl phosphinates (Scheme 12) [62, 65].

Several reports have dealt with the use of phosphinic pseudodipeptide building blocks in peptide couplings where both phosphinic and carboxylic acids are unprotected. Among the coupling reagents used successfully for such transformations are *N*,*N*′-carbonyldiimidazole (CDI) [66–68], (benzotriazol-1-yloxy)tris (dimethylamino)phosphonium hexafluorophosphate (BOP) [69, 70], (benzotriazol-1-yloxy)tripyrrolidinophosphonium hexafluorophosphate (pyBOP) [71, 72], and 1-ethyl-3-(3-dimethylaminopropyl)carbodiimide (EDC) [54, 73, 74]. Campagne et al. have studied both C- and N-terminus peptide elongation with BOP (or pyBOP) and verified that this process is epimerization-free and applicable to peptide synthesis in the solid phase [69]. In some case, EDC has also been used in the solid phase peptide synthesis using the Fmoc protocol [54, 75]. However, none of these protocols have ever been used in synthesis of phosphinic peptide libraries by automated synthetic methods.

Aiming at the facile embodiment of phosphinic pseudodipeptide building blocks into peptide structures using the Fmoc protocol, the quest for suitable protection for that purpose was pursued in very early works [76]. A major breakthrough toward this direction was the introduction of the 1-adamantyl (Ad) group by Yiotakis and Dive in 1996 [77]. The Ad group is totally compatible with the Fmoc-solid phase synthetic protocols and is easily removed under standard TFA-deprotection conditions. This approach has been validated in several reports by application to the development of large combinatorial libraries which allowed the identification of potent and selective inhibitors of peptidases [78–85]. The significance of this approach justifies the research efforts which followed, aiming at the optimization of building block preparation. In the original report, adamantylation of phosphinic dipeptide isosteres was performed by reaction of phosphinic acid 1-adamantyl bromide and silver oxide in refluxing chloroform [77]. Despite the high efficiency

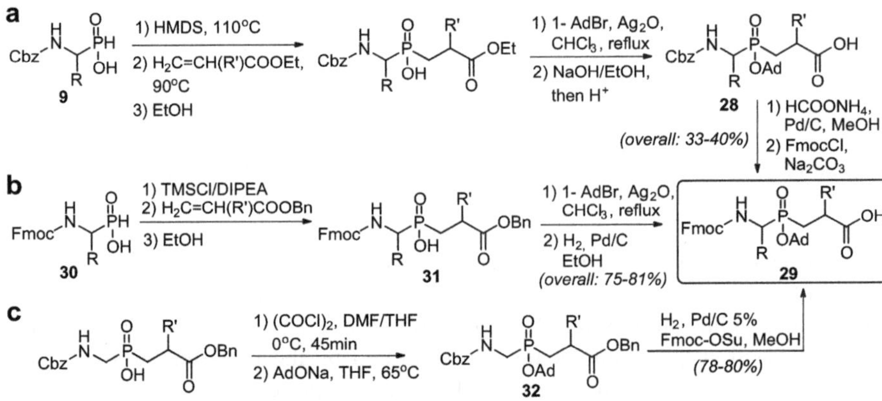

Scheme 13 Synthesis of Fmoc-protected pseudodipeptidic building blocks by (**a**) Yiotakis et al. [77], (**b**) Georgiadis et al. [88], and (**c**) Bhowmick et al. [86, 87]

of this protocol, other methods have been subsequently investigated, such as treatment of in situ-generated phosphinoyl chlorides with the sodium salt of adamantanol [81, 86, 87], but in most cases the AdBr/Ag_2O protocol proved superior [81, 87].

The first protocol reported for the synthesis of Ad-protected Fmoc building blocks **29** (Scheme 13a) involves five steps starting from Cbz-protected aminophosphinic acids **9** and ethyl acrylates [77]. Removal of the Cbz group of **28** by hydrogenolysis and reprotection of the amino group by Fmoc proved to be capricious, and an improved protocol was reported a few years later by Georgiadis et al. (Scheme 13b) [88]. In this protocol, Fmoc-protected aminophosphinic acids (**30**) and benzyl acrylates were employed in the Michael addition step using TMSCl/*N*,*N*-diisopropylethylamine (DIPEA) activation conditions. From this point, adamantylation of **31** and hydrogenolysis of the benzyl group afforded the target molecules **29** in almost double overall yield as compared to the prototype protocol. In a similar manner, Fields and co-workers prepared a GlyVal phosphinic isoster Fmoc building block by using an allyl acrylate in the Michael addition step and removing allyl group at the end of the synthesis with $CpRu(CH_3CN)_3PF_6$ (Cp: cyclopentadienyl) [85]. The same researchers observed that building blocks of type **32** can be transformed to the target building blocks **29** by hydrogenolysis and in situ re-protection using FmocOSu in a one-pot procedure (Scheme 13c) [86, 87].

Finally, Nasopoulou et al. presented a method to discriminate the two acidic groups of Fmoc diacids **25** by using the phenacyl (Pac) group, which can be selectively removed from the phosphinic moiety upon heating with TFA (Scheme 14) [89]. After Ad-protection, Pac carboxylic ester can be easily cleaved by metal reduction, furnishing the target molecules **29** in moderate overall yields.

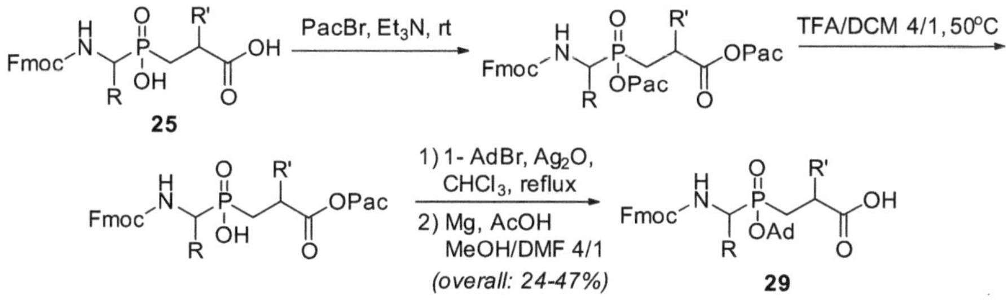

Scheme 14 Synthesis of Fmoc-protected pseudodipeptidic building blocks by Nasopoulou et al. [89]

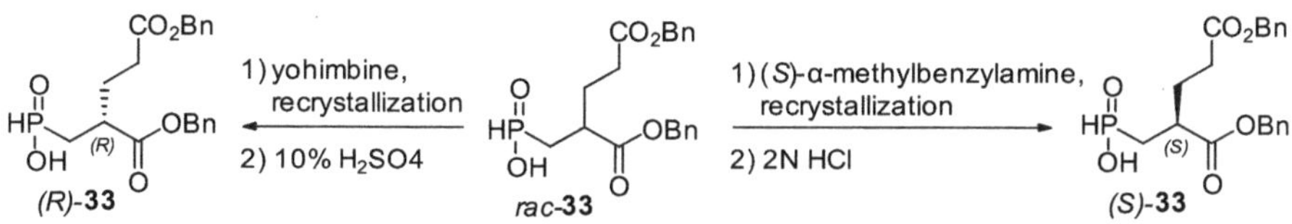

Scheme 15 Resolution of **33** by chiral amines [90]

3.5 Stereoselective Approaches

Control of phosphinic pseudodipeptide stereochemistry during their preparation is a subject of paramount importance, since active isomers are usually isolated in low yields from diastereoisomeric mixtures by chromatographic methods. Availability of enantiomerically pure aminophosphinic acids by applying the Baylis protocol in NP+C synthetic routes transposes the problem of finding ways to control P_1' stereochemistry. In particular, formation of a P_1' stereogenic center during *P*-Michael addition to acrylates limits opportunities for stereochemical control in this position. On the other hand, in N+PC synthetic protocols, it has been reported that enantioenriched phosphinic acids (i.e., **33**, Scheme 15) can be obtained by chiral resolution of racemic mixtures using (*S*)- or (*R*)-α-methylbenzylamines or yohimbine [71, 90]. Of course, this time the lack of stereoselectivity in amidoalkylation reactions produces two diastereoisomers which need to be separated by chromatography.

Regardless of the method used to construct the main pseudodipeptidic backbone, separation of different diastereoisomers has been achieved in several cases by selective crystallization at dipeptide [74, 91], tripeptide [74, 92, 93], or even tetrapeptide level [94]. The isomer corresponding to the natural peptide is usually crystallized out selectively; nevertheless, in some tripeptides it has been reported that non-natural peptide isomers are prone to selective crystallization in less polar solvents (Et_2O or $CHCl_3$), probably caused by differences in the intramolecular hydrogen bonding in these media (Scheme 16) [92, 93].

Early efforts to control P_1' stereochemical identity with only modest efficiency included enzymatic hydrolysis of *C*-terminal methyl carboxylates [18] and

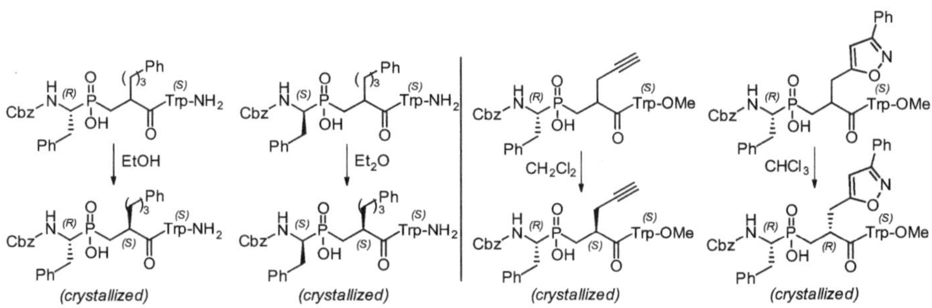

Scheme 16 Diastereoisomeric separation of phosphinic tripeptides by selective crystallization

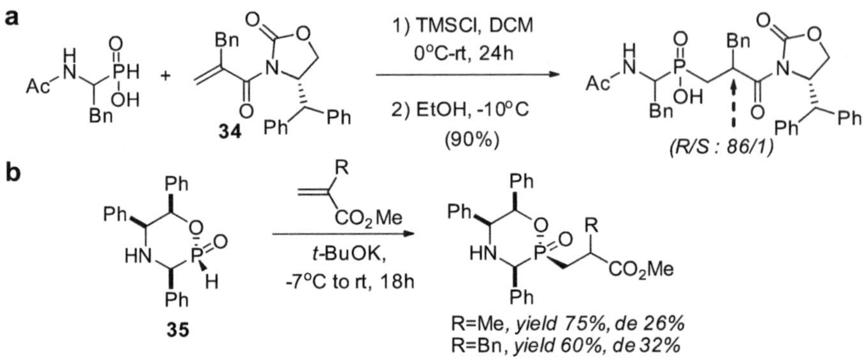

Scheme 17 Diastereoselective *P*-Michael additions by (**a**) Liu et al. [96] and (**b**) Cristau et al. [97, 98]

asymmetric hydrogenation of dehydroalanine pseudopeptides [95]. In 2002, the research group of Ebetino reported on the diastereoselective Michael addition of phosphinic or aminophosphinic acids to acrylates bearing Evans oxazolidinone-type auxiliaries (Scheme 17a) [96]. Under the developed conditions, up to 98% ee can be attained when DPM-substituted oxazolidinones (**34**) are employed. On the other hand, the Michael addition of 2-*H*-2-oxo-1,4,2-oxaza phosphinanes (**35**) to α-substituted acrylates has also been investigated, but poor stereoselectivities were observed (Scheme 17b) [97, 98]. However, the reaction can lead to high diastereoselectivities when β-substituted acrylates are employed.

Yamagishi et al. recently presented a novel method for the stereoselective synthesis of phosphinic pseudopeptide isosteres starting from an enantiopure α-aminoalkyl-*H*-phosphinate on both α-carbon and phosphorus atoms (**36**, Scheme 18) [99]. In particular, substrates **37** derived from a stereoretentive Michael addition of aminophosphinates **36** to *tert*-butyl acrylate, were diastereoselectively alkylated in the P_1' position with diastereoisomeric ratios (dr) ranging from 14:1 to 29:1. The stereochemical outcome of this process is highly controlled by the chirality at the phosphorus atom and the nature of the amino protecting group.

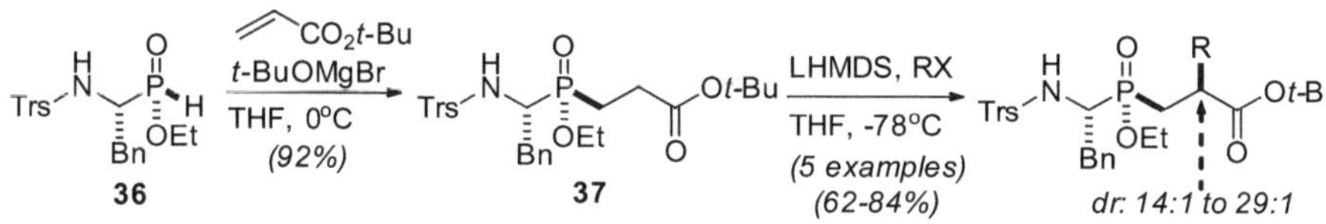

Scheme 18 Synthesis of phosphinic dipeptide isosteres through diastereoselective alkylation [99] (Trs: tritylsulfenyl)

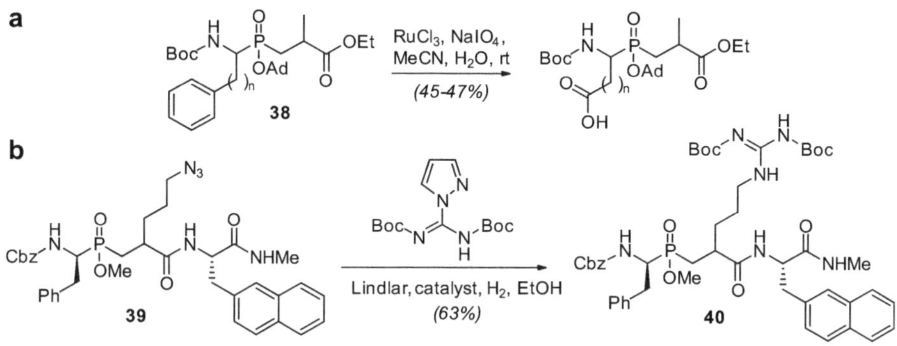

Scheme 19 Synthesis of (**a**) P_1 Asp and Glu [100] and (**b**) P_1' Arg [68] phosphinic dipeptide analogues by a post-modification approach

3.6 *Post-Modification of P_1 and P_1' Side Chains*

The need for post-modification protocols of phosphinic pseudodipeptide side chains stems from two main reasons. The first concerns the synthesis of phosphinic structures bearing side chains which cannot be present in the starting blocks necessary to construct the main backbone. With this aim, Georgiadis et al. devised a post-modification strategy for preparing pseudodipeptides with acidic residues in the P_1 position, since the classical NP+C strategy with aspartyl aminophosphinic acids failed [100]. This method is based on the preparation of phenylalanine and homophenylalanine derivatives (**38**) followed by the oxidative decomposition of phenyl groups (Scheme 19). In the same spirit, Kende et al. built Arg pseudopeptidic analogues **40** by introducing the guanidine group in suitable alkylazido P_1' side chains (**39**) [68].

The second and most important reason for pursuing post-modification is the possibility of widely diversifying pseudopeptide side chains at a late stage in the synthesis, thus enhancing screening options in medicinal projects. In an extension of an older report by Schoen and Parsons [49], Matziari et al. prepared a series of dehydroalanine pseudodipeptides by allylic displacement of 2-bromomethyl acrylate from aminophosphinic acid, and tested their reactivity in conjugate additions of

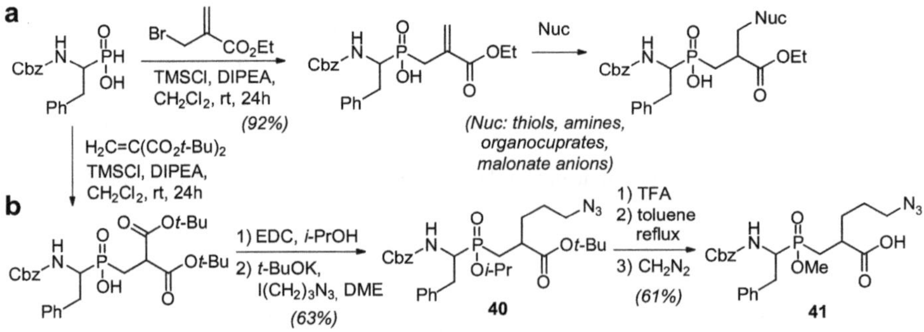

Scheme 20 Use of (**a**) dehydrodipeptides [101] and (**b**) malonic derivatives for post-modification [68] (DME: dimethoxyethane)

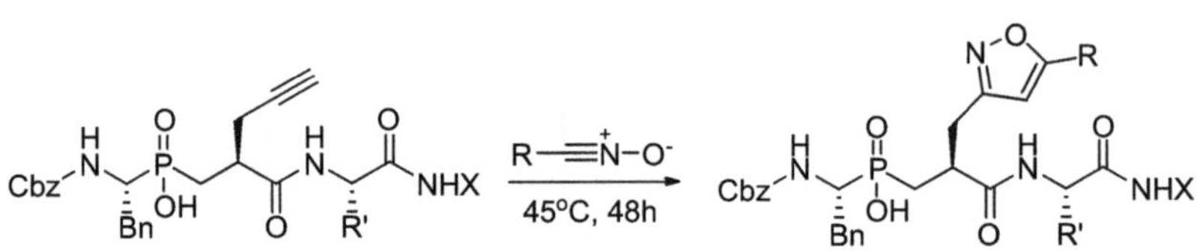

Scheme 21 A dipolar cycloaddition-based post modification approach [74]

various sulfur, nitrogen, or carbon nucleophiles (Scheme 20a) [101, 102]. Furthermore, in separate reports by the research groups of Yiotakis and Ebetino, malonic derivatives of types **22** or **40** were similarly synthesized and derivatized either by Knoevenagel-type reactions (Scheme 9b) or by alkylation/decarboxylation reactions (Scheme 20b), affording dehydro pseudodipeptides or pseudodipeptide **41**, respectively [52, 68].

One of the most powerful derivatization techniques of P_1' position was presented in 2003 by Makaritis et al. They prepared phosphinic pseudodipeptide precursors bearing a propargyl group in their P_1' side chain (**42**) and saw that these molecules can serve as excellent dipolarophiles in 1,3-dipolar cycloadditions with in situ-prepared nitrile oxides (Scheme 21) [74]. This approach can be easily applied in both solid phase and solution syntheses, and has been successfully used in the discovery of selective inhibitors of several Zn-metalloproteases [38, 84, 92, 103, 104].

Interesting work found in early patent literature deals with the use of dehydroalanine as an electrophile in the Michael addition of aminophosphinic acids, and diversification of the amino group appeared in the P_1' position by coupling reactions [105]. This concept was extended by Vassiliou et al. by allowing a similar precursor (**43**) to participate efficiently in benzylation diversification reactions (Scheme 22) [106].

An interesting approach to this direction was recently based on an Ireland–Claisen rearrangement triggered by the phospha-Michael addition of silyl

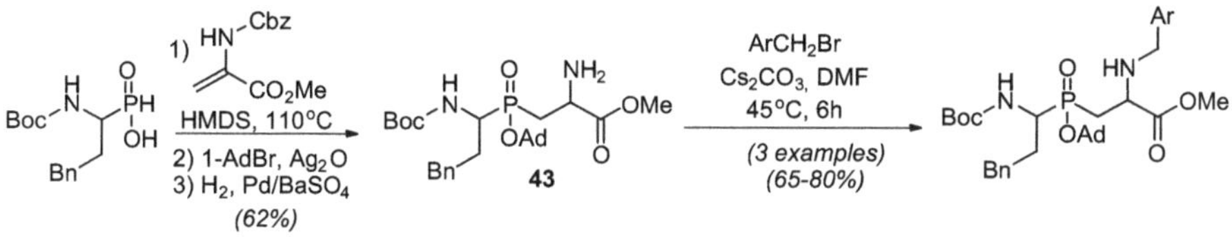

Scheme 22 Post-modification of phosphinic dipeptide precursors bearing an amino group [106]

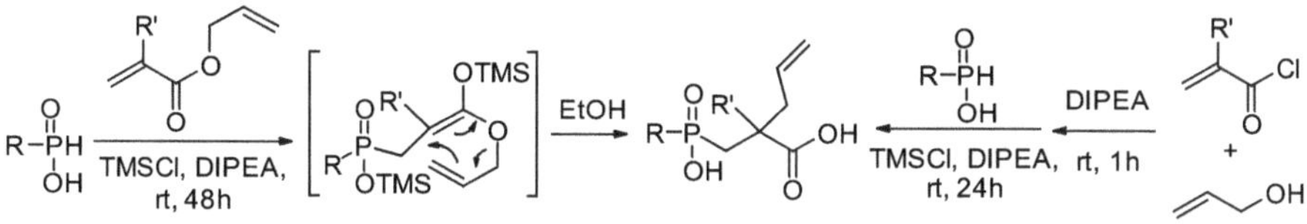

Scheme 23 Post-modification of phosphinic backbones based on an Ireland–Claisen rearrangement [107]

phosphonites to allyl acrylates (Scheme 23) [107]. This reaction was evolved in a one-pot three-component version between phosphinic acids, acryloyl chloride, and allylic alcohols, thus allowing the rapid synthesis of diverse phosphinic structures using simple starting materials in a single step (Scheme 23). The method can also lead to analogues with a quaternary center at the P_1' position of phosphinic structures, which offers an alternative route for obtaining conformational constrained phosphinic peptides.

3.7 *Conformationally Constrained Structures*

The introduction of rigid structural elements in phosphinic peptides which restrict their conformational mobility may serve as a promising technique for the identification of inhibitors with optimized inhibition profiles. In some cases, the synthesis of these molecules cannot be attained by standard protocols, and therefore several reports have focused on the construction of such structures by alternative routes. In such a report, Nasopoulou et al. have invented a three-center-two-component amidoalkylation annulation reaction between α,ω-carbamoylaldehydes **44** and phosphinic diacids **45** (Scheme 24a) [108]. Depending on the starting aldehyde, diverse phosphinic scaffolds **46** varying in the size of their rigidity element, the nature and stereochemistry of substituents, and the participation of heteroatoms in the azacyclic ring system can be obtained in one synthetic step and in high yields. In another report, a pyroglutamic ring at the P_1 position has been formed in the solid phase by base-catalyzed cyclization of glutamic analogue **47** (Scheme 24b) [75].

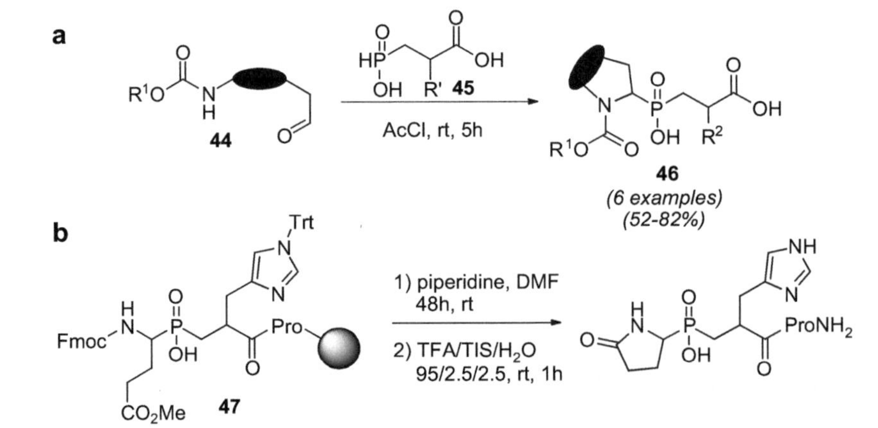

Scheme 24 Synthesis of conformationally constrained phosphinic structures at position P_1 by (**a**) Nasopoulou et al. [108] and (**b**) Matziari et al. [75] (TIS: triisopropylsilane, DMF: dimethylformamide)

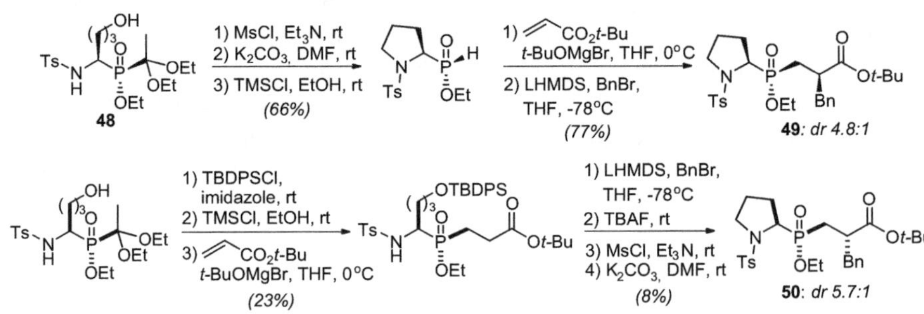

Scheme 25 Stereoselective synthesis of P_1 Pro analogues by Yamagishi et al. [99] (Ms: methanesulfonyl, TBDPS: *tert*-butyldiphenylsilyl, TBAF: tetrabutylammonium fluoride)

Recently, Yamagishi et al. utilized the chemistry of stereoselective alkylations developed in their laboratory to synthesize Phe-Pro pseudodipeptide isosteres (**49** and **50**, Scheme 25) [109]. By using the fully stereodefined phosphinate **48** as starting material (synthesized as in Scheme 4), the authors observed that the proline ring can easily be produced by an intramolecular substitution reaction and that the stereoselectivity of benzyl group introduction can be reversed, depending on whether this event follows or precedes the cyclization step.

As it will be discussed later, the introduction of a pseudo-Pro ring in the P_1' position has given rise to RXPA380, the most selective C-domain ACE inhibitor ever reported [110, 111]. Given the unsuccessful attempts to introduce pseudo-Pro ring to a phosphinic peptide isoster by a classical Michael addition approach (as proposed by earlier studies [112]), allylic substitution of acetate **51** by the Cbz-protected aminophosphinic analogue of phenylalanine was investigated (Scheme 26a) [111]. This route furnished dehydro pseudoproline analogue **52** in

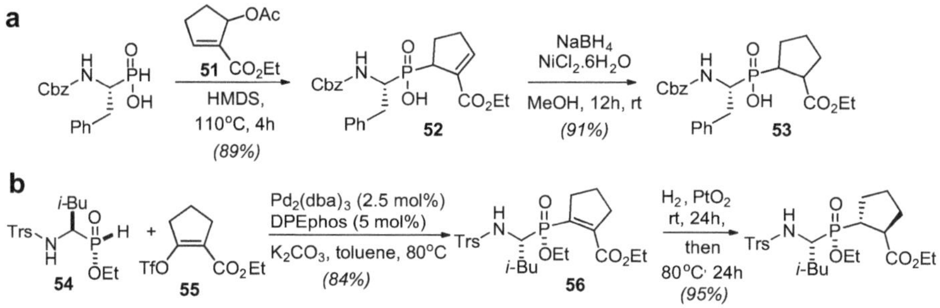

Scheme 26 Synthesis of pseudo-Pro phosphinic dipeptide isosteres (**a**) by Georgiadis et al. [111] and (**b**) by Yamagishi et al. [99] (dba: dibenzylideneacetone, DPEphos: bis-[2-(diphenylphosphino)phenyl]ether)

high yield and led to the target pseudo-Pro derivatives **53** after chemoselective nickel boride reduction. Yamagishi et al. recently proposed a stereoselective synthesis of P_1' pseudo-Pro phosphinic analogues by using a stereoretentive Pd-catalyzed cross-coupling of a fully stereodefined aminophosphonate (**54**) with triflate **55**, followed by a subsequent highly diastereoselective hydrogenation of the olefin moiety of **56** (Scheme 26b) [113]. It should be noted that the hydrogenation step involves an epimerization event of the initially *cis* disubstituted saturated product to the *trans* isomer which, according to the authors, is attributed to the chirality of the phosphorus atom.

4 Selected Examples of Potent Inhibitors of Zinc-Metalloproteases

Since the development of fosinoprilat, a number of potent phosphinic inhibitors have been reported toward various zinc-metalloproteinases. The examples discussed below which have been selected as crystal structures are available, and thus some insights on the phosphoryl binding mode with the zinc ion can be deduced and the value of phosphinic inhibitors in unveiling enzyme catalytic residues can be illustrated.

4.1 *Angiotensin-Converting Enzyme*

The central role played by ACE in cardiovascular pathologies such as hypertension and cardiac hypertrophy is well established. The action of ACE as the major mechanism in the biosynthesis of angiotensin II (AngII) has made it an excellent target for therapeutic intervention in the treatment of cardiovascular diseases [4]. While a number of ACE inhibitors have been developed and used effectively

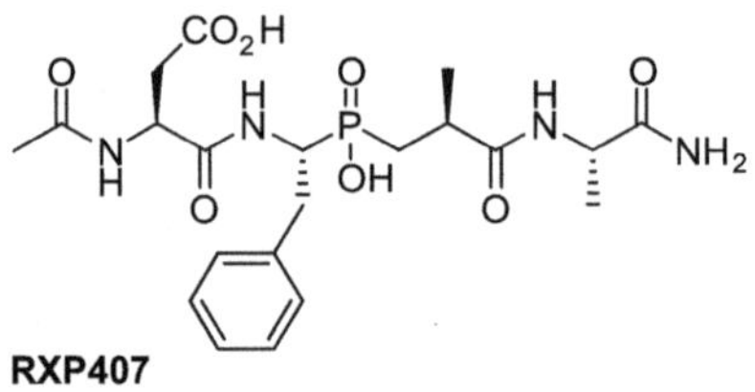

Ki (N) = 3 nM

Ki (C) = 2500 nM

RXP407

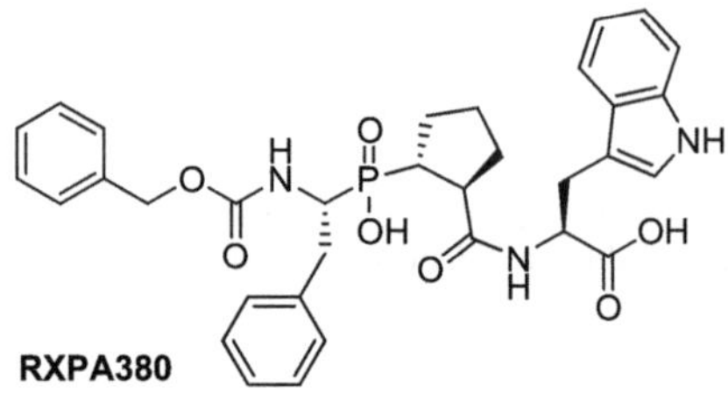

Ki (N) = 30 000 nM

Ki (C) = 6 nM

RXPA380

Fig. 5 Chemical structures of **RXP407** and **RXPA380** and inhibition constants for N- and C-domain active sites of ACE

for the treatment of hypertension [114, 115], adverse side effects such as persistent cough and angioedema are associated with ACE inhibition [116]. ACE is a type I transmembrane zinc dipeptidyl carboxypeptidase (EC 3.4.15.1), whose ectodomain is composed of two homologous domains (N and C) connected by an interdomain linker region. Each of the distinct domains of ACE harbors a catalytic active site, identified by the presence of the consensus zinc-binding motif HEXXH. Although the two domains of ACE are highly homologous, they display distinct substrate preferences. The C-domain of ACE is necessary and sufficient for the maintenance of proper basal blood pressure, and hence is viewed as the main site of Angiotensin II generation. On the other hand, the N-domain is more effective at cleaving renin-angiotensin system independent biological substrates of ACE, such as the anti-fibrotic hemoregulatory peptide acetyl-SDKP. In order to identify and understand the key components which distinguish N-domain activity from that of C-domain, inhibitors able to differentiate fully between the N- and C-domains have been devised. ACE inhibitors approved for clinical use are mixed inhibitors, exhibiting similar potency toward the N- and C-domains. By screening libraries of phosphinic peptides prepared by combinatorial chemistry, extremely selective inhibitors of the N-domain have been identified (Fig. 5) [80]. The ability to probe both the primed and the unprimed subsites of ACE N-domain (from S_2 to S_2' subsite) was critical to controlling the inhibitor selectivity, explaining why phosphinic peptides behaved as more selective inhibitors than "classical ACE inhibitors." In the RXP407 compound, the presence of the aspartic residue in the P_2 position was shown to play a key role in selectivity. X-Ray studies reveal that this role is explained by a single mutation between the N- and C-domains [117]. In the N-domain the aspartic side chain is pointing toward an arginine, while in the C-domain this residue is replaced by a glutamic residue, explaining why RXP407 is much less tolerated by the C-domain.

The extremely high selectivity of the RXPA380 compound has also been explained by the presence of residues probing the ACE active site from the S_2 to S_2' active sites [111, 118]. Here again, the corresponding crystal structure has illustrated that the simultaneous presence of a benzyl and tryptophan in the N- and C-terminal structure part of the inhibitor is critical for its ability to discriminate the C-domain from the N-domain. These inhibitors have been used as valuable pharmacological tools to advance our knowledge of the full functional role of both domains in vivo [110].

4.2 *Neprilysin*

Neutral endopeptidase (NEP, EC 3.4.24.11) is zinc-metallopeptidase involved in the metabolic inactivation of a number of bioactive peptides including the enkephalins, substance P, endothelin, bradykinin, and atrial natriuretic factor. Owing to the physiological importance of NEP in the modulation of nociceptive and pressor responses, considerable effort has been devoted to the development of potent and selective inhibitors as novel analgesics and antihypertensive agents [119]. Extremely potent inhibitors of NEP containing as a zinc-chelating group a thiol function have been successfully developed, such as the thiorphan compound [120, 121]. However, in the search for potent dual inhibitors of both NEP and aminopeptidase N (APN, EC 3.4.11.2), an amino zinc-exopeptidase, the presence of a free amine in the N-terminal position of the inhibitor was a strict requirement. In this case, the phosphinic peptide chemistry offers this possibility, leading to the design of the compound **57** (Table 1) [73]. The bulky biphenyl group in the P_1' position of the inhibitor has been selected as the S_1' subsite of NEP, which is known to tolerate large hydrophobic group. For this phosphinic inhibitor, Ki values of 1.2 nM and 5.6 nM have been reported toward NEP and APN, respectively.

4.3 *Thermolysin*

Thermolysin (TLN; EC 3.4.24.28), a thermostable bacterial protease isolated from *Bacillus thermoproteolyticus*, has been studied as the prototype of zinc-metallopeptidases at a time where no crystal structure was available for this class of proteases [122]. Crystallographic analysis of a number of TLN/inhibitor complexes has allowed an understanding of the binding mode of these inhibitors and allowed the mechanism of action of this protease to be determined [122]. These seminal studies have greatly inspired the development of NEP inhibitors, given the close structural relationship between TLN and NEP [123]. To examine further the structural relationships between these two peptidases, various phosphinic peptides were prepared. One of these compounds (**58**, Table 1) exhibits a Ki value of 26 nM toward thermolysin and 22 nM toward NEP [124].

Table 1 Selected examples of potent phosphinic peptide inhibitors of zinc-proteases

Zinc-protease name	Inhibitor structure	Ki (nM)
Neprilysin (NEP)	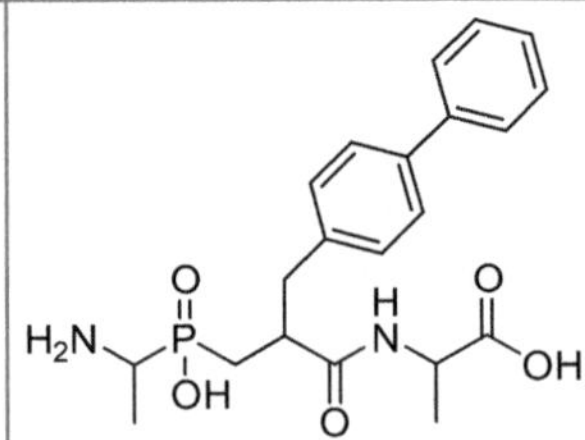 57	1.2 (5.6 nM for APN)
Thermolysin (TLN)	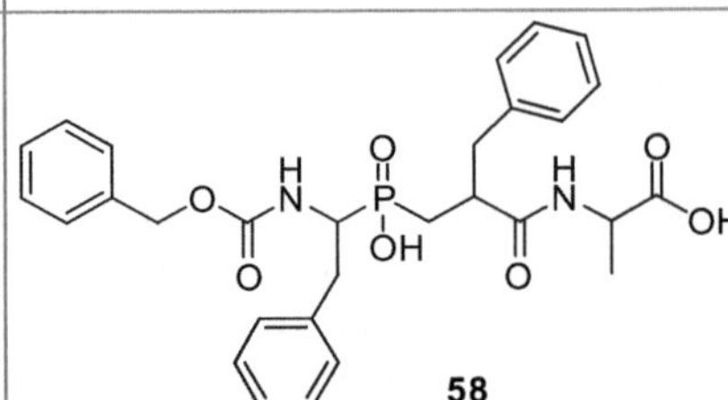 58	26
Leukotriene A4 hydrolase (LTA4H)	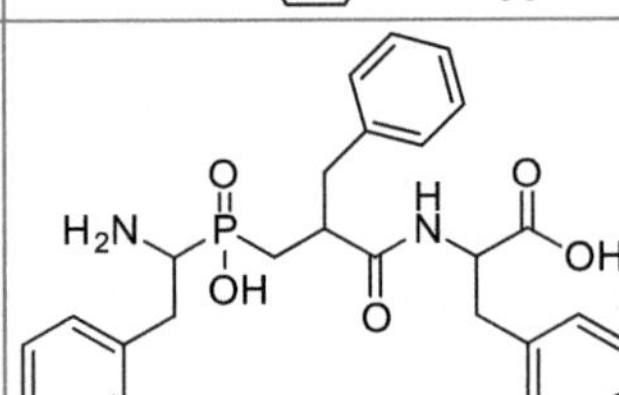 59	6.5
Endoplasmic reticulum (ER)-aminopeptidase (ERAP)	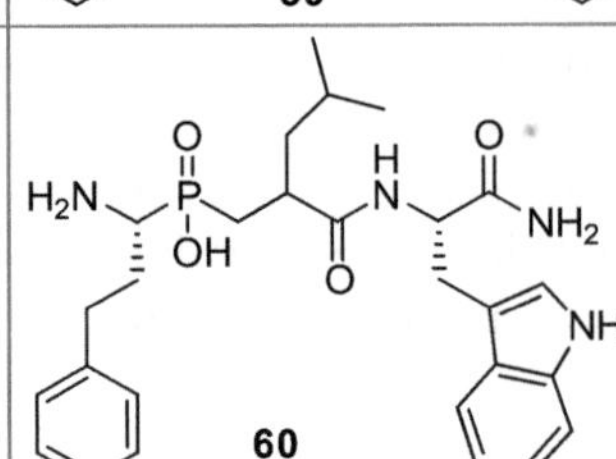 60	ERAP-1 = 33[a] ERAP-2 = 11[a]
Plasmodium falciparum amino-peptidases PfA-M17	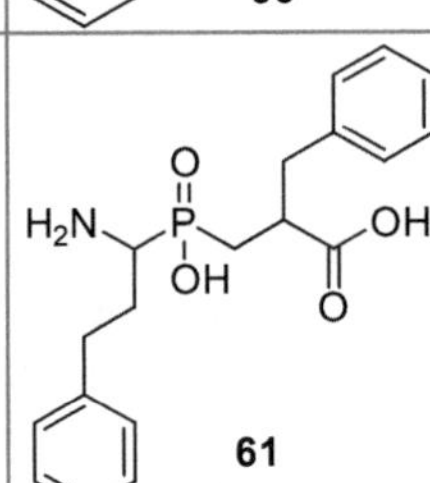 61	80

(continued)

Table 1 (continued)

Zinc-protease name	Inhibitor structure	Ki (nM)
Thrombin-activatable fibrinolysis inhibitor (TAFI)	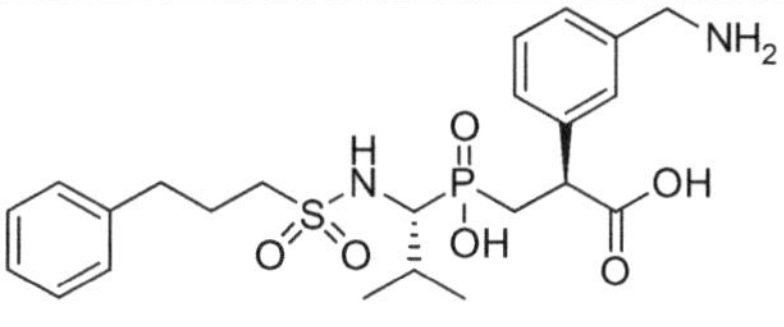 **62**	2
Glutamate carboxypeptidase II (GCPII)	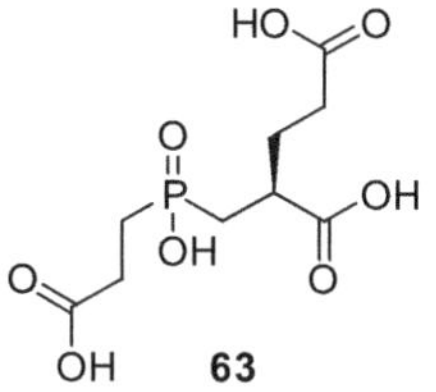 **63**	0.7

[a]Number referring to IC_{50} value

4.4 *Leukotriene A4 Hydrolase*

Leukotriene A4 hydrolase (LTA4H) is a bi-functional enzyme able to act as an aminopeptidase to remove selectively N-terminal arginine of various tripeptides and to catalyze the limiting step in the biosynthesis of LTB4 [125], a potent lipid chemoattractant involved in inflammation, immune responses, host defense against infection, platelet activating factor induced shock, and lipid homeostasis. The crystal structure of LTA4H has revealed that both activities take place in the same active site, which possesses the structural feature of zinc-aminopeptidases [126]. In line with this observation, compound **59** (Table 1) has been found to be a potent blocker of LTA4H with a Ki value of 6.5 nM [127].

4.5 *Endoplasmic Reticulum Aminopeptidases*

Endoplasmic reticulum (ER)-aminopeptidase ERAP-1 was first described as a leucine zinc-aminopeptidase, a growing family of mammalian zinc-containing aminopeptidases including membrane-bound placental leucine aminopeptidase (P-LAP), aminopeptidase A and N, and LTA4H. ERAP-1, and subsequently ERAP-2 were demonstrated to be final processing enzymes of the precursors of major histocompatibility complex (MHC) class I-presented antigenic peptides [128]. Their activity greatly affects the antigenic peptide repertoire presented to cytotoxic T lymphocytes and as a result can regulate cytotoxic cellular responses contributing to autoimmunity or immune evasion by viruses and cancer cells. Therefore, pharmacological regulation of their activity is a promising avenue for modulating the adaptive immune response, with possible applications in controlling autoimmunity, in boosting immune responses to pathogens, and in cancer immunotherapy. Potent inhibitors of ERAP-1 and ERAP-2 have recently been reported, exploiting phosphinic peptide chemistry. Compound **60** (Table 1) exhibits IC_{50} values of 33 and 11 nM towards ERAP-1 and ERAP-2, respectively, using a synthetic substrate in competition assays. These inhibitors were effective in cell presentation assays, inducing cell-surface antigen presentation and enhance cytotoxic T-cell responses [129].

4.6 Plasmodium Falciparum Aminopeptidases PfA-M17

The human malaria parasite *Plasmodium falciparum* expresses two neutral zinc-aminopeptidases, PfA-M1 and PfA-M17, which function in regulating the intracellular pool of amino acids required for parasite growth and development inside the red blood cell. These enzymes are essential for parasite viability and are validated therapeutic targets. As a result of the rapid emergence of drug-resistant parasites, blocking these peptidases with potent inhibitors has been proposed as a potential new antimalarial therapy. The simple phosphinic dipeptide **61** (Table 1) was reported to inhibit PfA-M17 with a Ki value of 80 nM [130].

4.7 Thrombin-Activatable Fibrinolysis Inhibitor

Under normal conditions, there is a balance between clot formation and clot dissolution. Thrombin-activatable fibrinolysis inhibitor (TAFI) is a zinc-metallocarboxypeptidase (MCP) which contributes to this balance by removing the C-terminal arginine and lysine residues from fibrin, thus reducing the formation of new plasmin and stabilizing the clot. TAFI is transformed through removal of its prodomain by thrombin-thrombomodulin into TAFIa, which is intrinsically unstable and has a short half-life in vivo. In vivo experiments indicate that selective inhibitors of TAFIa would be useful in the treatment of heart attacks and a useful adjunct to existing thrombolytic therapies [131]. BX 528 was reported as a potent inhibitor of TAFIa (2 nM) and has almost no measurable effect on carboxypeptidase N [132]. Compound **62** (Table 1) was designed to mimic the tripeptide Phe-Val-Lys. Given the rapid degradation of TAFIa in solution (also known as plasma carboxypeptidase B), the homologous protein porcine pancreatic carboxypeptidase B (pp-CpB) was used for crystallography studies to provide details on the binding mode interaction of this inhibitor [133].

4.8 Glutamate Carboxypeptidase II

The glutamate carboxypeptidase II (GCPII, E.C. 3.17.21), a zinc-dependent metallopeptidase, is characterized by the presence of two zinc ions in the enzyme active site, as compared to the aforementioned zinc-metallopeptidases which contain only a single zinc ion in their active site. Within the nervous system, GCPII hydrolyzes the natural substrate, *N*-acetyl-aspartyl-glutamate (NAAG), between aspartic and glutamic residues, participating directly in signal transmission via neural pathways [134]. As NAAG hydrolysis by GCPII in the nervous system leads to the increase in extracellular glutamate levels, the enzyme represents a therapeutic target for treatment of pathologies associated with dysregulated glutamatergic transmission. Highly selective and potent GCPII inhibitors were reported in the past and these showed efficacy in a variety of experimental models of neurological disorders, including neuropathic and inflammatory pain, stroke, diabetic neuropathy, amyotrophic lateral sclerosis, and schizophrenia. Moreover, GCPII is an excellent target for prostate cancer imaging and therapy because of its membrane localization and highly upregulated expression in prostate tumors and metastases [135]. Several reports have demonstrated the feasibility of imaging of GCPII-positive cells in experimental models of prostate cancer in vitro and in vivo, using low-molecular-weight GCPII ligands. The phosphinic analogue **63** (Table 1) of the Glu-Glu dipeptide exhibits a Ki value of 0.7 nM toward GCPII [136].

5 Insights from Crystal Structures

5.1 First Crystal Structure of a Zinc-Metalloprotease/ Phosphinic Inhibitor Complex

Astacin, a digestive zinc-endoprotease from the crayfish *Astacus astascus* L, is the prototype for the human astacin-like metalloproteases comprising seven proteases, one bone morphogenetic protein-1 (BMP-1), three tolloid-like proteinases (mTLD, mTll1 and mTll2), two meprins (α and β), and ovastacin [137]. The crystal structure of astacin in complex with a phosphinic peptide was the first 3D-stucture reported for this inhibitor family with a non-bacterial protease (PDB code, 1QJI) [138]. Interestingly, for this protease, the size of the inhibitor turned out to be a critical factor. Previous attempts to develop potent inhibitors harboring other zinc-chelating groups failed. Even for phosphinic peptides, only compounds comprising at least seven residues exhibit Ki values in the nanomolar range, while shorter inhibitors are much less potent [139]. As revealed by the X-ray structure (Fig. 6, left), the phosphinic pentapeptide (Z-Pro-Lys-Phe(PO_2-CH_2)Ala-Pro-OMe) is aligned into the active site, occupying the whole active site cleft with the phosphoryl group sitting above the zinc ion. The inhibitor backbone engages several hydrogen bonds with a β-strand of the protein active site (Fig. 6, right). The distances between the

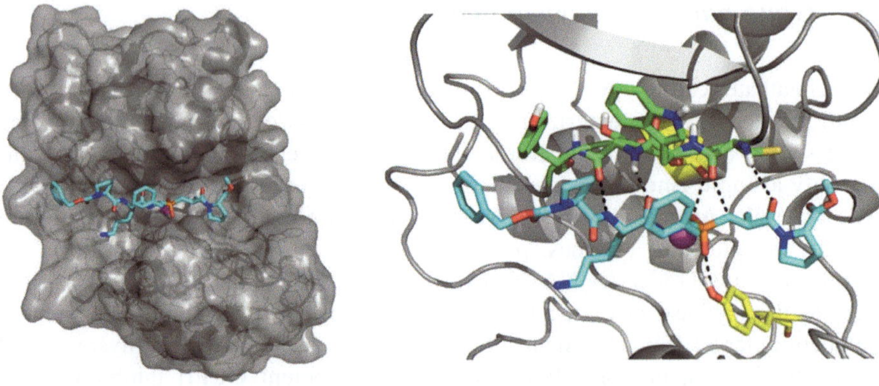

Fig. 6 *Left*: surface representation of astacin, with the inhibitor represented in ball and stick (*cyan* and *color coded* for atom); *right*: detail of the interaction between protein (*green* and *yellow*) and inhibitor (*cyan*); zinc (*purple sphere*); astacin backbone is represented as *gray cartoon*

two phosphoryl oxygens and the zinc ion are 1.8 and 2.7 Å, respectively (Fig. 7). Moreover, the oxygen atoms of the phosphoryl group also interact, respectively, with a glutamic and a tyrosine residue (yellow stick, Fig. 6). Inspection of this complex does not reveal the occurrence of other specific contacts between inhibitor and protein atoms. This feature may explain why the length of the inhibitor, by providing several hydrogen bond opportunities, is critical for inhibitor affinity.

The interaction between the tyrosine residue and the phosphoryl group was not obvious to predict, as in the free enzyme this residue occupied another position. Thus, between the free and bound states, this residue is moving to interact with the phosphoryl group. Finally, the carbon atom of the phosphinic methylene is a hydrogen bond distance to a protein carbonyl. Based on this result, it can be predicted that, with a peptide substrate, the position occupied by the methylene will be replaced by an amide group, placing this NH hydrogen in an optimal position to form a hydrogen bond with the same carbonyl group. The phosphoryl group of the inhibitor, which mimics a water-attacked carbonyl group of the scissile peptide bond, thus highlights an important role played by the glutamic and tyrosine residues in the catalytic process. From this structure, it can be suggested that the glutamic will act a proton shuttle or a general base and the tyrosine will stabilize the transition-state structure [140].

5.2 Detail of the Interaction Between the Active Site Zinc Ion and the Phosphoryl Group

Interactions seen in the astacin complex between two oxygen atoms borne by the phosphoryl group and the glutamic residue were expected to be observed in other zinc-metalloproteases. Indeed, this glutamic is contained in the highly conserved

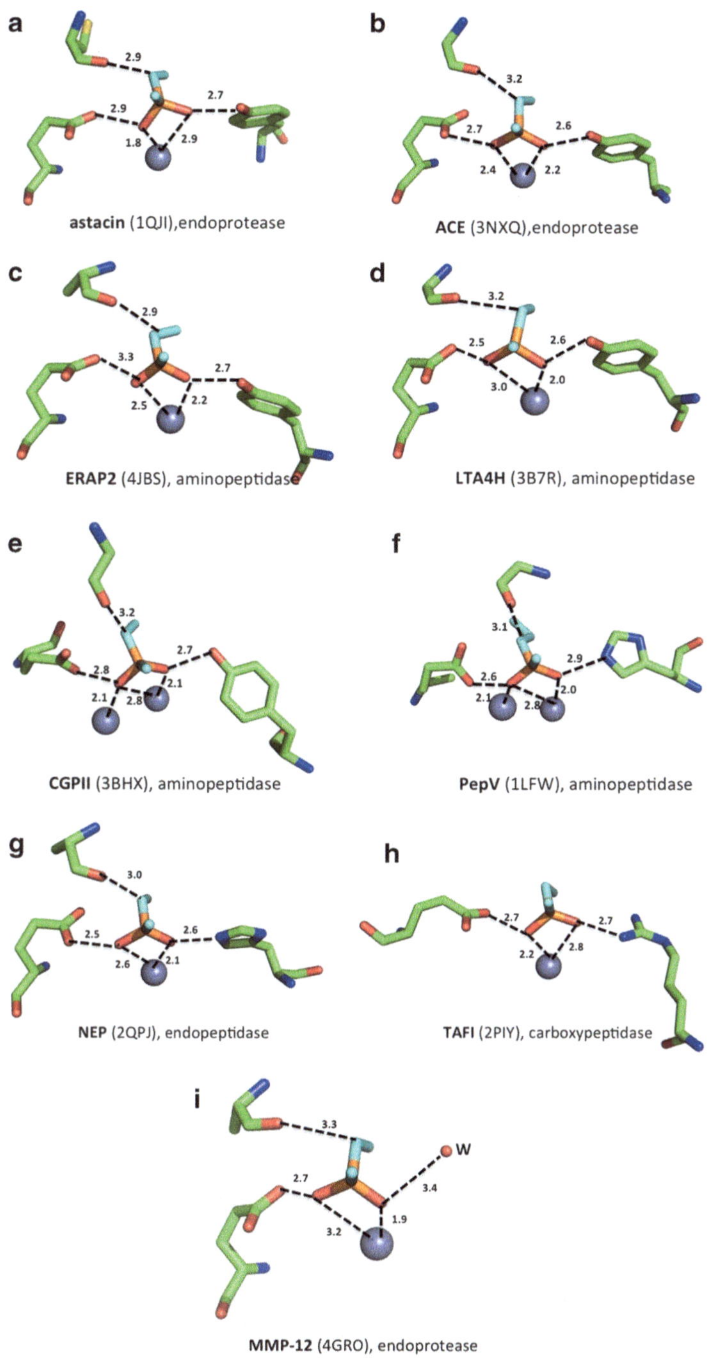

Fig. 7 Detail of the interactions between the phosphoryl group of phosphinic peptide inhibitors and active site residues in crystal structure of various zinc-proteases

signature sequence "His-Glu-X-X-His" present in most zinc-metalloproteases. Among the 111 crystal structures of zinc-metallopeptidases in complex with phosphinic peptide inhibitors archived in the protein data bank, we have selected a set of examples illustrating the content of structural information revealed by these complexes and the value of phosphinic peptide inhibitors to highlight key interactions.

Comparing astacin and ACE crystal structure (PDB code 3NXQ) in the complex with phosphinic peptide inhibitors at the active site zinc ion level (Fig. 7a, b) reveals marked conserved structural features. The interactions seen in astacin are conserved in ACE, with only small differences in distances between inhibitor and protein atoms. As suggested above, the interaction with the glutamate was expected, as this residue is conserved and present in the signature sequence of the zinc-protease family (HEXXH). In contrast, the interaction with the tyrosine residue was less obvious and less difficult to predict, as this residue is not present in a stretch of residues highly conserved in zinc-proteases. It is worth noting that these two proteins have extremely different protein folding, and astacin is cleaving proteins, while ACE hydrolyzes only peptides. Despite these differences, these two complexes reveal clearly the conservation of interactions which should be critical for the catalytic process, involving in this example the same residues in the zinc ion vicinity. The same interactions and residue nature are observed in the Pz-peptidase A endopeptidase, from the thermophilic bacterium *Geobacillus collagenovorans,* which hydrolyzes peptides containing a collagen-specific tripeptide sequence (−Gly-Pro-*X*-), but does not act on collagen proteins themselves (PDB code 3AHN) [141]. In the next two examples, two aminopeptidases have been selected, ERAP-2 and LTA4H (PDB code 4JBS and 3B7R, respectively), and here, again, the same structural features are observed, with the participation of a tyrosine residue and the presence of a carbonyl group in hydrogen bond distance to one proton of the phosphoryl methylene group (Fig. 7c, d). These interactions are also strictly conserved in two other zinc-aminopeptidases, e.g., (1) a malarial neutral aminopeptidase (PfA-M1, PDB code 3EBI) involved in the terminal stages of hemoglobin digestion, an essential step for the provision of amino acids used for parasite growth and development within the erythrocyte [142], and (2) aminopeptidase N from *E. Coli* (PDB code 2ZXG), a peptidase expressed in a broad spectrum of species (bacteria, plants, insects and mammals) in which it plays an important role in the final digestion of peptides. This peptidase is also widely distributed in mammalian tissues, including the central nervous system, and is particularly abundant in the kidney, the intestine, and the lung [142, 143]. As this enzyme is involved in the catabolism of enkephalins in association with neprilysin, mixed inhibitors have been developed to sustain nociception mediated by enkephalin in brain and peripheral organs.

Some zinc-exopeptidases are characterized by the presence of two zinc ions in their active sites. Protein folding of these exopeptidases is extremely different from the one observed in the aforementioned mono-zinc proteases. Despite such a major difference in the overall protein structure, in the GCPII carboxypeptidase (Fig. 7e, PDB code 3BHX) the structural elements mentioned above are present [136]. Thus, for GCPII, the use of phosphinic inhibitor helps to uncover relationships between these two-zinc exopeptidases and more classical mono zinc-proteases. The presence of similar structural features between these enzymes argues for a conserved catalytic mechanism in these enzymes, involving the same nature of residues.

PepV from *lactobacillis delbrueckii*, a dinuclear zinc peptidase, has been characterized as an unspecific amino dipeptidase. Crystal structure of PepV in complex with a phosphinic dipeptide (PDB code 1LFW) reveals the positioning of the phosphinic moiety in the active site of this peptidase [144]. As shown in Fig. 7f, the topology of the glutamate residue and of the carbonyl group, interacting with the phosphoryl methylene group, is equivalent to that seen in GCPII; however, in PepV a histidine is replacing the tyrosine of GCPII. The involvement of a histidine in the transition-state stabilization, instead of a tyrosine, has been observed in thermolysin (PDB code 1OS0) and NEP (PDB code 2QPJ) (Fig. 7g). In TAFI, a carboxypeptidase, the crystal structure reveals two unusual features – the presence of an arginine residue interacting with phosphoryl group and the absence of a protein carbonyl group near to the phosphoryl methylene group (PDB code 2PIY) [133]. MMP-12 crystal structure is another example in which the stabilization of the transition-state is unusual, as while a glutamate is interacting with one oxygen of the phosphoryl group, as expected, the other phosphoryl oxygen is not interacting with a protein residue, but instead with water (PDB code 4GR0) [145].

From the examples discussed above and the 111 structures of zinc-protease/phosphinic inhibitor complexes archived in the PDB, it follows that the two phosphoryl oxygens interact with the zinc ion in asymmetrical fashion in most cases. The precise orientation of the phosphoryl group with respect to the zinc ion is probably related to the interactions of these oxygens with other atoms from the protein active site. Among these, in all structures, one oxygen atom from a glutamate is systematically involved at hydrogen bond distance to one phosphoryl oxygen. This structural feature is related to the strict conservation of the HEXXH sequence signature in zinc-proteases. With the exception of MMP-12, the other oxygen atom of the phosphoryl group interacts with a protein atom, a proton from the hydroxyl group of a tyrosine residue in most cases. As mentioned for the astacin, the involved tyrosine is moving upon inhibitor binding; thus the use of phosphinic peptide inhibitor reveals the exact orientation taken by this tyrosine residue in stabilizing the transition-state structure.

The proximity between one oxygen atom of the phosphoryl group and an oxygen borne by a glutamic residue suggests, as mentioned above for the astactin complex, that a proton should be taken upon complex formation to avoid repulsive effect between two negatively charged residue (PO_2^- and COO^-). To address this issue explicitly, the MMP-12 interaction with the RXP470.1 selective inhibitor has been studied by isothermal titration calorimetry at several pH values and in different buffers [145]. This study has confirmed that in solution the formation of the enzyme/inhibitor complex involves a proton uptake from the buffer by the glutamate active site residue. This proton uptake explains why the affinity of the inhibitor increases at lower pH, passing at pH 7 from a Ki value of 0.4 nM to a Ki of 45 pM at pH 5.5. The more the glutamic MMP-12 active site residue population is protonated in solution at a given pH, the higher the inhibitor affinity. This dependence of the affinity on pH value can be of some importance, as in many pathological tissues the local pH is often shifted toward more acidic values. According to our results, this should lead to increase inhibitory potency. It will be of some importance to determine experimentally whether this expected proton uptake is occurring in other zinc-proteases with similar pH dependence. Indeed, the contribution of this proton uptake and the interaction between the protonated active site glutamic residue and the phosphoryl group may depend on the distance between these residues. This distance is determined by the exact positioning of the phosphoryl group with respect to the zinc ion and the influence of other protein atoms, as with the proton of the tyrosine hydroxyl group.

6 In Vivo Studies

As far as the targeting of zinc proteases in vivo is concerned, it is important to underline that most of the known zinc proteases are expressed outside the cell, either in a membrane-bound form located at the membrane surface, or as a soluble form that can interact with the different proteins of the extracellular matrix or with a receptor. Thus, as compared to targets located inside the cell, zinc proteases are rather accessible, as their targeting does not require crossing of the cell membrane. Because of the presence of a strong negative charge borne by the phosphoryl group at neutral pH, such compounds cannot cross the intestinal barrier, thus limiting their oral administration. In the case of the fosinoprilat ACE inhibitor, oral formulation has relied on a non-permanent masking of the phosphoryl charge by introducing an ester group on the phosphoryl (Fosinopril, Fig. 8). Under this ester form, intestinal crossing was possible and the fosinoprilat was subsequently liberated in plasma through the action of an esterase [146]. This lack of oral availability can be overcome by intraperitoneal (i.p.) or intravenous (i.v.) administration of inhibitor to animals. Despite the presence of peptide bonds in phosphinic peptide structure, several in vivo studies have revealed the remarkable stability of this class of compounds [110, 147–151]. As no major metabolites were detected in these

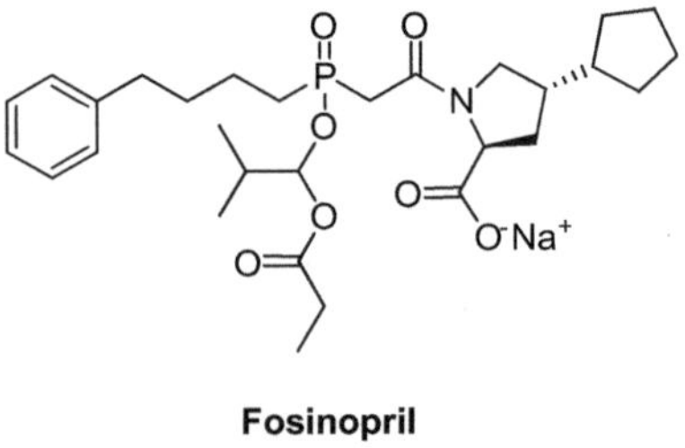

Fosinopril

Fig. 8 Structure of Fosinopril

studies, this also suggests that these compounds are protected from modifications by cytochrome P450 enzymes of the liver. This remark particularly holds for non-polar phosphinic peptides, which are mainly eliminated by hepato-biliary excretion. Possible explanations for the unexpected absence of rapid metabolism of phosphinic peptides under physiological conditions could be the strong negative charge of the phosphinic group. This negative charge may prevent both the interaction of these pseudo-peptides with nonspecific peptidases and their uptake by hepatocytes. The in vivo stability observed for some phosphinic peptide inhibitors made possible their use in probing the function of their targets in animal models. Selective ACE inhibitors have been used to probe the function of the N- and C-domain of somatic ACE towards the hydrolysis of their natural substrates. This study has suggested the occurrence of some interactions between these two domains of ACE and the impact of these interactions on domain selectivity in cleaving substrates [110]. Exploiting the in vivo stability of phosphinic peptides, a radiolabeled broad-spectrum inhibitor of MMPs has been prepared and used to detect by radioimaging the presence of MMP active forms in tumors [148]. The same inhibitor has also been evaluated in a rat model of liver ischemia, a model in which most hydroxamate inhibitors turned out to induce strong hepatotoxicity. Hepatic ischemia, occurring during liver transplantation or cardiogenic shock, induces the expression of several MMPs, an event suspected to be associated with increased necrosis of hepatic cells. In agreement with this proposal, treatment with a broad-spectrum phosphinic inhibitor of MMPs was observed to reduce injury triggered by ischemia in rat liver [149, 152]. The development of a highly selective inhibitor of MMP-12 has made it possible to test this compound in a mice model of atherosclerosis and viral infection. In atherosclerosis, the selective inhibition of MMP-12 was shown to block fully plaque growth and rupture [150]. Based on these observations, selective MMP-12 inhibitors are currently being developed to predict vulnerable plaques in patients by SPECT or PET imaging. In viral infection, inhibition of MMP-12 was observed in a mice model to increase the blood level of interferon-α, leading to reduce viral replication [151]. All these studies demonstrate that phosphinic peptides are useful tools for performing pharmacological studies and testing functional hypotheses. Further efforts may lead in the near future to the delivery of inhibitors for medical applications, a long time period after the pioneering research on the fosinoprilat drug.

References

1. Turk B (2006) Nat Rev Drug Discov 5:785–799
2. Puente XS, Sanchez LM, Overall CM, Lopez-Otin C (2003) Nat Rev Genet 4:544–558
3. Lopez-Otin C, Matrisian LM (2007) Nat Rev Cancer 7:800–808
4. Zaman MA, Oparil S, Calhoun DA (2002) Nat Rev Drug Discov 1:621–636
5. Peyman A, Budt KH, Spanig J, Ruppert D (1993) Angew Chem Int Ed Engl 32:1720–1722

6. Sheng XC, Casarez A, Cai R, Clarke MO, Chen X, Cho A, Delaney Iv WE, Doerffler E, Ji M, Mertzman M, Pakdaman R, Pyun HJ, Rowe T, Wu Q, Xu J, Kim CU (2012) Bioorg Med Chem Lett 22:1394–1396
7. Ondetti MA, Condon ME, Reid J, Sabo EF, Cheung HS, Cushman DW (1979) Biochemistry 18:1427–1430
8. Holmes MA, Matthews BW (1981) Biochemistry 20:6912–6920
9. Petrillo EW Jr, Ondetti MA (1982) Med Res Rev 2:1–41
10. Devel L, Czarny B, Beau F, Georgiadis D, Stura E, Dive Vincent V (2010) Biochimie 92:1501–1508
11. Christianson DW, Lipscomb WN (1988) J Am Chem Soc 110:5560–5565
12. Holden HM, Tronrud DE, Monzingo AF, Weaver LH, Matthews BW (1987) Biochemistry 26:8542–8553
13. Hanson JE, Kaplan AP, Bartlett PA (1989) Biochemistry 28:6294–6305
14. Mookhtiar KA, Marlowe CK, Bartlett PA, Van Wart HE (1987) Biochemistry 26:1962–1965
15. Dive V, Yiotakis A, Nicolaou A, Toma F (1990) Eur J Biochem 191:685–693
16. Yiotakis A, Lecoq A, Nicolaou A, Labadie J, Dive V (1994) Biochem J 303:323–327
17. Merz KM Jr, Kollman PA (1989) J Am Chem Soc 111:5649–5658
18. Morgan BP, Scholtz JM, Ballinger MD, Zipkin ID, Bartlett PA (1991) J Am Chem Soc 113:297–307
19. Yiotakis A, Georgiadis D, Matziari M, Makaritis A, Dive V (2004) Curr Org Chem 8:1135–1158
20. Mucha A (2012) Molecules 17:13530–13568
21. Thottathil JK, Ryono DE, Przybyla CA, Moniot JL, Neubeck R (1984) Tetrahedron Lett 25:4741–4744
22. McKittrick BA, Stamford AW, Weng X, Ma K, Chackalamannil S, Czarniecki M, Cleven RM, Fawzi AB (1996) Bioorg Med Chem Lett 6:1629–1634
23. Miller DJ, Hammond SM, Anderluzzi D, Bugg TDH (1998) J Chem Soc. Perkin Trans 1:131–142
24. Baylis EK, Campbell CD, Dingwall JG (1984) J Chem Soc. Perkin Trans 1:2845–2853
25. Cowart M, Kowaluk EA, Kohlhaas KL, Alexander KM, Kerwin JF Jr (1996) Bioorg Med Chem Lett 6:999–1002
26. Lecoq A, Yiotakis A, Dive V (1994) Synth Commun 24:2877–2882
27. Khomutov AR, Osipova TI, Khurs EN, Alferov KV, Khomutov RM (1996) Russ Chem Bull 45:1961–1964
28. Belyankin AV, Khomutov AR, Zhukov YN, Kartasheva ON, Khomutov RM (1997) Russ Chem Bull 46:133–136
29. Zhukov YN, Khomutov AR, Osipova TI, Khomutov RM (1999) Russ Chem Bull 48:1348–1351
30. Alferov KV, Zhukov YN, Khurs EN, Osipova TI, Khomutov RM (2001) Russ Chem Bull 50:316–318
31. Liboska R, Pícha J, Hančlová I, Buděšínský M, Šanda M, Jiráček J (2008) Tetrahedron Lett 49:5629–5631
32. Jiao XJ, Verbruggen C, Borloo M, Bollaert W, De Groot A, Dommisse R, Haemers A (1994) Synthesis 23–24
33. Li S, Whitehead JK, Hammer RP (2007) J Org Chem 72:3116–3118
34. Olszewski TK, Boduszek B (2011) Synthesis 437–442
35. Borloo M, Jiao XY, Wojtowicz H, Rajan P, Verbruggen C, Augustyns K, Haemers A (1995) Synthesis 1074–1076
36. Boyd EA, Chan WC, Loh VM Jr (1996) Tetrahedron Lett 37:1647–1650
37. Hatam M, Martens J (1995) Synth Commun 25:2553–2559
38. Mores A, Matziari M, Beau F, Cuniasse P, Yiotakis A, Dive V (2008) J Med Chem 51:2216–2226

39. Yamagishi T, Mori JI, Haruki T, Yokomatsu T (2011) Tetrahedron: Asymmetry 22:1358–1363
40. Haruki T, Yamagishi T, Yokomatsu T (2007) Tetrahedron: Asymmetry 18:2886–2893
41. McCleery PP, Tuck B (1989) J Chem Soc. Perkin Trans 1:1319–1329
42. Yao Q, Yuan C (2013) J Org Chem 78:6962–6974
43. Cristau HJ, Coulombeau A, Genevois-Borella A, Pirat JL (2001) Tetrahedron Lett 42:4491–4494
44. Cristau HJ, Coulombeau A, Genevois-Borella A, Sanchez F, Pirat JL (2002) J Organomet Chem 643–644:381–391
45. Buchardt J, Meldal M (2000) J Chem Soc Perkin Trans 1:3306–3310
46. Dorff PH, Chiu G, Goldstein SW, Morgan BP (1998) Tetrahedron Lett 39:3375–3378
47. Manzenrieder F, Kessler H (2009) Adv Exp Med Biol 611:11–12
48. Vassiliou S, Tzouma E (2013) J Org Chem 78:10069–10076
49. Schoen WR, Parsons WH (1988) Tetrahedron Lett 29:5201–5204
50. Gurulingappa H, Buckhaults P, Kumar SK, Kinzler KW, Vogelstein B, Khan SR (2003) Tetrahedron Lett 44:1871–1873
51. Gurulingappa H, Buckhalts P, Kinzler KW, Vogelstein B, Khan SR (2004) Bioorg Med Chem Lett 14:3531–3533
52. Matziari M, Nasopoulou M, Yiotakis A (2006) Org Lett 8:2317–2319
53. Chen S, Coward JK (1996) Tetrahedron Lett 37:4335–4338
54. Matziari M, Yiotakis A (2005) Org Lett 7:4049–4052
55. Rozhko LF, Ragulin VV (2005) Amino Acids 29:139–143
56. Dmitriev ME, Ragulin VV (2010) Tetrahedron Lett 51:2613–2616
57. Dmitriev ME, Ragulin VV (2012) Tetrahedron Lett 53:1634–1636
58. Ross FC, Botting NP, Leeson PD (1996) Bioorg Med Chem Lett 6:2643–2646
59. Bertenshaw SR, Rogers RS, Stern MK, Norman BH, Moore WM, Jerome GM, Branson LM, McDonald JF, McMahon EG, Palomo MA (1993) J Med Chem 36:173–176
60. Štrancar K, Boniface A, Blanot D, Gobec S (2007) Arch Pharm (Weinheim) 340:127–134
61. Chackalamannil S, Chung S, Stamford AW, McKittrick BA, Wang Y, Tsai H, Cleven R, Fawzi A, Czarniecki M (1996) Bioorg Med Chem Lett 6:1257–1260
62. Reiter LA, Rizzi JP, Pandit J, Lasut MJ, McGahee SM, Parikh VD, Blake JF, Danley DE, Laird ER, Lopez-Anaya A, Lopresti-Morrow LL, Mansour MN, Martinelli GJ, Mitchell PG, Owens BS, Pauly TA, Reeves LM, Schulte GK, Yocum SA (1999) Bioorg Med Chem Lett 9:127–132
63. Goulet JL, Kinneary JF, Durette PL, Stein RL, Harrison RK, Izquierdo-Martin M, Kuo DW, Lin TY, Hagmann WK (1994) Bioorg Med Chem Lett 4:1221–1224
64. Georgiadis D, Dive V, Yiotakis A (2001) J Org Chem 66:6604–6610
65. Reiter LA, Jones BP (1997) J Org Chem 62:2808–2812
66. Krapcho J, Turk C, Cushman DW, Powell JR, DeForrest JM, Spitzmiller ER, Karanewsky DS, Duggan M, Rovnyak G, Schwartz J, Natarajan S, Godfrey JD, Ryono DE, Neubeck R, Atwal KS, Petrillo EW Jr (1988) J Med Chem 31:1148–1160
67. Yiotakis A, Lecoq A, Vassiliou S, Raynal I, Cuniasse P, Dive V (1994) J Med Chem 37:2713–2720
68. Kende AS, Dong HQ, Liu X, Ebetino FH (2002) Tetrahedron Lett 43:4973–4976
69. Campagne JM, Coste J, Guillou L, Heitz A, Jouin P (1993) Tetrahedron Lett 34:4181–4184
70. Chen H, Noble F, Coric P, Fournie-Zaluski MC, Roques BP (1998) Proc Natl Acad Sci U S A 95:12028–12033
71. Reiter LA, Mitchell PG, Martinelli GJ, Lopresti-Morrow LL, Yocum SA, Eskra JD (2003) Bioorg Med Chem Lett 13:2331–2336
72. Manzenrieder F, Frank AO, Huber T, Dorner-Ciossek C, Kessler H (2007) Bioorg Med Chem 15:4136–4143
73. Chen H, Noble F, Mothé A, Meudal H, Coric P, Danascimento S, Roques BP, George P, Fournié-Zaluski MC (2000) J Med Chem 43:1398–1408

74. Makaritis A, Georgiadis D, Dive V, Yiotakis A (2003) Chem Eur J 9:2079–2094
75. Matziari M, Bauer K, Dive V, Yiotakis A (2008) J Org Chem 73:8591–8593
76. Lloyd J, Schmidt JB, Hunt JT, Barrish JC, Little DK, Tymiak AA (1996) Bioorg Med Chem Lett 6:1323–1326
77. Yiotakis A, Vassiliou S, Jiráček J, Dive V (1996) J Org Chem 61:6601–6605
78. Jiracek J, Yiotakis A, Vincent B, Lecoq A, Nicolaou A, Checler F, Dive V (1995) J Biol Chem 270:21701–21706
79. Jirácek J, Yiotakis A, Vincent B, Checler F, Dive V (1996) J Biol Chem 271:19606–19611
80. Dive V, Cotton J, Yiotakis A, Michaud A, Vassiliou S, Jiracek J, Vazeux G, Chauvet MT, Cuniasse P, Corvol P (1999) Proc Natl Acad Sci U S A 96:4330–4335
81. Buchardt J, Ferreras M, Krog-Jensen C, Delaissé JM, Foged NT, Meldal M (1999) Chem Eur J 5:2877–2884
82. Buchardt J, Schiødt CB, Krog-Jensen C, Delaissé JM, Foged NT, Meldal M (2000) J Comb Chem 2:624–638
83. Christensen C, Groth T, Schiødt CB, Foged NT, Meldal M (2003) QSAR Comb Sci 22:737–744
84. Devel L, Rogakos V, David A, Makaritis A, Beau F, Cuniasse P, Yiotakis A, Dive V (2006) J Biol Chem 281:11152–11160
85. Fields GB, Lauer-Fields J, Brew K, Lauer-Fields J, Hammer RP, Li S, Whitehead JK (2007) J Am Chem Soc 129:10408–10417
86. Bhowmick M, Sappidi RR, Fields GB, Lepore SD (2011) Biopolymers 96:1–3
87. Bhowmick M, Fields GB (2012) Int J Pept Res Ther 18:335–339
88. Georgiadis D, Matziari M, Yiotakis A (2001) Tetrahedron 57:3471–3478
89. Nasopoulou M, Matziari M, Dive V, Yiotakis A (2006) J Org Chem 71:9525–9527
90. Vitharana D, France JE, Scarpetti D, Bonneville GW, Majer P, Tsukamoto T (2002) Tetrahedron: Asymmetry 13:1609–1614
91. Chen H, Noble F, Roques BP, Fournié-Zaluski MC (2001) J Med Chem 44:3523–3530
92. Jullien N, Makritis A, Georgiadis D, Beau F, Yiotakis A, Dive V (2010) J Med Chem 53:208–220
93. Matziari M, Dellis D, Dive V, Yiotakis A, Samios J (2010) J Phys Chem B 114:421–428
94. Yiotakis A, Makaritis A, Georgiadis D, Matziari M, Dive V (2002) Peptides 2002:330–331
95. Parsons WH, Patchett AA, Bull HG, Schoen WR, Taub D, Davidson J, Combs PL, Springer JP, Gadebusch H, Weissberger B, Valiant ME, Mellin TN, Busch RD (1988) J Med Chem 31:1772–1778
96. Liu X, Hu XE, Tian X, Mazur A, Ebetino FH (2002) J Organomet Chem 646:212–222
97. Cristau HJ, Pirat JL, Virieux D, Monbrun J, Ciptadi C, Bekro YA (2005) J Organomet Chem 690:2472–2481
98. Monbrun J, Dayde B, Cristau HJ, Volle JN, Virieux D, Pirat JL (2011) Tetrahedron 67:540–545
99. Yamagishi T, Ichikawa H, Haruki T, Yokomatsu T (2008) Org Lett 10:4347–4350
100. Georgiadis D, Matziari M, Vassiliou S, Dive V, Yiotakis A (1999) Tetrahedron 55:14635–14648
101. Matziari M, Georgiadis D, Dive V, Yiotakis A (2001) Org Lett 3:659–660
102. Matziari M, Beau F, Cuniasse P, Dive V, Yiotakis A (2004) J Med Chem 47:325–336
103. David A, Steer D, Bregant S, Devel L, Makaritis A, Beau F, Yiotakis A, Dive V (2007) Angew Chem Int Ed Engl 46:3275–3277
104. Bregant S, Huillet C, Devel L, Dabert-Gay AS, Beau F, Thai R, Czarny B, Yiotakis A, Dive V (2009) J Proteome Res 8:2484–2494
105. Aminoethylphosphinic acid derivatives (1988) EP282219,
106. Vassiliou S, Xeilari M, Yiotakis A, Grembecka J, Pawełczak M, Kafarski P, Mucha A (2007) Bioorg Med Chem 15:3187–3200
107. Rogakos V, Georgiadis D, Dive V, Yiotakis A (2009) Org Lett 11:4696–4699

108. Nasopoulou M, Georgiadis D, Matziari M, Dive V, Yiotakis A (2007) J Org Chem 72:7222–7228
109. Yamagishi T, Kinbara A, Okubo N, Sato S, Fukaya H (2012) Tetrahedron: Asymmetry 23:1633–1639
110. Georgiadis D, Beau F, Czarny B, Cotton J, Yiotakis A, Dive V (2003) Circ Res 93:148–154
111. Georgiadis D, Cuniasse P, Cotton J, Yiotakis A, Dive V (2004) Biochemistry 43:8048–8054
112. Demange L, Dugave C (2001) Tetrahedron Lett 42:6295–6297
113. Yamagishi T, Tashiro N, Yokomatsu T (2011) J Org Chem 76:5472–5476
114. Anthony CS, Masuyer G, Sturrock ED, Acharya KR (2012) Curr Med Chem 19:845–855
115. Dive V, Chang CF, Yiotakis A, Sturrock ED (2009) Curr Pharm Des 15:3606–3621
116. Redelinghuys P, Nchinda AT, Sturrock ED (2005) Ann N Y Acad Sci 1056:160–175
117. Anthony CS, Corradi HR, Schwager SLU, Redelinghuys P, Georgiadis D, Dive V, Acharya KR, Sturrock ED (2010) J Biol Chem 285:35685–35693
118. Corradi HR, Chitapi I, Sewell BT, Georgiadis D, Dive V, Sturrock ED, Acharya KR (2007) Biochemistry 46:5473–5478
119. Roques BP, Noble F, Dauge V, Fournie-Zaluski MC, Beaumont A (1993) Pharmacol Rev 45:87–146
120. Patey G, De La Baume S, Schwartz JC, Gros C, Roques B, Fournie-Zaluski MC, Soroca-Lucas E (1981) Science 212:1153–1155
121. Roques BP, Lucas-Soroca E, Chaillet P, Costentin J, Fournie-Zaluski MC (1983) Proc Natl Acad Sci U S A 80:3178–3182
122. Matthews BW (1988) Acc Chem Res 21:341–347
123. Tiraboschi G, Jullian N, Thery V, Antonczak S, Fournie-Zaluski MC, Roques BP (1999) Protein Eng 12:141–149
124. Selkti M, Tomas A, Gaucher JF, Prange T, Fournie-Zaluski MC, Chen H, Roques BP (2003) Acta Crystallogr D Biol Crystallogr 59:1200–1205
125. Samuelsson B, Dahlen SE, Lindgren JA, Rouzer CA, Serhan CN (1987) Science 237:1171–1176
126. Thunnissen MM, Nordlund P, Haeggstrom JZ (2001) Nat Struct Biol 8:131–135
127. Tholander F, Muroya A, Roques BP, Fournie-Zaluski MC, Thunnissen MM, Haeggstrom JZ (2008) Chem Biol 15:920–929
128. Tanioka T, Hattori A, Masuda S, Nomura Y, Nakayama H, Mizutani S, Tsujimoto M (2003) J Biol Chem 278:32275–32283
129. Zervoudi E, Saridakis E, Birtley JR, Seregin SS, Reeves E, Kokkala P, Aldhamen YA, Amalfitano A, Mavridis IM, James E, Georgiadis D, Stratikos E (2013) Proc Natl Acad Sci U S A 110:19890–19895
130. McGowan S, Oellig CA, Birru WA, Caradoc-Davies TT, Stack CM, Lowther J, Skinner-Adams T, Mucha A, Kafarski P, Grembecka J, Trenholme KR, Buckle AM, Gardiner DL, Dalton JP, Whisstock JC (2010) Proc Natl Acad Sci U S A 107:2449–2454
131. Wang YX, Zhao L, Nagashima M, Vincelette J, Sukovich D, Li W, Subramanyam B, Yuan S, Emayan K, Islam I, Hrvatin P, Bryant J, Light DR, Vergona R, Morser J, Buckman BO (2007) Thromb Haemost 97:45–53
132. Adler M, Bryant J, Buckman B, Islam I, Larsen B, Finster S, Kent L, May K, Mohan R, Yuan S, Whitlow M (2005) Biochemistry 44:9339–9347
133. Adler M, Buckman B, Bryant J, Chang Z, Chu K, Emayan K, Hrvatin P, Islam I, Morser J, Sukovich D, West C, Yuan S, Whitlow M (2008) Acta Crystallogr Sect D Biol Crystallogr 64:149–157
134. Zhou J, Neale JH, Pomper MG, Kozikowski AP (2005) Nat Rev Drug Discov 4:1015–1026
135. Bostwick DG, Pacelli A, Blute M, Roche P, Murphy GP (1998) Cancer 82:2256–2261
136. Barinka C, Hlouchova K, Rovenska M, Majer P, Dauter M, Hin N, Ko YS, Tsukamoto T, Slusher BS, Konvalinka J, Lubkowski J (2008) J Mol Biol 376:1438–1450
137. Gomis-Ruth FX, Trillo-Muyo S, Stocker W (2012) Biol Chem 393:1027–1041

138. Grams F, Dive V, Yiotakis A, Yiallouros I, Vassiliou S, Zwilling R, Bode W, Stocker W (1996) Nat Struct Biol 3:671–675
139. Yiallouros I, Vassiliou S, Yiotakis A, Zwilling R, Stocker W, Dive V (1998) Biochem J 331:375–379
140. Yiallouros I, Grosse Berkhoff E, Stocker W (2000) FEBS Lett 484:224–228
141. Kawasaki A, Nakano H, Hosokawa A, Nakatsu T, Kato H, Watanabe K (2010) J Biol Chem 285:34972–34980
142. McGowan S, Porter CJ, Lowther J, Stack CM, Golding SJ, Skinner-Adams TS, Trenholme KR, Teuscher F, Donnelly SM, Grembecka J, Mucha A, Kafarski P, Degori R, Buckle AM, Gardiner DL, Whisstock JC, Dalton JP (2009) Proc Natl Acad Sci U S A 106:2537–2542
143. Fournie-Zaluski MC, Poras H, Roques BP, Nakajima Y, Ito K, Yoshimoto T (2009) Acta Crystallogr D Biol Crystallogr 65:814–822
144. Jozic D, Bourenkow G, Bartunik H, Scholze H, Dive V, Henrich B, Huber R, Bode W, Maskos K (2002) Structure 10:1097–1106
145. Czarny B, Stura EA, Devel L, Vera L, Cassar-Lajeunesse E, Beau F, Calderone V, Fragai M, Luchinat C, Dive V (2013) J Med Chem 56:1149–1159
146. Morrison RA, Singhvi SM, Peterson AE, Pocetti DA, Migdalof BH (1990) Drug Metab Dispos 18:253–257
147. Junot C, Gonzales MF, Ezan E, Cotton J, Vazeux G, Michaud A, Azizi M, Vassiliou S, Yiotakis A, Corvol P, Dive V (2001) J Pharmacol Exp Ther 297:606–611
148. Dive V, Andarawewa KL, Boulay A, Matziari M, Beau F, Guerin E, Rousseau B, Yiotakis A, Rio MC (2005) Int J Cancer 113:775–781
149. Defamie V, Laurens M, Patrono D, Devel L, Brault A, Saint-Paul MC, Yiotakis A, Barbry P, Gugenheim J, Crenesse D, Dive V, Huet PM, Mari B (2008) Hepatology 47:177–185
150. Johnson JL, Devel L, Czarny B, George SJ, Jackson CL, Rogakos V, Beau F, Yiotakis A, Newby AC, Dive V (2011) Arterioscler Thromb Vasc Biol 31:528–535
151. Marchant DJ, Bellac CL, Moraes TJ, Wadsworth SJ, Dufour A, Butler GS, Bilawchuk LM, Hendry RG, Robertson AG, Cheung CT, Ng J, Ang L, Luo Z, Heilbron K, Norris MJ, Duan W, Bucyk T, Karpov A, Devel L, Georgiadis D, Hegele RG, Luo H, Granville DJ, Dive V, McManus BM, Overall CM (2014) Nat Med 20:493–502
152. Cursio R, Mari B, Louis K, Rostagno P, Saint-Paul MC, Giudicelli J, Bottero V, Anglard P, Yiotakis A, Dive V, Gugenheim J, Auberger P (2002) FASEB J 16:93–95

Top Curr Chem (2015) 360: 39–114
DOI: 10.1007/128_2014_566

Published online: 3 December 2014

Synthesis and Biological Applications of Phosphinates and Derivatives

David Virieux, Jean-Noël Volle, Norbert Bakalara, and Jean-Luc Pirat

Abstract This review first outlines general considerations on phosphinic acids and derivatives as bioisosteric groups. The next sections present key aspects of phosphinic acid-based molecules and include a brief description of the biological pathways involved for their activities. The synthetic aspects and the biological activities of such compounds reported in the literature between 2008 and 2013 are also described.

Keywords Bioisostere · Phosphinate · Phosphinic acid · Tetrahedral transition state

Contents

D. Virieux (✉), J.-N. Volle, and J.-L. Pirat (✉)
UMR 5253 - ICG Montpellier, Equipe AM2N, ENSCM, 8, rue de l'Ecole, Normale, 34296 Montpellier Cedex 5, France
e-mail: david.virieux@enscm.fr; jean-luc.pirat@enscm.fr

N. Bakalara
INSERM U-1051. Institut des neurosciences de Montpellier, 80 rue Augustin Fliche, 34091 Montpellier, France

Abbreviations

ABCN	1,1′-Azobis(cyclohexanecarbonitrile)
acac	Acetylacetonate
Ac	Acetyl
ADP	Adenosine diphosphate
AIBN	2,2′-Azobisisobutyronitrile
anhyd	Anhydrous
Ar	Aryl
ATP	Adenosine triphosphate
AZT	Azidothymidine
BHMT	Betaine-homocysteine methyltransferase
Bn	Benzyl
Boc	*tert*-Butoxycarbonyl
BP	Bis(phosphonate)
BSA	*N*,*O*-(Bistrimethylsilyl)acetamide

Bu	Butyl
Bz	Benzoyl
Cbz	Benzyloxycarbonyl
CNS	Central nervous system
d	Day(s)
DABCO	1,4-Diazabicyclo[2.2.2]octane
DAST	Diethylaminosulfur trifluoride
DBU	1,8-Diazabicyclo [5.4.0]undec-7-ene
DCC	*N,N*-Dicyclohexylcarbodiimide
DEAD	Diethyl azodicarboxylate
DIAD	Diisopropyl azodicarboxylate
DIEPA	Diisopropylethylamine
DMAP	4-(Dimethylamino)pyridine
DMF	Dimethylformamide
DMSO	Dimethyl sulfoxide
DOTA	1,4,7,10-Tetraazacyclododecane-1,4,7,10-tetraacetic acid
dppp	Diphenylphosphinopropane
EC_{50}	Half maximal effective concentration
EDC	1-Ethyl-3-(3-dimethylaminopropyl)carbodiimide
Et	Ethyl
FDA	Food and drug administration
Fmoc	9-Fluorenylmethoxycarbonyl
GABA	γ-Aminobutyric acid
GPCRs	G-Protein-coupled-receptors
GR	Glutamate racemase
h	Hour(s)
HATU	1-[Bis(dimethylamino)methylene]-1*H*-1,2,3-triazolo[4,5-*b*]-pyridinium 3-oxid hexafluorophosphate
HBTU	*N,N,N′,N′*-Tetramethyl-*O*-(1*H*-benzotriazol-1-yl) uroniumhexafluorophosphate
HCV	Hepatitis C virus
HIV	Human Immunodeficiency Virus
HMDS	Hexamethyldisilazane
HMPA	Hexamethylphosphoric triamide
HOBt	Hydroxybenzotriazole
i-Pr	Isopropyl
LDA	Lithium diisopropylamide
LogP	Partition-coefficient
LTMP	Lithium 2,2,6,6-tetramethylpiperidide
m-CPBA	*m*-Chloroperoxybenzoic acid
MDCK	Madin-Darby canine kidney
Me	Methyl
mGluR	Metabotropic glutamate receptors
MIC	Minimum inhibitory concentration

mol	Mole(s)
NA	Neuraminidase
NBS	*N*-Bromosuccinimide
NNM	*N*-Methyl morpholine
NNRTI	Non-nucleoside reverse transcriptase inhibitor
Np	Naphthyl
NRTI	Nucleoside reverse transcriptase inhibitor
PALA	*N*-Phosphonoacetyl-L-aspartate
Ph	Phenyl
Pr	Propyl
PTM	Post-translational modification
rt	Room temperature
RT	Reverse transcriptase
s-Bu	*sec*-butyl
TBDMS	*tert*-Butyldimethylsilyl
t-Bu	*tert*-Butyl
TEBAC	Triethylbenzylammonium chloride
THF	Tetrahydrofuran
TLESR	Transient lower oesophageal sphincter relaxation
TMG	Tetramethylguanidine
TMS	Trimethylsilyl
TMS-N_3	Trimethylsilyl azide
TS	Transition state
UTP	Uridine 5′-triphosphate

1 Introduction

Phosphates play significant roles in nature and they pervade the living world [1, 2]. By contrast, naturally reduced phosphorus derivatives (i.e. phosphonates or phosphinates) are rarely encountered in living organisms and they still represent an underused functional group for the development of bioactive compounds [3, 4]. The last few years have seen the revival of the work, originally started early in the 1970s, relating to the growing information on the genomic and metabolic pathways of bacteria [5, 6]. Interestingly, the few synthetic or naturally occurring molecules which contain a phosphinate function play key roles in many different areas of life science. This review first outlines general considerations on phosphinic acids and derivatives as bioisosteric groups. The next sections present key aspects of phosphinic acid-based molecules and include a brief description of the biological pathways involved for their activities. Phosphinopeptides and related metalloprotease inhibitors are presented in a dedicated chapter by Vincent Dive and Athanasios Yiotakis. The synthetic aspects and the biological activities of such compounds reported in the literature between 2008 and 2013 are also described.

2 Phosphinic Acids and Derivatives as Bioisosteric Groups

In order to improve efficacy and selectivity or to modify the sensitivity to metabolic transformations by alteration of chemo and physical properties, the design of new medicines or crop protection chemicals benefited from the concept of bioisosterism [7]. In life science, bioisosteres are substituents or functional groups which induce similar biological response. By contrast, a bioisostere can have different shape or physical/chemical properties when compared to the original group or substituent. Bioisosterism is a useful strategy to address a number of aspects associated with the development of drug candidates such as improving potency or selectivity, altering physical properties (pK_a, solubility), changing metabolic transformation and for industrial purposes relating to acquiring novel intellectual property.

2.1 *Phosphinates as Phosphate Bioisosteres*

Phosphinic acids are probably, with the phosphonic acids, the simplest phosphate bioisosteres [8]. Because of their tetrahedral geometry, phosphinic acids remain quite similar in shape and are approximately isosteric with phosphates. The main difference is the presence of two phosphorus–carbon bonds (phosphinic acids) or one phosphorus–carbon bond and a phosphorus–hydrogen bond (*H*-phosphinic acids). One intrinsic character of the P–C bond is its chemical and enzymatic stability when compared to P–O or P–N bonds [9]. As a consequence, phosphinates (as well as phosphonates) are often used when hydrolysis becomes a bottleneck for the activity. The number of acidic functions but also pK_a values are different between phosphate and phosphinic acids [10–12]. Indeed, the replacement of oxygen atoms by hydrogens in hypophosphorous acid slightly changed the pK_a when compared to the first acidity of phosphoric acid (Fig. 1). However, the presence of an alkyl group directly bonded to the phosphorus results in an increase

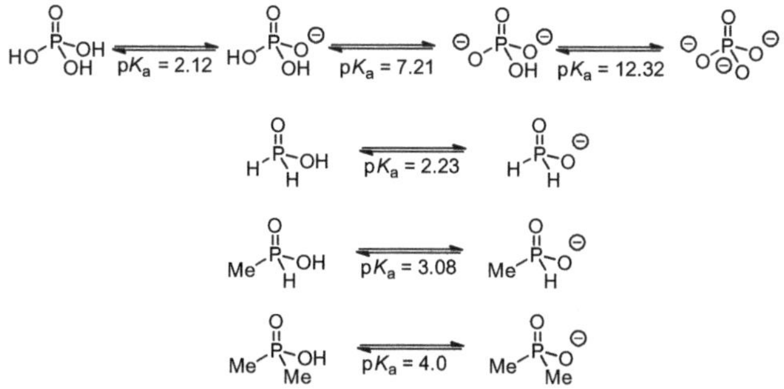

Fig. 1 pK_a of phosphinic acids compared to phosphoric acid

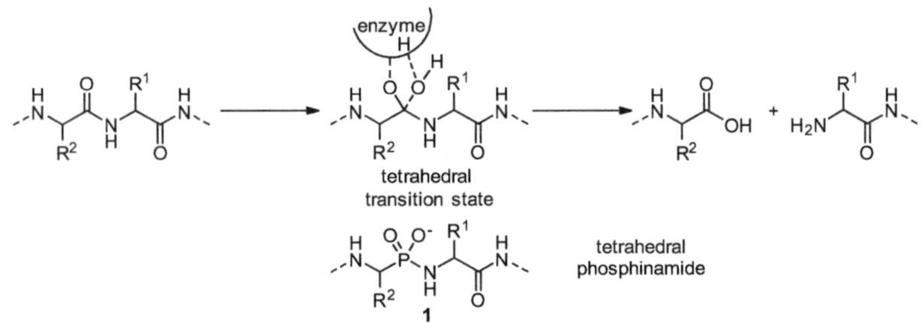

Fig. 2 Analogy between TS of amide bond cleavage and phosphinamide

of the pK_a from 2.23 to 3.08 and then 4.0 in connection with the positive inductive effect of the alkyl group.

2.2 *Phosphinic Acids as Bioisosteres of Carboxylic Acids and Their Derivatives*

As carboxylic acids, phosphinic acids are monoacidic, but replacement of a carboxylic group by a phosphinic one modulates down the pK_a of the resulting compounds by one to three units. Moreover, owing to the Csp^2 nature of carboxylic acids, their trigonal shape strongly differs from the tetrahedral geometry observed for phosphinic acids. Consequently, phosphinic acids are more polar than their carboxylic counterpart, lowering the partition coefficient (Log*P*), and are often considered under their anionic form in living organisms. Another structural feature of phosphinic acids is the facility for cation complexation [13, 14].

Phosphinic acids are often cited for the inhibition of proteolytic enzymes (i.e. protease, peptidase). It has been suggested that they interfere inside the active site of the enzyme directly with the amino acid residues or cofactors which bind the natural substrate. The pivotal step is the determination of the transition state structure of substrate and then the elaboration of structurally similar inhibitors [15]. Peptidases or proteases cleave the C–N peptidic bond via the formation of a tetrahedral transition state (TS). It is commonly admitted that phosphinamides **1** are structurally close to the TS in terms of shape and electronic repartition (Fig. 2).

2.3 *Other Bioisosteric Relationships*

Although the biological properties of phosphinic acids as carboxylic or phosphoric acids are well reported, their uses as bioisosteres of other functionalities remain relatively unexplored. Some year ago, Hall highlighted the analogy between phosphine oxide and an alcohol [16, 17]. The phosphine oxide pharmacophore **3** was

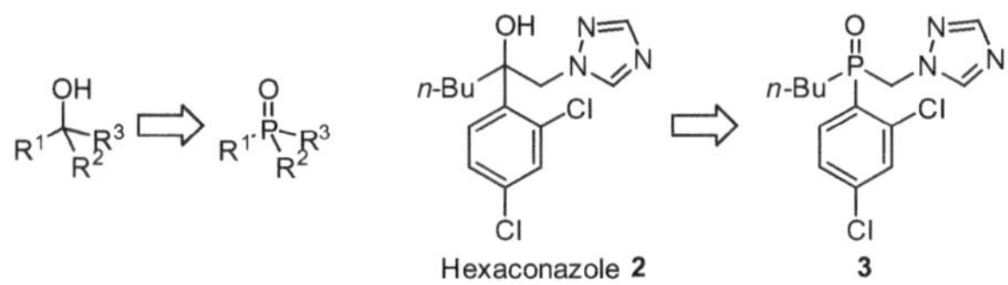

Fig. 3 Hexaconazole **1** and its phosphine oxide analogue **2**

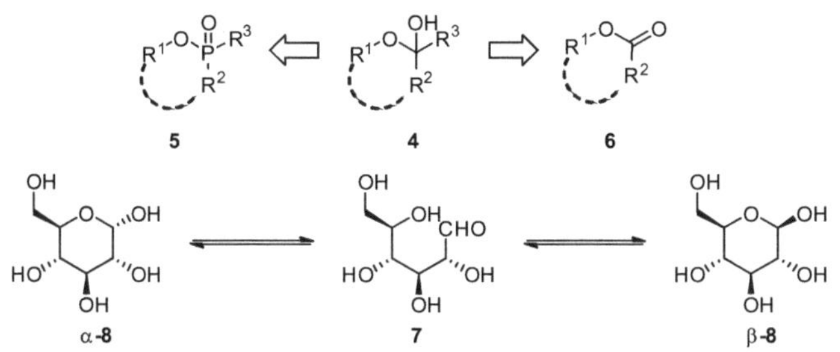

Fig. 4 Phosphinolactone as lactol bioisostere

compared to the parent compound, hexaconazole **2**. Interestingly, if the fungicide activity of **3** remains lower than **2**, this approach demonstrated the potential of the P=O bond to mimic a tertiary alcohol function (Fig. 3).

In a broadened extension of this concept, Pirat et al. exemplified this analogy to lactol **4** and phosphinolactone **5** functions (see Sect. 4.5). The cyclic hemiacetals are widely represented in biologically active compounds, particularly in sugar derivatives. The lactol group **7** is in equilibrium with the corresponding hydroxyaldehyde (**α-8** and **β-8**) or hydroxyketone and this dynamic equilibrium is sometimes directly connected to the activity. However, in the development of new drug, it becomes important to suppress this possibility of interconversion. Lactol **4** is then often replaced by robust lactone **6** (Fig. 4). The sp^2 hybridized ester group of lactone ring **6** possesses a partial analogy with lactol **4** and therefore it can be considered only as an imperfect bioisostere. From a structural point of view, the tetrahedral geometry of the phosphinolactone group **5** can be directly addressed as a mime of the hemiketal function.

3 Phosphinic Acids as Antivirals

3.1 Reverse Transcriptase Inhibitors of HIV

The emergence of the Human Immunodeficiency Virus (HIV) in the early 1980s pushed the discovery of new drugs and new modes of action to cure virus-induced illnesses. Early anti-HIV drugs focused on the inhibition of the reverse transcriptase

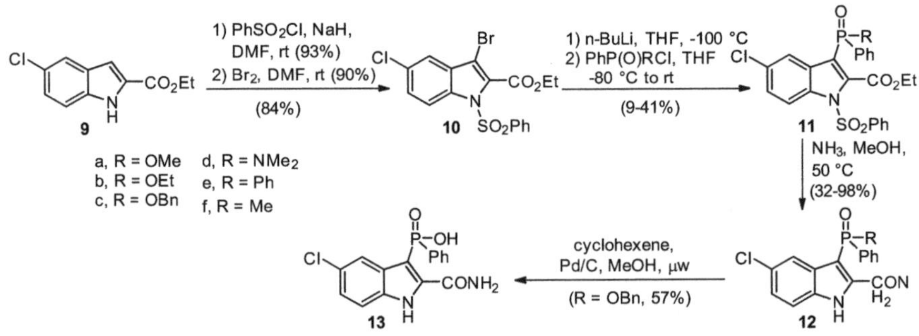

Scheme 1 Synthesis of phosphinates **12** and **13** as NNRTIs

(RT). This viral protein is the target of narrow spectrum but rather specific inhibitors with reduced toxicity profiles. Nucleoside reverse transcriptase inhibitors (NRTIs) are nucleoside analogues competing with the natural substrate of RT. Three consecutive phosphorylations are required for the activity of NRTIs, but the first is generally considered as the most difficult and prompted the development of mono-phosphorylated analogues (i.e. nucleotide derivatives). If phosphate and phosphonate derivatives are well represented, the phosphinates are still rarely encountered as phosphate surrogate in this series.

In parallel, non-nucleoside reverse transcriptase inhibitors (NNRTIs) inhibit RT by a non-competitive binding to an allosteric site closely located to the catalytic site [18]. The short distances separating the allosteric and the binding sites suggest that NNRTIs alter the function of RT and directly disturb the interactions between the nucleotide natural substrates and the active site.

First generation of NNRTI was impeded by the rapid appearance of resistance when they were used as monotherapy. Y181C and K103N are probably the most concerning virus mutants. Fortunately, in spite of conformational changes induced by amino acid modifications of the mutant-RT, the recently developed inhibitors targeted interactions with the conserved residues (such as tryptophan W229) of RT. They also introduced increased conformational flexibility, allowing the adaptation of the inhibitor to the modification of spatial arrangement induced by the mutations.

Searching for a new inhibitor of RT, Idenix developed compounds **12** and **13** (Scheme 1) [19]. 2-Ethoxycarbonyl indole **9** was protected at the nitrogen atom by deprotonation using sodium hydride followed by the reaction of phenylsulfonyl chloride and was consecutively brominated at position 3, affording the bromoindole **10** in 84%. The racemic phosphine oxides or phosphinates **11a–f** were obtained by lithiation from *n*-butyllithium and the reaction of the appropriate phenylphosphinyl chloride or phenylphosphonochloridate in yields of 9–41%. Finally, the *N*-protecting group was removed and the esters were transformed into carboxamide **12a–f** by reaction of ammonia in methanol. The free phosphinic acid **13** (R = Ph) is obtained by hydrogen cleavage of the benzyl group.

Table 1 EC50 on cell-based assays (in nM) of **12** and **13** phosphorus based NNRTI

	R	WT	K103N	Y181C	K103N/Y181C
12a	OMe	0.7	6.8	10.1	446.6
12b	OEt	1.2	16.4	13.7	>1,250
12c	OBn	33.7	>1,250	>1,250	>1,250
12d	NMe_2	13.9	>1,250	467	>1,250
12e	Ph	>1,250	>1,250	>1,250	>1,250
12f	Me	0.8	1,382	266	>1,250
13	OH	>1,250	> 1,250	> 1,250	>1,250
(*Rp*)-**12a**	OMe	0.1	1.2	3.6	137.4
(*Sp*)-**12a**	OMe	181.7	>1,250	>1,250	>1,250

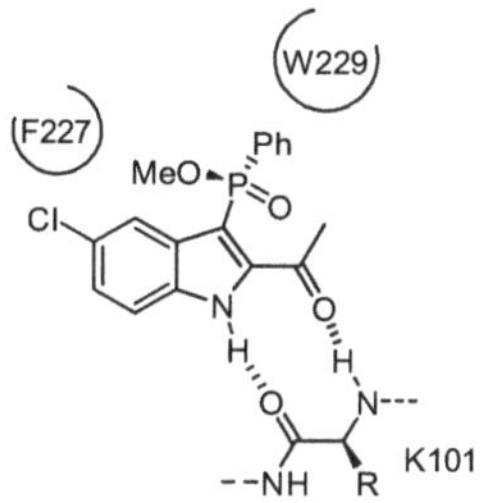

Fig. 5 Binding mode of phosphinate (Rp)-**12a**

The antiviral activities were evaluated in a cell-based assay against the wild type HIV-1 virus and NNRTI resistant strains K103N, Y181C and the double-mutant K103N-Y181C (Table 1). Large R substituents (**12c**: R = OBn, **12e**: R = Ph) or free phosphinic acid **13** induced weak activities. Only phosphinates bearing small neutral group showed low to subnanomolar activities (**12a**: R = OMe, **12b**: R = OEt). By contrast, when compared to phosphinates **12a** and **12b**, the phosphine oxide **12f** was only active on the wild type HIV-1 virus.

Chirality at phosphorus deeply affected the activity of the most efficient methyl phosphinate **12a**. The enantiomer (*Rp*)-**12a** was 1,800-fold more active than the corresponding (*Sp*)-**12a** on the wild type HIV-1 virus and exhibited the same behaviour on the resistant mutant strains.

The (*Rp*)-**12a** was an unprecedented scaffold and its binding mode to the reverse transcriptase of HIV-1 is similar to Efavirenz. Biological results and docking experiments were in accordance with hydrogen bonding interactions with lysine residue K101 of the reverse transcriptase (Fig. 5). In addition, the methoxy group and the phenyl ring overlaid with lipophilic pockets. (*Rp*)-**12a** reached the phase 2b clinical trials but this drug has been placed on "clinical hold".

Boranophosphate analogues, in which a borane group (BH_3) replaces one of the non-bridging phosphoryl oxygen atoms, are bioisosteres of phosphate. The borane group is isoelectronic with the oxygen atom but the distribution of charge density is different and boranophosphate analogues are generally less hydrophilic, less sensitive to nuclease when compared to normal nucleic acids, and demonstrate

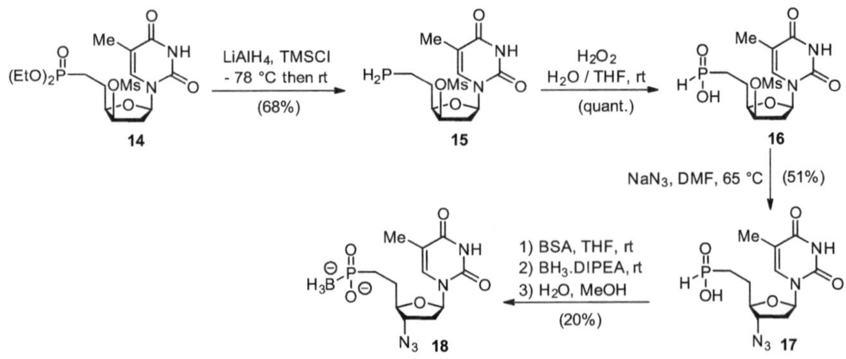

Scheme 2 Synthesis of boranophosphinate **18** analogue of AZT

improved substrate properties. As a consequence, boranophosphates and their analogues were shown as promising alternatives in antiviral therapy when nucleoside reverse transcriptase inhibitors were involved [20, 21].

In this context, Alvarez et al. synthesized a boranophosphinate **18**, analogue of AZT (Scheme 2) [22]. The formation of the phosphorus-boron motif required six steps, starting from the stable thymine phosphonate **14**. A full reduction by lithium aluminium hydride followed by partial oxidation mediated by hydrogen peroxide afforded the *H*-phosphinic acid **16**. After the introduction of the azide group, azido-*H*-phosphinate **17** undergoes a borylation and desilylation to yield to the boranophosphinate **18**.

The boranophosphinate **18** was screened in vitro for antiviral activity against HIV-1, HIV-1(III_B) and HIV-2(ROD) but no activity was observed, even at 200 μM concentrations. Similar results were reported for hepatitis C virus (HCV) and herpes simplex virus.

3.2 *Hepatitis C Virus NS3/4A Protease Inhibitors*

By infecting 3% of world's population, Hepatitis C virus (HCV) infection is a significant public health problem [23]. The standard treatment consists of combining interferon-α (Peg-Intron® and Pegasys®) with ribavirin. Nevertheless, only 50% of patients respond to this therapy [24], which is also poorly tolerated. For these reasons, new therapeutic pathways targeting viral enzymes have emerged, and a particular attention has been focused on NS3 serine protease complex. The NS3 complex cleaves at various regions in the viral polyprotein [25] and it has been found that negative modulation of NS3 activity inhibits viral replication [26]. In addition, NS3 complex can also shut down the host interferon response induced to protect the cell against viral infection [27]. Recently, three NS3 protease inhibitors were evaluated in clinic: Ciluprevir [28], Telaprevir [29] and Boceprevir [30], and the latter two were approved by FDA. A combination of Telaprevir with

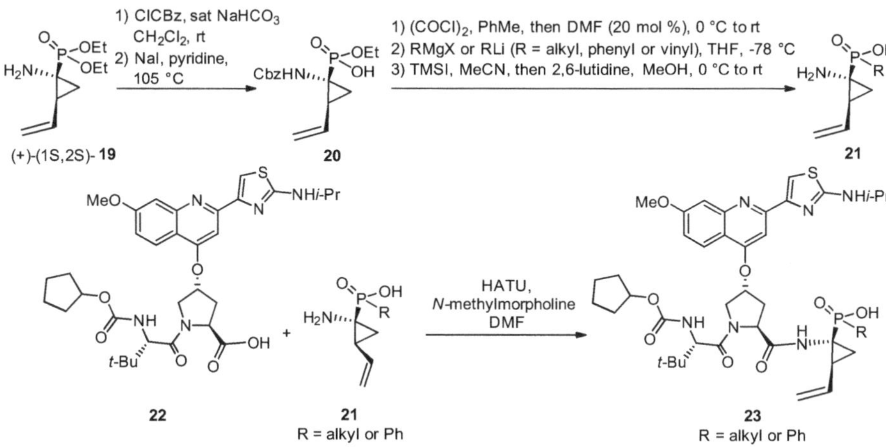

Scheme 3 Preparation of acyclic phosphinic acids **23** analogues to Ciluprevir

peginterferon/ribavirin was revealed as more efficient compared with peginterferon/ribavirin alone [31]. However, adverse effects (i.e. rash or anaemia) appeared with the Telaprevir group. In order to overcome these drawbacks, a molecular design modification was made. Clarke et al. reported synthesis and biological evaluation of a series of acyclic phosphinic acids **23** and their macrocyclic analogues **29** and **29′** structurally close to Ciluprevir [32, 33], whereas Pompei et al. [34] reported a phosphinic acid analogue **32** of clinical evaluated compound MK-7009 [35].

Preparation of the acyclic phosphinic analogues **23** to Ciluprevir consisted of a peptide coupling between α-aminophosphinic acid **21** [32] and dipeptide **22** [36]. The α-aminophosphinic acid **20** was accessible from enantiopure precursor **19** (synthesis described by Pyun et al. [37] or Sheng et al. [38]), applying a synthetic sequence of five steps (Scheme 3). After coupling, a series of compounds **23** harbouring various R groups (Ph, Me, Et, *n*-Bu, *i*-Pr, Bn and mono- and disubstituted benzyl derivatives) were obtained [32].

Macrocyclic phosphinic acid analogues **29** and **29′** were prepared by a 15- to 16-step synthesis (Scheme 4) from a series of α-aminophosphinic acids **21′** which were also prepared from starting materials **19**. Available protected 4-hydroxy proline **24** was coupled with the aminoacid **25**, and after basic hydrolysis of the resulting ester, the carboxylic acid was transformed into lactone by intramolecular esterification, and then the lactone was opened by introduction of **21′** to produce **26**. After transformation of **26** into sulfonyl esters, ring-closing metathesis process and introduction of compound **27**, structure macrocycle phosphinates **28** were obtained. Hydrolysis of phosphinate ester, combined with cleaving the protecting Boc group and introduction of a new carbamate function, led to **29** (Ar = Ph, *o*-Cl-Ph). Saturated series **29′** (Ar = Ph, *o*-mono- and *o*,*o′*-disubstituted phenyl derivatives) were obtained by diimide reduction of **29**.

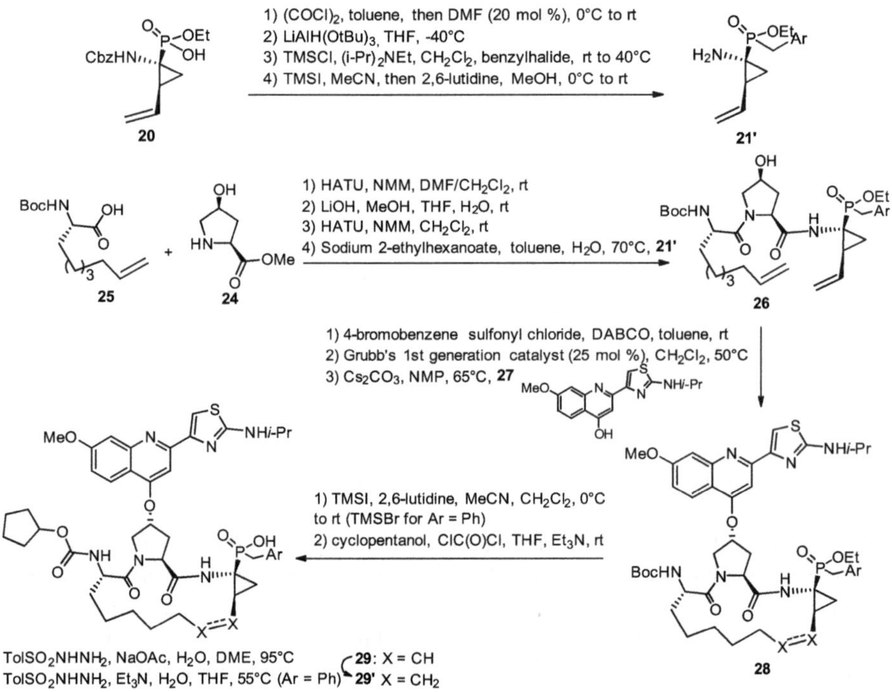

Scheme 4 Preparation of macrocyclic phosphinic acids **29** and **29′** analogues to Ciluprevir

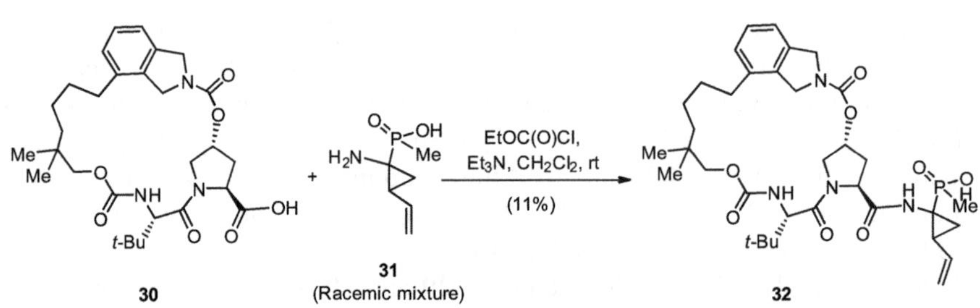

Scheme 5 Preparation of macrocyclic phosphinic acid **32** analogue to MK-7009

The synthesis of phosphinate derivatives **32** described by Pompei was performed by coupling of the key macrocyclic acid **30** [35, 39], with racemic phosphinic acid **31** (Scheme 5) [34]. This coupling occurred by converting the carboxylic acid of pyrolidine ring with ethyl chloroformate into an anhydride intermediate which was subsequently trapped by the amino function of phosphinic acid derivatives.

Compounds **23** (R = alkyl, aryl), **29** and **29′** exhibited a high inhibitor activity against HCV NS3 protease. Their IC_{50} values against the full length of HCV-NS3/4A protease activity ranged between 6 and 86 nM for compounds **23** and between 3 and 11 nM for compounds **29** and **29′**. Their EC_{50} values for HCV replication in

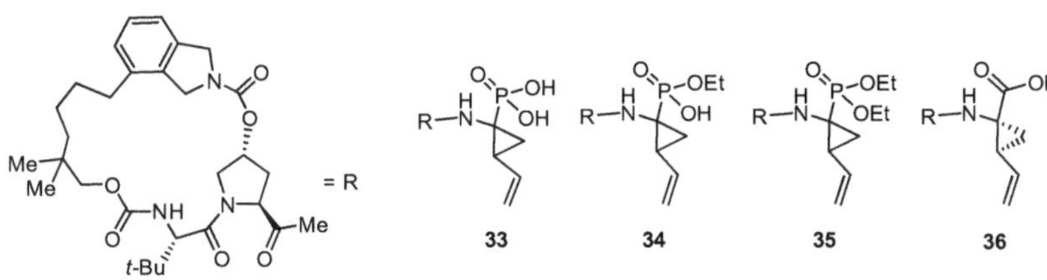

Fig. 6 Analogues **33–36** to macrocyclic phosphinic acid **32**

HuH7 cells ranged from 47 to 3,080 nM for compounds **23** and from 5 to 10 nM for compounds **29** and **29′**. Most values measured were similar to Ciluprevir ($IC_{50} = 1$ nM and $EC_{50} = 1$ nM) [32, 33] and those of their phosphonic analogues ($IC_{50} \leq 1$ nM and $EC_{50} < 100$ nM) [32]. For macrocyclic phosphinic acid **32**, inhibitory evaluation over HCV-NS3 protease activity ($K_i = 2.4$ nM) was slightly better than its analogues phosphonic acid **33** and monoethyl phosphonate **34**, with, respectively, inhibition constant values of 70 and 16 nM (Fig. 6). On the other hand, a convincing result has been found for **32** in comparison with diethyl phosphonate **35** and carboxylic acid **36** derivatives, which showed a K_i of 2,900 and 2,800 nM, respectively [34].

3.3 Influenza A (H3N2) Neuraminidase Inhibitors

Influenza A virus, the causative agent of recurrent worldwide pandemic flu, is responsible for substantial public health (morbidity and mortality) and economic problems. Vaccination raises an immune response against a common conserved region of the viral surface hemagglutinin HA. Unfortunately, influenza virus undergoes genetic recombination leading to antigenic shifts. These changes allow the virus to escape the immune response. One can consider that viral chemotherapy provides “a cost-effective stockpiling option for reducing the impact of fast spreading pandemy” [40]. Only four small molecule drugs are available to minimize illness and excess deaths. This small library of drugs includes zanamivir and oseltamivir as neuramidase inhibitors and amantadine and rimantadine as ion channel blockers. Unfortunately, a high level of resistance of influenza A virus to adamantine-type M2 ion channel blocker developed, and oseltamivir-resistance to seasonal and pandemic strains are emerging [40, 41]. For these reasons, intensive research is underway to discover new antiviral molecules in order to overcome resistance phenomena. In this perspective, neuraminidase (NA), an essential enzyme cleaving host sialic acid, is used as the drug target. Hydroxyphosphonates and the parent α-hydroxyphosphonic acids were disclosed as antiviral activities [42–46]. In 2013, Khorshin and Pozdeev reported the preparation of α-γ-dihydroxyphosphinates as a new class of protective agent against influenza A subtype H3N2 [47]. The molecules which targeted **38a–c** were produced by

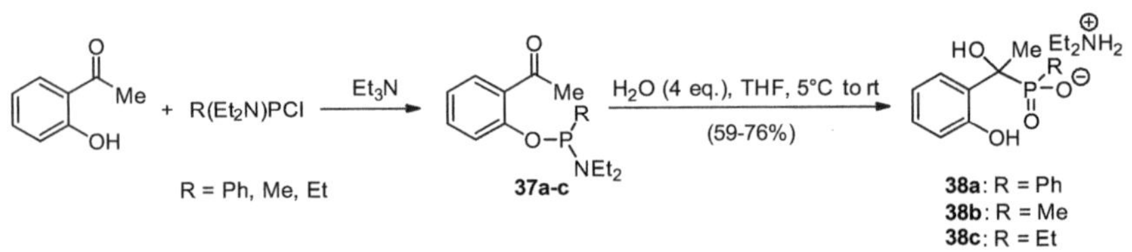

Scheme 6 Synthesis of α-γ-dihydroxyphosphinates **38a–c**

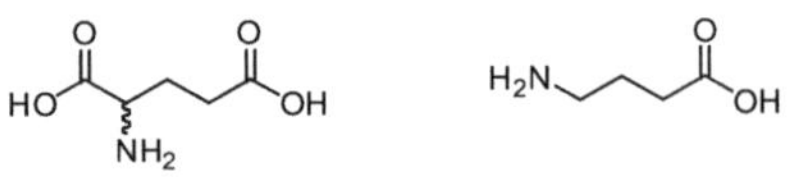

Fig. 7 Structure of Glutamate **39** and GABA **40** neurotransmitters

hydrolysis/intramolecular Abramov hydrophosphinylation of readily available amidophosphonites **37a–c** (Scheme 6) [48].

Products **38a–c** were tested in vitro for their antiviral activity through Madin-Darby canine kidney (MDCK) cells, infected by influenza A virus strain St Petersburg/34/72 (H3N2) [47]. The antiviral activity of **38b** was weaker ($EC_{50} = 28$ μM) than rimantadine ($EC_{50} = 5.0$ μM), the reference compound. Both phosphonate analogues (R = EtO and *n*-PrO) also gave disappointing results. Alternatively, in vivo tests performed after inoculation of influenza virus A/Aichi/2/68 (H3N2) in mice were more convincing. Phosphinates **38a–c** exhibited significant anti-influenza protective activities. The survival rate of mice reached seven out of ten after 8 days, but dropped to three out of ten after 14 days. Nevertheless, these promising results were not as good as expected when compared to rimantadine, which afforded a survival rate of nine out of ten after 8 days and seven out of ten after 14 days.

4 Glutamate and GABA-Based CNS Therapeutics

In the mammalian central nervous system (CNS), glutamate **39**, the main excitatory neurotransmitter and γ-aminobutyric acid **40** (GABA), the main inhibitory neurotransmitter, achieve together the proper functioning of the brain [49, 50] (Fig. 7). Glutamate and GABA are released from nerve terminals in high concentration modifying the membrane potential of the receptive neuron, generating an excitatory (EPSP) or an inhibitory post-synaptic potential (IPSP). Ligand-gated ion channels, G-Protein-Coupled-Receptors (GPCRs), transporters and anabolic and catabolic enzymes are the components of the glutamate and GABA neurotransmitter system.

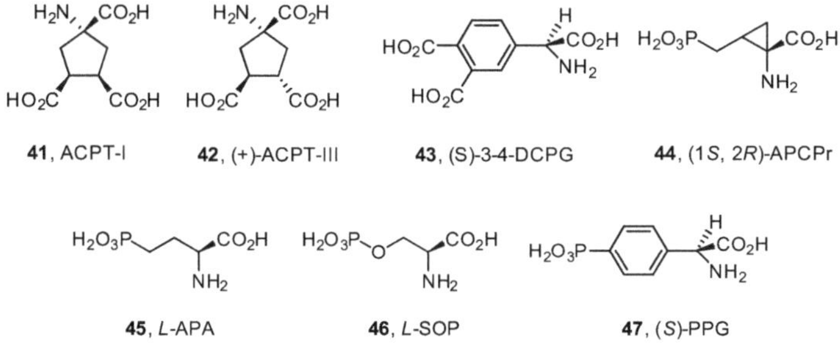

Fig. 8 Structure of the glutamate receptor agonists

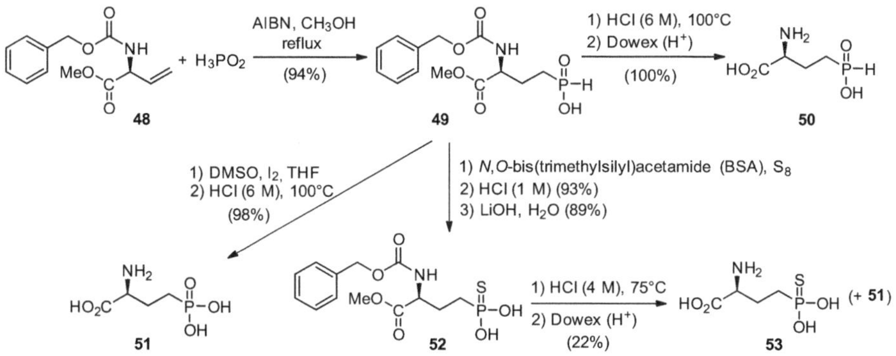

Scheme 7 Phosphinic potent agonists of group III metabotropic glutamate receptors **50**

4.1 Selective Agonists of Metabotropic Glutamate Receptor

Many studies have demonstrated that metabotropic glutamate receptors (mGluR) can serve as new therapeutic targets for a wide panel of brain disorders [51] (schizophrenia [52], ischemia after-effects, convulsions, Parkinson's disease [53], pain…). This G-Protein-Coupled-Receptors (GPGCR) family is subdivided into three groups according to their sequence similarity, transduction mechanism and pharmacological profiles. Consequently, a large number of glutamate analogues (41–47) have been synthesized and assayed in order to identify selective ligands, some of which have proved to be successful and even reached advanced clinical phases [49] (Fig. 8).

In order to activate or enhance the activity of group III of the metabotropic glutamate receptors, new and potent phosphinic agonists were synthesized by Acher et al. in 2007 [54]. Indeed, these presynaptic receptors inhibit the adenylate cyclase and the release process. γ-Phosphinic acid derivative **49** is the key intermediate for the formation of compounds **51**, **52** and **53** (Scheme 7). Starting from a radical addition of hypophosphorous acid to the *N*-CBz protected vinylglycine methyl ester **48**, the phosphinic derivative compound **49** was obtained in 94%

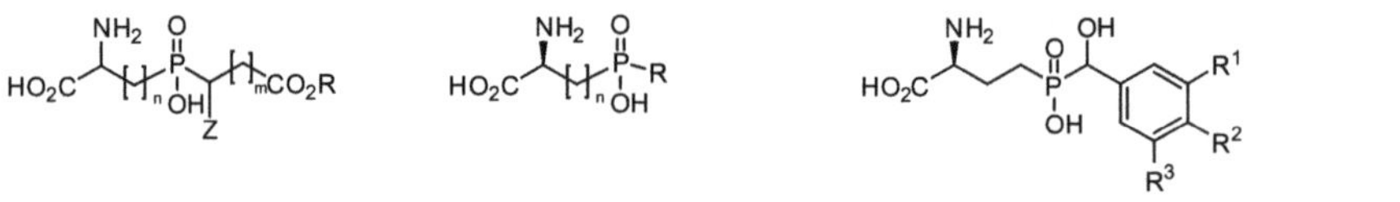

54a: n = 2, m = 1, Z = H **55a:** n = 2, R = $CH_2CH(CH_2CO_2H)CO_2H$ **56**: LSP1-2111 ($R^1 = NO_2$, $R^2 = OH$, $R^3 = OMe$)
54b: n = 1, m = 1, Z = CO_2H **55b:** n = 2, R = $CH_2CH[(CH_2)_2OH]CO_2H$ **57**: LSP4-2022 ($R^1 = OCH_2CO_2H$, $R^2 = R^3 = H$)
54c: n = 1, m = 3, Z = CO_2H **55c:** n = 2, R = $CH_2CH(Cl)CO_2H$

Fig. 9 New metabotropic glutamate receptor agonists

yield. Oxidation of the P–H and deprotection of the amine function performed in HCl (6 M) gave **51** in almost quantitative yield. *H*-Phosphinic derivative **50**, was obtained in quantitative yield using HCl (6 M) under reflux starting from **49**. The thiophosphonic derivative **53** was obtained after silylation of compound **49**, followed by a smooth oxidation to the protected thiophosphonate **52** (in good yield) and a final deprotection in the presence of compound **51** in a 7:3 ratio. A cation exchange resin column afforded pure thiophosphonic acid **53** in 22% yield.

Acher demonstrated that replacing the phosphonate by a phosphinic group in L-AP4 resulted in loss of activity, although changing the phosphonate to a thiophosphonate had the opposite effect. These results confirm the critical role of the additional acidic function and its negative charge in glutamate analogues, which probably provides stronger binding to specific basic residues and then stabilizes the active conformation of the receptor.

More recently, Acher et al. described a new series of phosphinic metabotropic glutamate receptor agonists **54–57** [55–60]. These compounds selectively activate subtype receptors of group III ($mGlu_4$, $mGlu_6$, $mGlu_7$, $mGlu_8$) (Fig. 9).

The strategy for the synthesis of phosphinate **54a** or **54b** involves the activation of the phosphorus atom into the P^{III} reactant, followed by successive reactions with a Michael acceptor (ethyl acrylate or diethyl maleate), dibromoethane, and an esterification using triethyl orthoformate. The incorporation of the amino acid group accomplished by nucleophilic substitution of the anion of the diethylacetamidomalonate yielded the branched phosphinates in 50–77% yield. The final deprotected compounds were obtained in 11–21% yield after acidic deprotection and decarboxylation and they were purified by ion exchange chromatography. Compound **54b** was synthesized by alkylation of the phosphonite intermediate with acetamidoacrylic acid, followed by acidic deprotection and ion exchange chromatography purification (Scheme 8).

Another strategy was performed for phosphinic acid **54c**. First, a radical addition of H_3PO_2 on unactivated diethylallylmalonate afforded regioselectively the terminal adducts **58** and **58′**. The next steps were identical to the first method giving a yield inferior to 5% (Scheme 9).

The synthesis of compounds **55** started from *H*-phosphinic acid **49b**. The formation of the second P–C bond was achieved by in situ generation of the phosphonite species (TMSCl or BSA) followed by a conjugate addition of the appropriate Michael acceptor, or alkyl halides (silyl-Arbuzov reaction), or

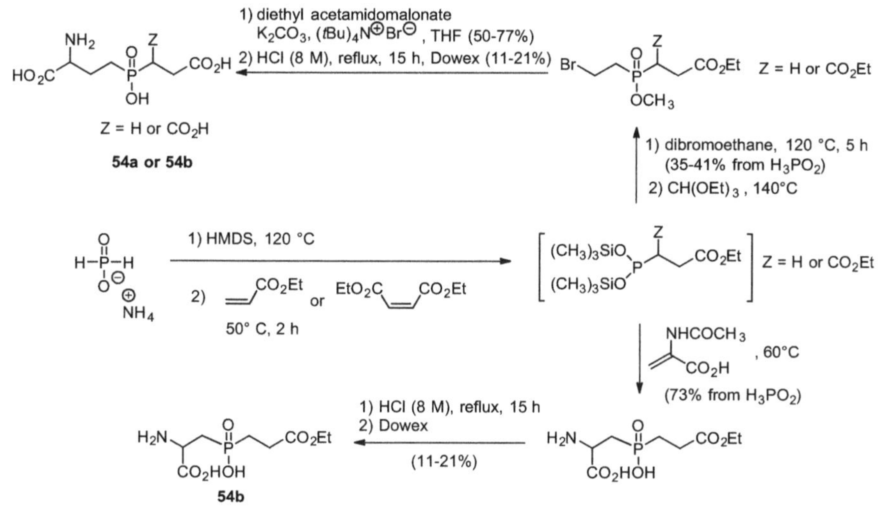

Scheme 8 Synthesis of metabotropic glutamate receptor agonists **54a, b**

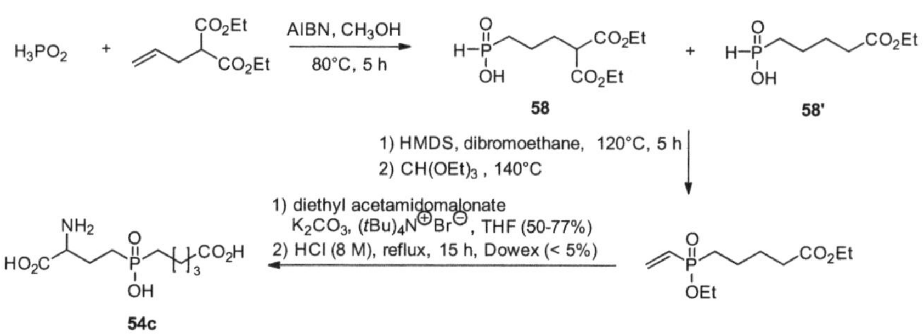

Scheme 9 Synthesis of metabotropic glutamate receptor agonist **54c**

aldehydes (silyl-Abramov reaction) in 29–95% yields. Acidic deprotection followed by ion exchange chromatography purification afforded the desired phosphinates **55** in 3–69% yield (Scheme 10).

Two aminophosphinates **55a** and **55b** were show to be the most potent agonists in this new series of (*R*)-PCEP (Fig. 10).

Hydroxymethyl phosphinates **56** and **57** were obtained starting from phosphinic acid **49a**. According to the previous methodology, the silyl-Abramov reaction on various aldehydes led to the formation of the addition products. Acidic deprotection and ion exchange chromatography purification afforded the desired compounds **56** or **57** (Scheme 11).

The biological assay on **56** showed that it preferentially activates mGlu4 receptor. Hydroxymethyl phosphinic acid **57** was also highly potent at $mGlu_4$ and was shown to be selective towards $mGlu_8$. These two compounds had good oral bioavailability, blood–brain-barrier penetration, and efficacy in vivo in rodents.

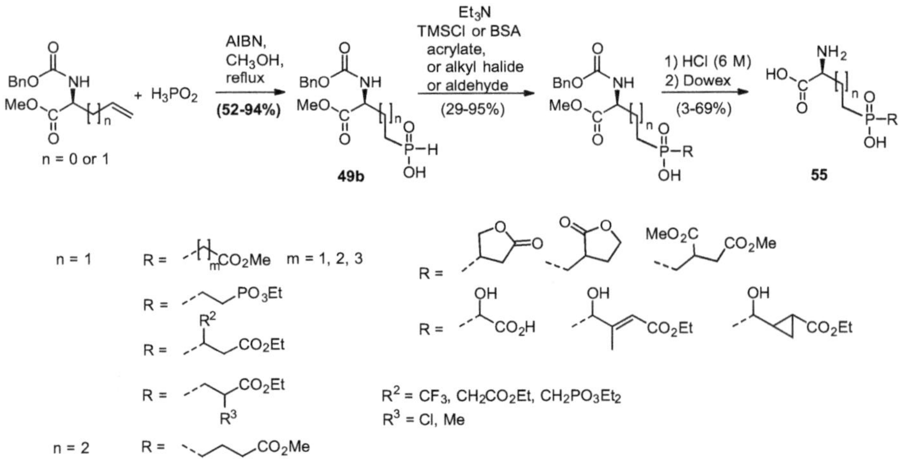

Scheme 10 Synthesis of metabotropic glutamate receptor agonists **55**

55a **55b**

Fig. 10 New potent agonists **55a, b** in the (*R*)-PCEP series

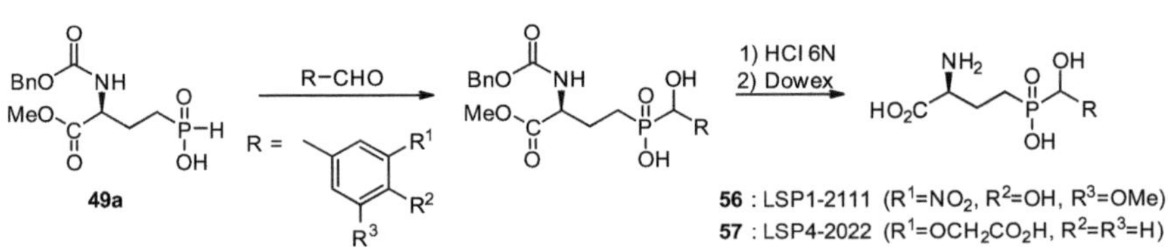

Scheme 11 Synthesis of metabotropic glutamate receptor agonists **56** and **57**

They showed high potency with individual group-III receptors and they are in the pipeline of clinical drug candidates for Parkinson's disease and, possibly, for further human neurodegenerative disorders.

4.2 Selective Agonists or Antagonists of *γ*-Aminobutyric Acid 40 (GABA)

Perturbation of the GABAergic inhibition is responsible for neurological disorders such as epilepsy, anxiety disorders, and schizophrenia [61]. $GABA_A$, $GABA_C$ ligand-gated ion channel and $GABA_B$, a single class of GPCRs, are of enormous medical significance and have various effects, including those on cognition [62],

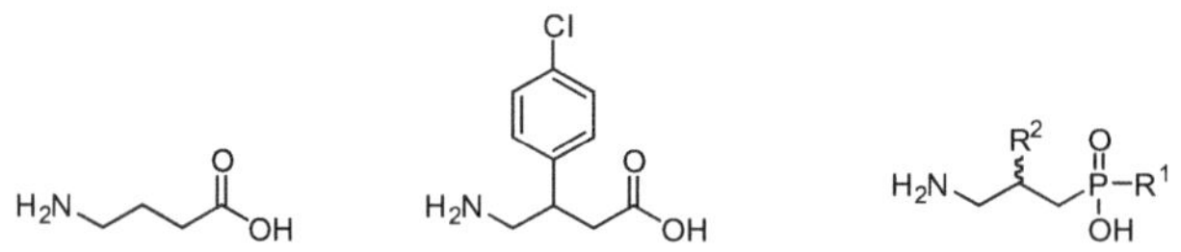

40, GABA   Baclofen   **59**, R^1 = H, Me; R^2 = H, OH, F, =O

Fig. 11 Structure of GABA and $GABA_B$ agonists

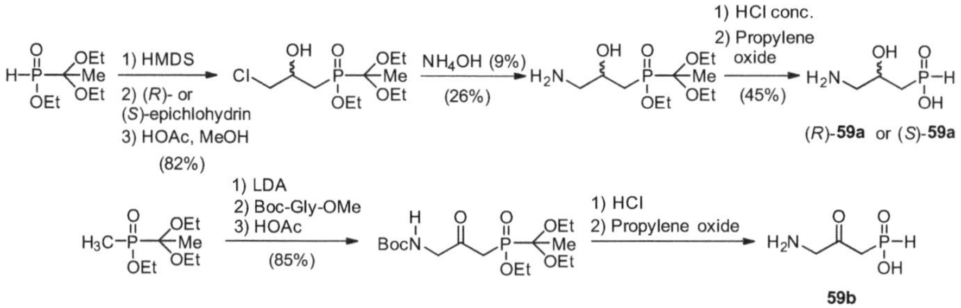

Scheme 12 Synthesis of GABA analogues **59a, b**

nociception [63], vision [64] and circadian rhythms [65]. To design potent and selective ligands, series of compounds were generated via the bioisosteric replacement of the carboxylic acid group by a phosphinic acid.

4.2.1 GABA-Like Analogues

Elebring demonstrated that $GABA_B$ agonists such as baclofen, previously exploited for gastroesophageal reflux disease (GEDR) and transient lower esophageal sphincter relaxations (TLESRs), could, at high doses, have central nervous system related effects [66]. For that purpose, he decided to explore derivatives of 3-aminopropylphosphinic acid **59** ($R^1 = R^2 = H$), which has been demonstrated to be a potent and selective $GABA_B$ agonist [67] (Fig. 11).

The optically pure 2-hydroxy analogues **59a** were prepared from ethyl (diethoxyethyl)-*H*-phosphinate. The corresponding silyl phosphonite intermediate, obtained from the reaction of HMDS with (*R*)- and (*S*)-epichlorhydrin, afforded, after deprotection, (*R*)- and (*S*)-**59a**, respectively. The 2-oxo analogue **59b** was prepared starting from the reaction of metallated methylphosphinate with *N*-Boc-glycine methyl ester followed by deprotection (Scheme 12).

Replacement of one hydrogen or a hydroxyl group by fluorine has been classically used to modulate metabolism or off-target activity. In such a way, fluorine can modulate the pK_a of amino and phosphinic acid groups. The preparation of the racemic fluoro analogues **59c–d** was achieved starting from (diethoxyethyl) phosphinate and (methyl)phosphinate. After transformation into the corresponding silylphosphonites, they were subjected to the addition to ethyl 2-fluoroacrylate

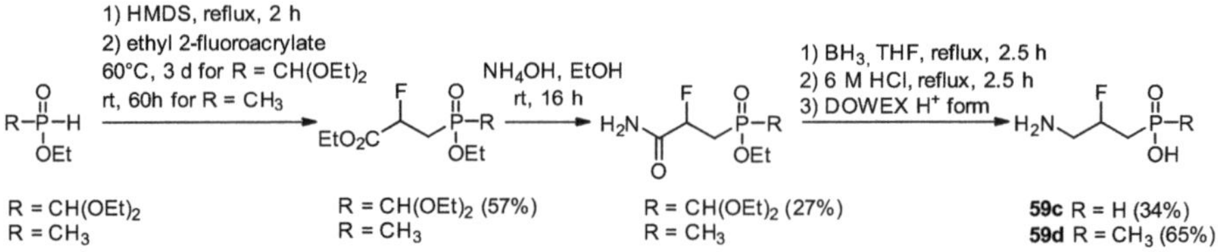

Scheme 13 Synthesis of 2-fluoro GABA analogues **59c, d**

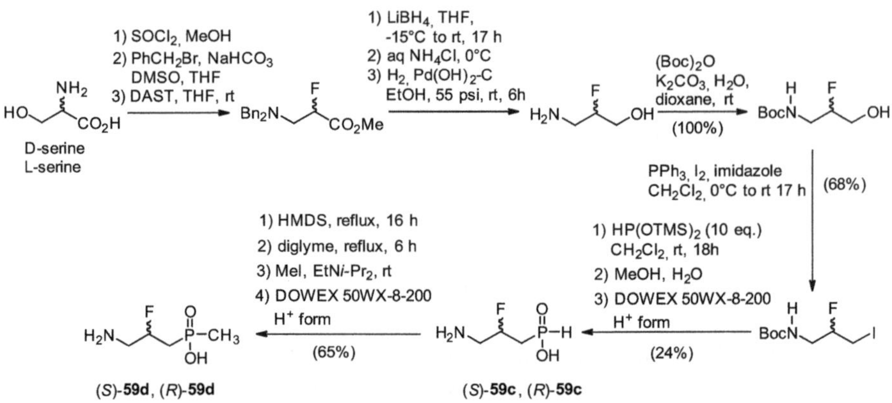

Scheme 14 Synthesis of pure 2-fluoro-analogues (*S*)- and (*R*)-**59c** and (*S*)- and (*R*)-**59d**

affording 4-phosphino-2-fluoropropanoates. The carboxylic acid was then converted to amide and reduced by treatment with borane (Scheme 13).

The syntheses of enantiopure 2-fluoro-analogues employed either D-serine or L-serine as starting materials. The stereospecific transformations of the β-hydroxy-α-amino acid into the α-fluoro-β-amino acid derivatives were performed in accordance with specific rearrangement developed by Somekh and Shanzer [68]. This method involved fluorination of the *N*,*N*-dibenzyl derivative of the β-hydroxy-α-amino acid ester with DAST and then reduction of the ester followed by the removal of the benzyl group. The amino group was Boc-protected, and the hydroxyl group was replaced by an iodine. Two successive alkylations through the silyl-Arbuzov reaction led, respectively, to the formation of the functional alkyl chain and the methyl-affording (*S*)-**59d** and (*R*)-**59d,** respectively (Scheme 14).

Binding affinities and agonistic properties of the compounds were measured in detail and compared to baclofen. In general, P–H derivatives have in vitro potency in the same range at the $GABA_B$ receptor for the P-methyl derivatives [(S)-59d and (R)-59-d], but they behave very differently in animal studies with regard to inhibition of transient lower oesophageal sphincter relaxations (TLESRs) as well as to possible side effects. Introducing a fluorine atom leads to $GABA_B$ agonists with potencies which exceed those of any previously known $GABA_B$ agonist. Compound (*R*)-**59c** has to be evaluated in humans with respect to inhibition of TLESR.

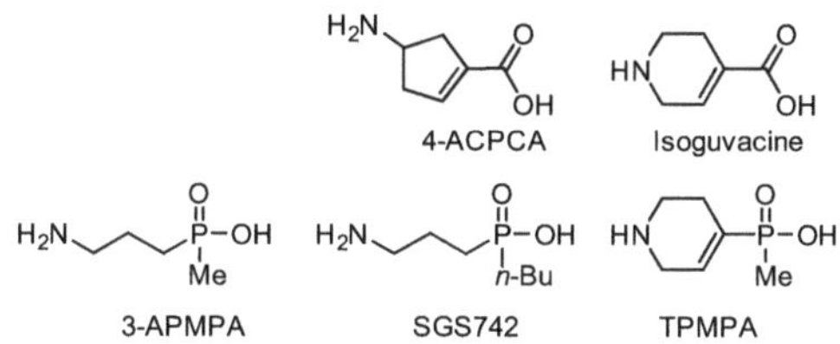

Fig. 12 Active compounds at GABA receptors

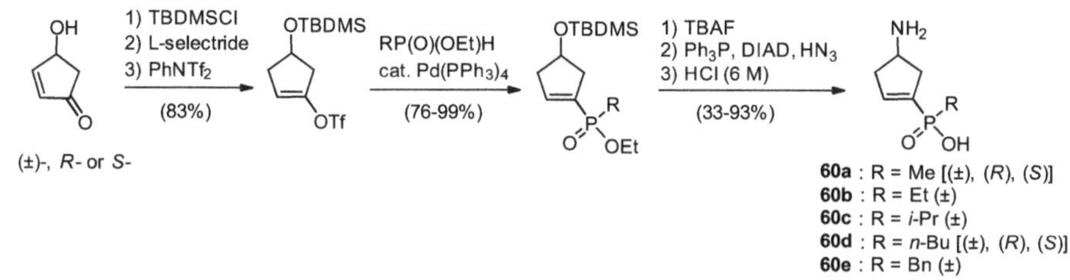

Scheme 15 Synthesis of pure 4-ACPCA phosphinic acid analogues **60a–e**

4.2.2 Conformationally Restrained Isogavacine and 4-ACPCA Phosphinic Analogues

Because GABA has a flexible backbone, it is able to adopt a number of conformations when interacting with various macromolecular targets. This characteristic of GABA can be utilized to provide selective ligands through the generation of conformationally restricted analogues [69]. In 2008, Hanrahan described the synthesis, the pharmacological activity and structure-activity relationships of restricted phosphinic acid analogues of γ-aminobutyric acid and investigated the three major GABA receptor subtypes [70]. The approach was to replace the carboxylic acid by a bioisostere, the phosphinic acid, known to reduce activity at $GABA_A$ receptors (Fig. 12). The first targets were Isogavacine and 4-ACPCA phosphinic acid analogues.

After protection of 4-hydroxycyclopentenone with TBDMSCl and subsequent reduction and trapping of the enolate intermediate as the triflate, Kumar et al. formed the P–C bond using a pallado-catalyzed coupling reaction for a variety of alkyl-*H*-phosphinates with 2.5 mol% of $Pd(PPh_3)_4$ [70]. The alkylphosphinate esters were isolated in 76–99% yields using mild conditions. After deprotection using TBAF, the amino function was generated from a one-pot Mitsunobu–Staudinger reaction. Finally, the phosphinate ester was hydrolyzed using aqueous HCl and the crude product purified via ion exchange chromatography and recrystallisation to give the free amino derivatives **60a–e** (Scheme 15).

Four-membered ring cyclobutane derivatives were also investigated. The synthesis proceeded from the Boc-protected 3-amino cyclobutanone [71]. The phosphorus–carbon bond formation was achieved using the silyl-Abramov reaction in an approximate 1:4 ratio of *cis*- and *trans*-isomers [72]. The two isomers can be

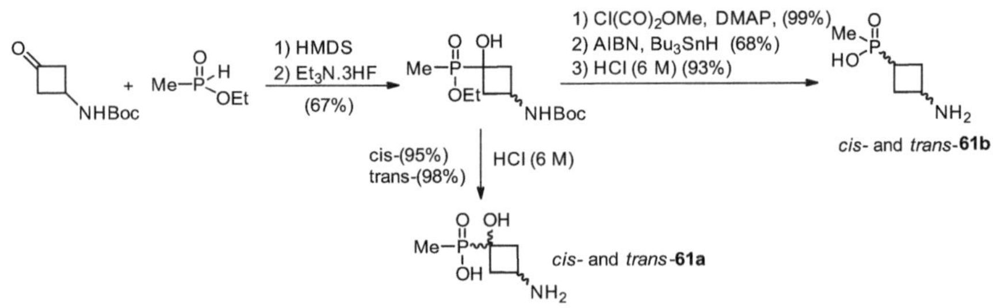

Scheme 16 Synthesis of cyclobutyl phosphinic analogues **61a, b**

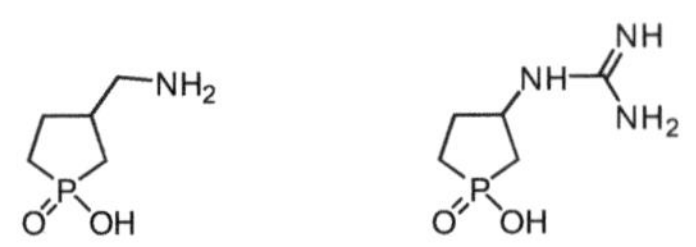

Fig. 13 Aminophospholanes **70** and **74** as GABA analogues

separated by chromatography and deprotected with HCl affording *cis*- and *trans*-**61a**. Compounds **61b** were obtained by a modified Barton deoxygenation, which proceeds with racemization, and deprotection with HCl (Scheme 16).

The activity of these conformationally restricted analogues has been investigated at the three major GABA gated ion channels ($GABA_A$, $GABA_C$ and $GABA_B$). The pharmacological data shows that the nature of the alkyl substituent, the stereochemistry and the existence of an alkene either α or β to the acid functionality are important in determining optimal antagonist activity and selectivity. Compound (*S*)-**60d** is currently under investigation in animal system as an antagonist candidate for elucidating the role of $GABA_C\rho_1$.

4.2.3 Cyclic Phosphinic Analogues of $GABA_C\rho$ Receptor Antagonists

In 2010, Chebib et al. investigated γ-aminobutyric acid analogues based on cyclic phosphonic and phosphinic acids in order to obtain selective $GABA_C$ receptor antagonists [73]. In the series, two cyclic phosphinic acids. 3-AMOHP **70** and 3-GOHP **74**, showed selectivity at ρ_1 $GABA_C$ receptors (Fig. 13).

Esterification of hypophosphorous acid was achieved by reaction with tetraisopropyl orthosilicate **62**, which, when treated with methyl acrylate **63**, afforded the triester **65**, which was cyclized to **66** in the presence of *n*-butyllithium. Selective hydrolysis of compound **66** with hydrochloric acid (3 M) in the presence of isopropyl alcohol gave, in moderate yield, the six-membered phosphaketone **67**. The phospholane-3-carboxylic **68** was obtained from ring contraction mediated by thallium(III) along with oxidation reaction in one step. The carboxylic acid function

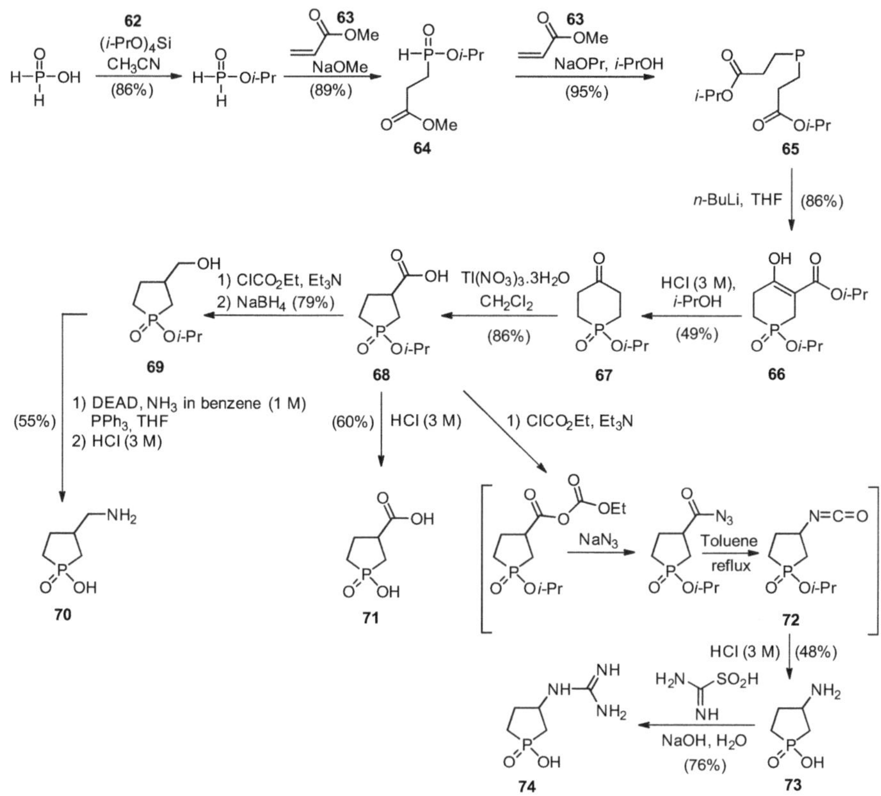

Scheme 17 Synthesis of aminophospholanes as GABA analogues

was transformed into primary alcohol **69** using $NaBH_4$ through the in situ formation of mixed carbonic-carboxylic acid anhydrides. Finally, a one-pot Mitsunobu–Staudinger reaction, followed by phosphinate ester hydrolysis, using aqueous HCl, and purification using ion exchange chromatography gave, after recrystallization, the free aminomethyl phospholane **70**. The diacid **71** was obtained from ester hydrolysis product **68** using aqueous HCl, in 60% yield. The amino phosphinate **73** was obtained using a modified Curtius rearrangement of the acid **68**, followed by the hydrolysis of the isocyanate **72**. Guanylation of **73** using formamidine sulfinic acid in the presence of sodium hydroxide afforded 3-(guanidino)-1-oxo-1-hydroxyphospholane **74** (3-GOHP) in 76% yield (Scheme 17).

All the compounds synthesized were investigated against the three major GABA ion channel and receptor families. 3-AMOHP **70** and 3-GOHP **74** are potent and selective $GABA_C$ receptor antagonists. These results offer new knowledge of the architecture of the $GABA_C$ receptor ligands for future studies.

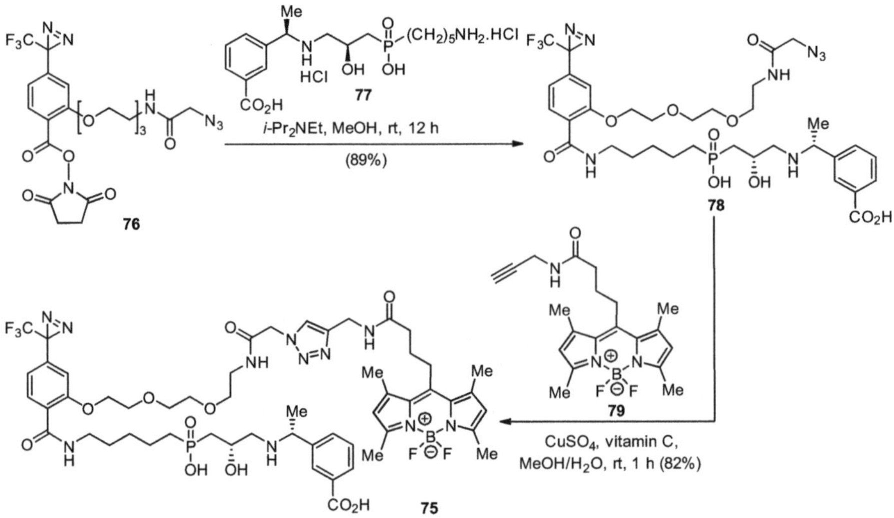

Scheme 18 Synthesis of fluorescent probe **75**

4.2.4 Selective Fluorescent Probes for the GABA Receptors

The development of fluorescent probes specifically targeting GABA ion channels and GPCRs represents a major challenge. In 2008, Nan described the design, the synthesis, and the biological evaluation of small fluorescent phosphinic probes specifically targeting unmodified or native $GABA_B$ receptors.

The synthesis of probe **75**, via procedures similar to those reported previously with some modifications, is outlined in Scheme 18 [74]. Coupling compound **76**, already described in the literature [75–77], with the bioactive ligand **77** gave compound **78**, which was subsequently reacted with the fluorophore **79** to give probe **75** employing the copper-catalyzed [3 + 2] azide-alkyne cycloaddition [78].

Pharmacological studies showed that the probe **75** conserved reasonable antagonist potency for $GABA_B$ receptor. Subsequent photoaffinity labelling and different fluorescent microscopy experiments also showed that the probe labels the functional $GABA_B$ receptor on the cell surface with high specificity. These experiments performed in living cells suggest useful applications for this method.

In 2012, Johnsson described the first fluorescent sensor for GABA, a synthetic fluorophore and a fluorescent $GABA_B$ receptor antagonist [79]. Phosphinic molecules **80** and **81** were used as starting reagents for the generation of tethered fluorescent ligands **82** and **83** using methods described in detail in the above-mentioned Johnsson article (Fig. 14).

Hanrahan reported in 2013 the synthesis and the pharmacological studies of a series of fluorescent ligands containing different-sized linkers and fluorophores based around (*S*)- and (*R*)-4-ACPBPA (Scheme 19) as selective $GABA_C$

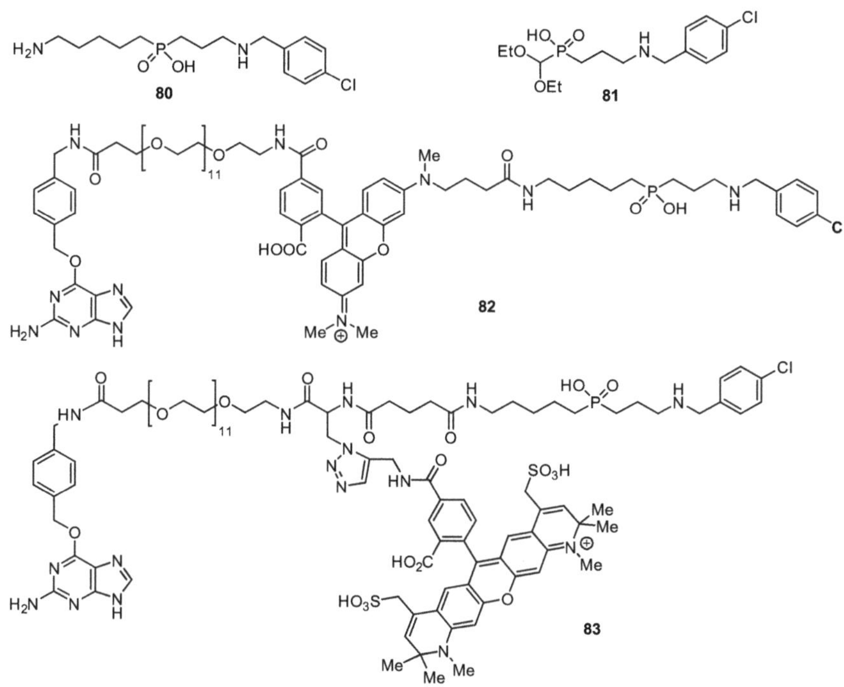

Fig. 14 Fluorescent sensors for GABA receptors

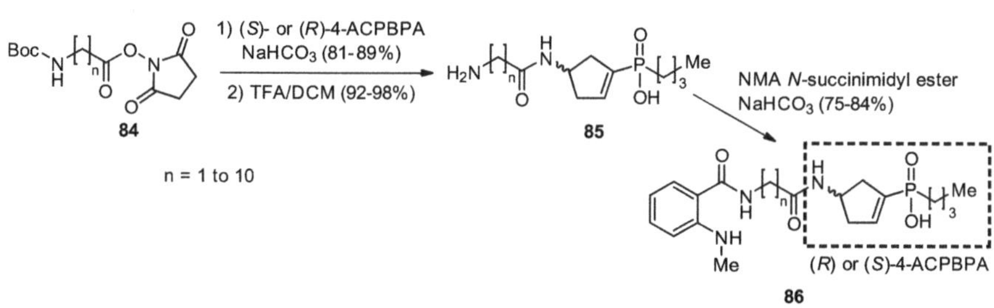

Scheme 19 Synthesis of NMA-fluorescent ligands ρ1 $GABA_C$ antagonists

antagonists [80]. The purpose was to develop more potent and selective fluorescent probes for studying the localization and function of $GABA_C$ receptors in living cells.

Previously reported selective potent $GABA_C$ antagonists (*S*)- and (*R*)-4-ACPBPA were used for the design of the fluorescent $GABA_C$ antagonists. To introduce molecular flexibility into the conjugated fluorescent ligand, Hanrahan et al. chose a linker size that varied from zero to ten carbon atoms and three different fluorophores [NMA (*N*-methylanthranilic acid), 4,4-difluoro-4-bora-3*a*,4*a*-diaza-*s*-indacene (BODIPY) and 7-nitrobenz-2-oxa-1,3-diazol-4-yl chloride

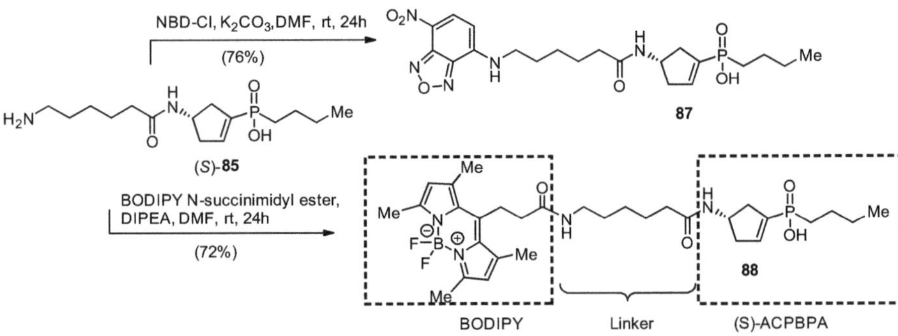

Scheme 20 Synthesis of biotinylated ρ1 $GABA_C$ antagonists **87** and **88**

(NBD), known to have different excitation wavelengths and distinctive properties. The activated *N*-succinimidyl esters **84** were reacted with the $GABA_C$ antagonist (*S*)- and (*R*)-4-ACPBPA in an aqueous sodium carbonate solution and removal of the Boc group using TFA in CH_2Cl_2 to afford free amino compounds **85** in good yields. Target compounds **86** were obtained by amine coupling of free aminoderivatives **86** with the NHS-activated fluorophore NMA in aqueous sodium carbonate solution (Scheme 19).

NBD fluorescent probe **87** was prepared from condensation between free amino derivative (*S*)-**85** ($n=5$) and NBD chloride in 76% yield. BODIPY derivative **88** was obtained starting from a condensation between the compound (*S*)-**85** ($n=5$) and BODIPY *N*-succinimidyl ester in 72% yield (Scheme 20).

These compounds are the first fluorescent and biotinylated probes designed as selective antagonists for ρ_1 $GABA_C$. Only compound **88** displayed moderate potency ($IC_{50}=58.6$ μM compared to (*S*)-4-ACPBPA $K_{B(\rho 1)}=9.76$ μM) and selectivity (more than five times greater) for ρ_1 over $GABA_A$ receptor.

5 Phosphinic Acids for Cancer Treatments

5.1 *Inhibitors of Folylpoly-γ-Glutamyl Synthetase*

Folate derivatives differ from each other according to their reduction state, presence and nature of one carbon substituents, and occurrence of poly-γ-glutamyl chains linked to the intrinsic glutamate of the folic acid structure. These folates play a central role for a diversity of one-carbon transfer of metabolic processes (anabolic and catabolic), including thymidylate, purine, serine, glycine and methionine biosynthesis [81]. Among them, the poly-γ-glutamyl folates are essential for cell growth and survival [81, 82]. Indeed, the poly-γ-glutamyl metabolites retain folate within the cell because only monoglutamate is substrate for the folate efflux systems and their diffusion through the plasma membrane is prevented by their

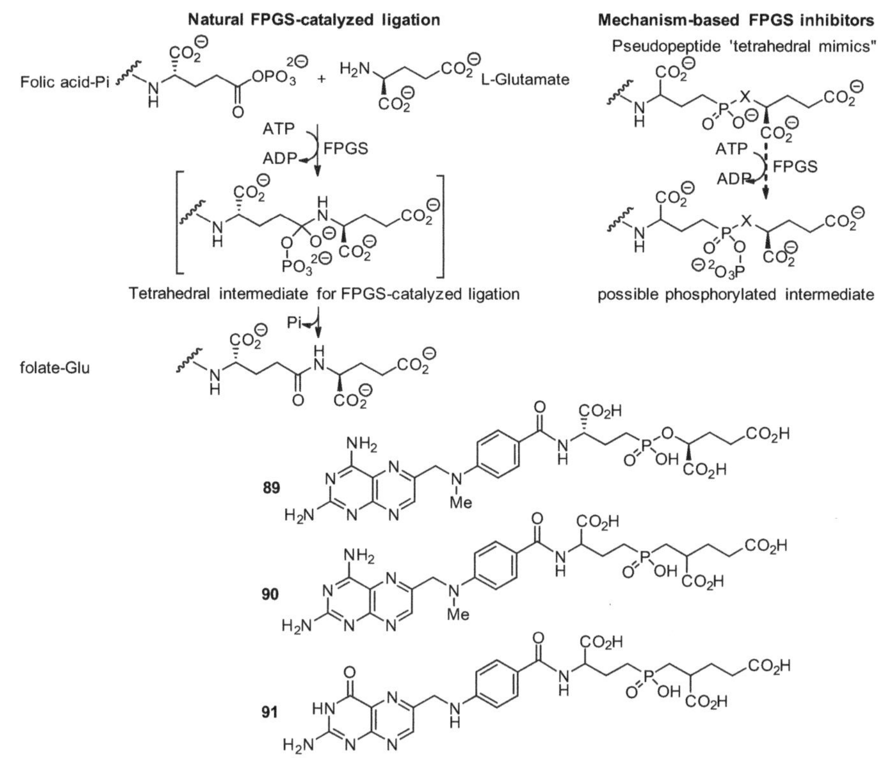

Fig. 15 Natural FPGS-catalyzed ligation for formation of folylpoly-γ-glutamate, mechanism-based FPGS inhibitors, prior phosphonic and phosphinic acids inhibitors

high negative charge. In addition, poly-γ-glutamyl metabolites have a better affinity towards folate-dependant enzymes. Their crucial role was demonstrated by mutational deletion of folylpoly-γ-glutamate synthetase (FPGS, EC 6.3.2.17). Indeed, expression of this mutated enzyme, in the absence of end-products of the folate metabolism (thymidine, purines, serine, etc.), was shown to be lethal [82, 83].

Folylpoly-γ-glutamate synthetase (FPGS, EC 6.3.2.17) is an enzyme which catalyzes the ATP-dependent addition of glutamate moieties to folate (folic acid anion) and their derivatives (Fig. 15). Because of the high dividing rate of cancer cells, FPGS has been considered as a potential target in cancer chemotherapy [84, 85]. In this context, Coward et al. elaborated various phosphorus-containing pseudopeptides as FPGS inhibitors, such as diastereomerically pure phosphonate **89** [86] or biologically preferred phosphinates **90** and **91** (mixture of four diastereomers) [87, 88] able to simulate the postulated transient tetrahedral intermediate of FPGS-catalyzed reaction.

In order to increase potency and even specificity of FPGS inhibitors, Coward et al. considered the preparation of analogues based on three different kind of modifications: the nature of the stereogenic centres bearing the carboxylic function

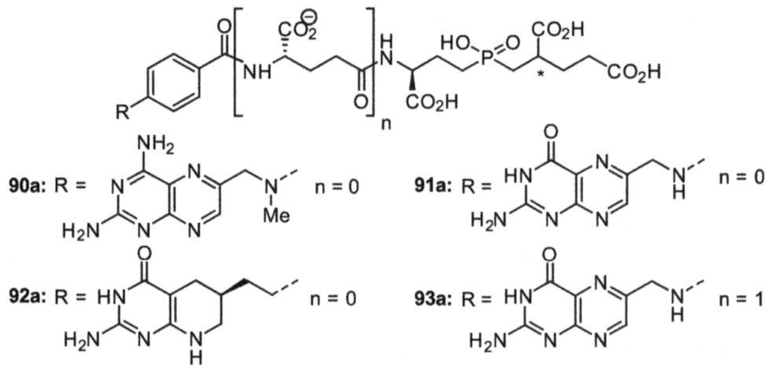

Fig. 16 Modified phosphinic acid FPGS inhibitors

(**90a** and **91a**), the nature of the heterocyclic R-group **92a** and the number of glutamate motifs **93a** (Fig. 16). For the preparation of targeted molecules, each phosphinate diastereomer has been prepared individually and was coupled with three different heterocycles: 4-amino-4-deoxy-10-methylpteroate [89], pteroate [90] and 5,10-dideazatetrahydropteroate [90].

For example, both diastereomers under salt form **90a1** and **90a2** (Scheme 21) [89] were prepared by coupling pure diastereomer pseudopeptide **94a** or **94b** with acyl azide **95**, AMPte-N_3, derived from 4-amino-4-deoxy-10-methylpteroic acid [91], using the strategy described [87]. Both pseudopeptides **94a** and **94b** were obtained according a multistep sequence from the corresponding protected phenyl selenide **96** (Scheme 21) [89]. The latter was transformed into protected α-vinylglycine **97** by concomitant selenium oxidation and β-elimination of phenyl selenium oxide formed. Introduction of phosphinic acid residue on vinylglycine was performed by radical reaction of ammonium *H*-phosphinate salt with triethylborane/air, affording **98** with a high conversion rate. Introduction of the glutamate moiety on **98**, has been led by a Michael addition process with Evans's olefin **99** to give only two separable diastereomers, **100a** and **100b** (combined yield 72%). Acrylate derivative **99** was accessible in two steps, affording acrylic derivative **101** (overall 75%). Compound **101** was obtained from ethyl malonate and *tert*-butyl acrylate **102** in three steps (overall 49%). The Evans chiral auxiliary of each protected pseudopeptide **100a** and **100b** was removed by treatment with lithium hydroperoxide in a mixture of THF/water. The free carboxylic acids were then esterified by diazomethane. The complete deprotection of the esters **103a** and **103b** was carried out under strong acid conditions to afford the pseudopeptides **94a** and **94b** in excellent yields (Scheme 21) [90].

Both diastereomers **91a1** and **91a2**, as well as both diastereomers **92a1** and **92a2**, were prepared by coupling of (6*R*)-5,10-dideaza-5,6,7,8-tetrahydropteroyl azide with pseudopeptide **94a** or **94b** in DMSO and triethylamine. For both isomers **93a1** and **93a2**, they were accessible by mixing pteroyl azide **95** and Glu-γ-Glu-γ[ψ{P(O)(OH)-CH_2}glutarate phosphinic acid pseudotripeptide (pure isomer) [89].

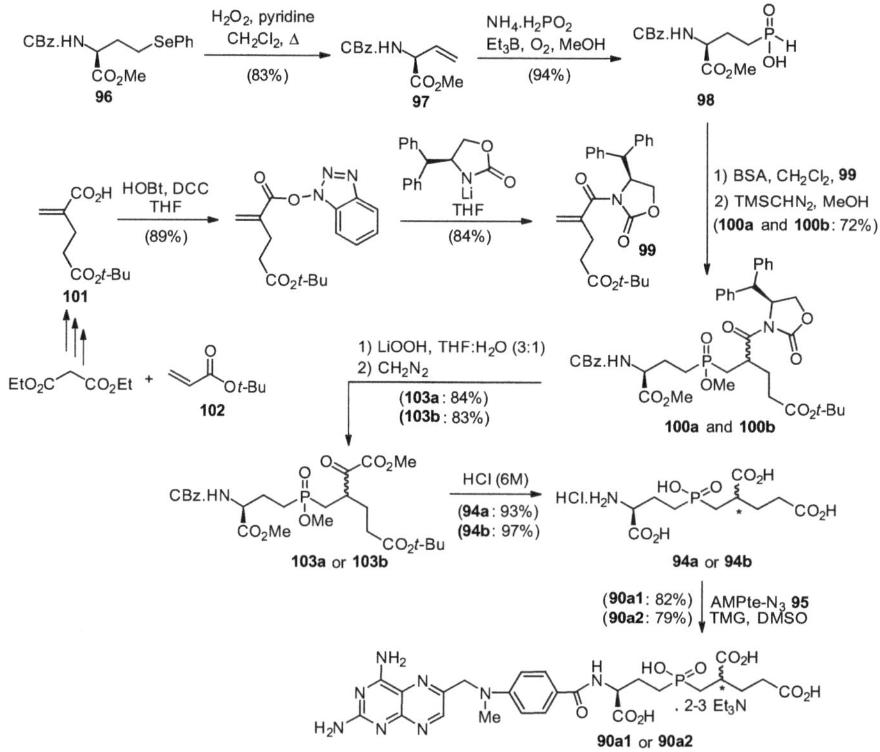

Scheme 21 Preparation of a phosphinic acid FPGS inhibitor

Among separate diastereomeric pairs of **90a**, **91a**, **92a** and **93a**, diastereomers having a high *R*f (HPLC) gave a better inhibition of the recombinant human cytosolic FPGS [90]. For instance, the compound **90a1** (high *R*f) was about 13-fold ($IC_{50} = 3.5 \pm 0.3$ nM) more potent than the diastereomer **90a2** with low *R*f ($IC_{50} = 45 \pm 0.3$ nM), and about fourfold more efficient than original mixture of the four diastereomers **90** ($IC_{50} = 13.6 \pm 1.3$ nM). These results suggest – with current knowledge of FPGS stereospecificity – that the high *R*f species is likely analogous to the natural L-Glu-ψ-L-Glu configuration. Nevertheless, the best inhibition observed with the phosphinic acid mimics series, was supported by the isomer **92a1** (high *R*f) with an IC_{50} of 1.7 ± 0.2 nM [90].

5.2 *Inhibitors of Aspartate Transcarbamoylase*

Aspartate transcarbamoylase (ATCase, EC 2.1.3.2) is an allosteric enzyme which controls the first step of pyrimidine de novo biosynthesis. It catalyzes the condensation of L-aspartic acid with carbamyl phosphate to produce carbamoyl-L-aspartate

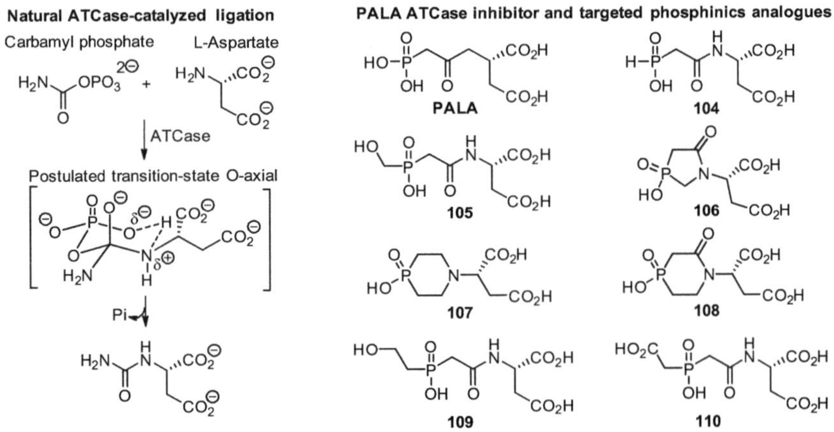

Fig. 17 Mechanism ATCase-catalyzed ligation, PALA ATCase inhibitor and targeted phosphinic analogues **104–110**

and orthophosphate (Fig. 17) [92, 93]. This irreversible reaction is a key step regulating the pathway of pyrimidine de novo biosynthesis. *N*-Phosphonoacetyl-L-aspartate (PALA) has been reported as a potent inhibitor of ATCase ($K_i = 16$ nM) and has been evaluated in cancer chemotherapy [94–96]. Combination of PALA with dipyridamole revealed some effectiveness against various resistant cancer cells [97]. Nevertheless, insufficient bioavailability of PALA, coupled to ATCase activity over-expression does not block the pyrimidine biosynthesis. Probably for this reason, PALA failed to pass the clinical trials for use as monotherapy. In 2009, Montchamp et al. designed phosphinate derivatives **104–107** as analogues of PALA, with an aim of a more potent and selective inhibitor of ATCase (Fig. 17) [98, 99]. Phosphinic monoacid rather than phosphonic diacid should improve the bioavailability compared to PALA.

H-Phosphinate **104** was synthesized according to two synthetic sequences [98, 99]. For instance, cinnamyl alcohol and hypophosphorous acid were coupled using a palladium-catalyzed allylation reaction, affording cinnamyl-*H*-phosphinic acid **111** [100, 101], which was protected using orthoacetate. Ozonolysis of the protected phosphinate **112**, followed by oxidation, gave the carboxyphosphinate **113**. A peptidic coupling between **113** and L-aspartic acid dibenzyl ester formed the protected phosphinate **114**. Finally, deprotective reactions, consisting of catalytic hydrogenation and acidic hydrolysis, yielded carbamoylmethyl-*H*-phosphinate **104** (Scheme 22).

Hydroxymethyl phosphinate **105** was prepared in three steps from hydroxymethyl-*H*-phosphinic acid **115** [102]. *H*-Phosphinic acid **115** was silylated to give the P^{III} intermediate, which reacted in a silyl-Arbusov reaction with the bromoacetyl derivative of aspartic acid to give **116**. A deprotective step using hydrogen afforded phosphinic acid **105**. From intermediate **116**, cyclic phosphinic acid **106** was also accessible in two steps. First, a 5-exo-tet cyclization under

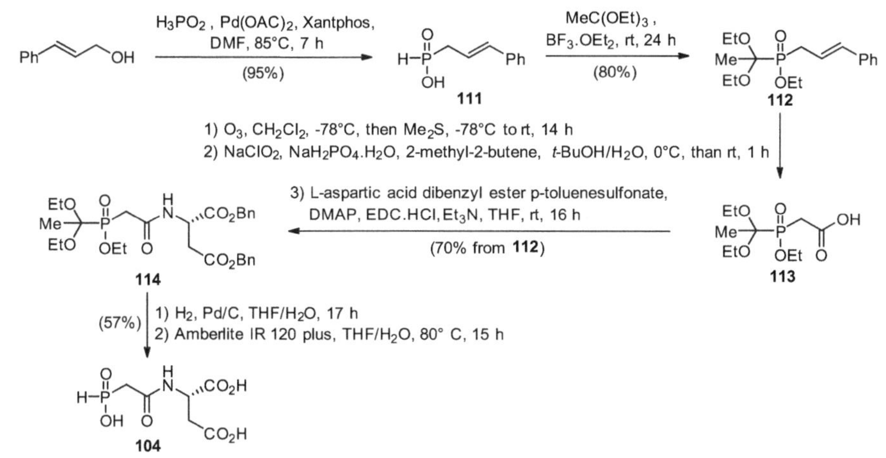

Scheme 22 Synthesis of PALA analogue **104**

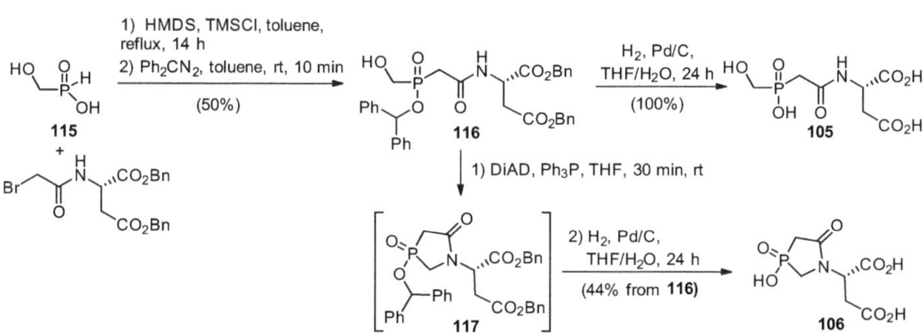

Scheme 23 Synthesis of hydroxmethylphosphinic acid **105** and phosphapyrrolidone **106**

Mitsunobu conditions generated 4-phosphapyrrolidone **117**, which gave phosphinic acid **106** after cleavage by hydrogen of the diphenylmethyl group (Scheme 23).

4-Azaphosphinane **107** was obtained from bis(cinnamyl)phosphinate **118** [103]. The reaction of the dialdehyde, obtained by a double ozonolysis of **118**, with dibenzyl aspartate afforded the cyclic benzyl phosphinate **119**. A pallado-catalyzed hydrogenolysis produced 4-azaphosphinane **107** in 85% yield (Scheme 24).

Other analogues of PALA were also synthesized. A chemical sequence using isoprenylphosphinate **121** as common precursor to three targeted structures **108**–**110** has been elaborated. Phosphinic acid precursor **120** was obtained by palladium-catalyzed hydrophosphinylation from isoprene and hypophosphorous acid [104]. The phosphinate **121** was formed by alkylation of isoprenyl-*H*-phosphinic acid **120** using *N*-bromoacetyl aspartic acid dibenzyl ester under silyl-Arbusov conditions. The resulting phosphinic acid was esterified using benzylbromide in the presence of silver oxide [105]. In one way, an ozonolysis of **121** followed by a

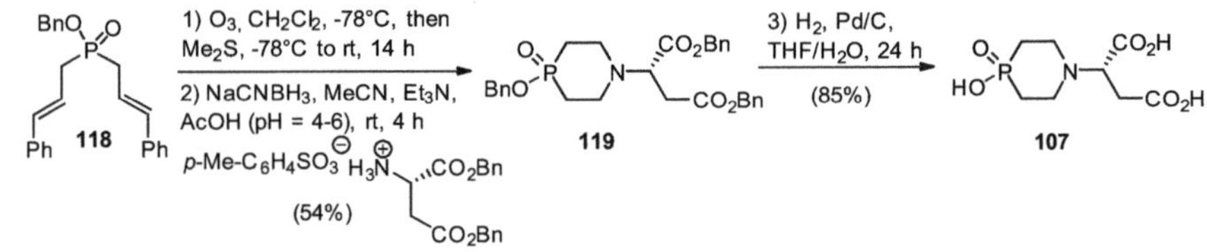

Scheme 24 Preparation of 4-azaphosphinane **107**

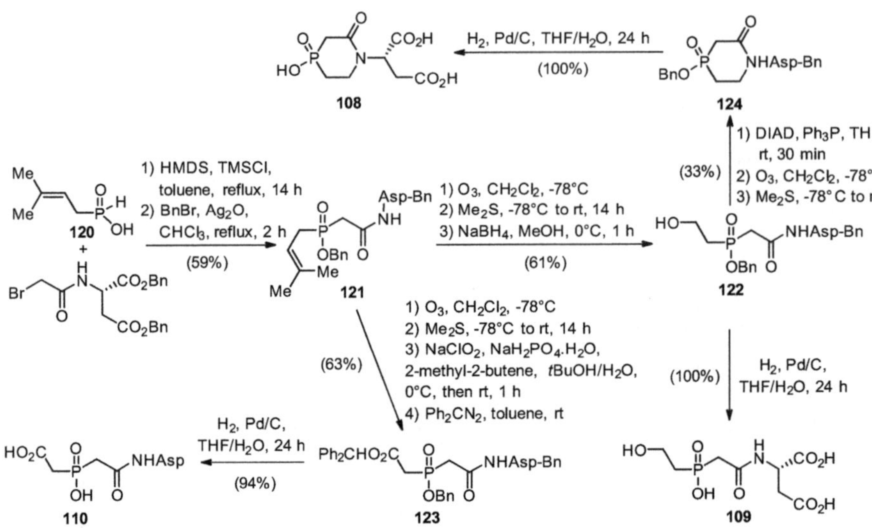

Scheme 25 Preparation of 4-azaphosphinanone **108** and phosphinic acids **109** and **110**

reduction of the aldehyde function produced alcohol **122**, whereas an oxidation/esterification sequence led to 2-carbamyl-2′-carboxyphosphinate **123**. From **122** and **123**, a debenzylation furnished the targeted compounds **109** and **110**, respectively. Finally, an intramolecular Mitsunobu reaction converted the alcohol **122** into azaphosphinane-2-one **124**. The cleavage of the protecting groups afforded quantitatively the free acid **108** (Scheme 25).

All the substances **104–110** were evaluated as ATCase inhibitors, using *E. coli* Histidine-tagged C3 ATCase [100]. All inhibition constants ranging from 190 to 3,600 nM were at least ten times higher than the K_i value of 16 nM of the reference molecule **PALA**. The best inhibition constant was found for the α-hydroxyphosphinic acid **105** ($K_i = 190$ nM). The cyclic azaphosphinanes **106** and **107** were shown to be inactive. Two others structures belonging to the phosphonic acid class, **125** and **126**, and one phosphoric acid derivative **127** were also evaluated (Fig. 18). Contrary to phosphonic acid **126**, product **125** was inactive, whereas phosphate analogue **127** presented a close biological response to hydroxymethyl phosphinate **105**. Most of tested compounds were revealed to be competitive inhibitors, except for compound **109** which was non-competitive. Two

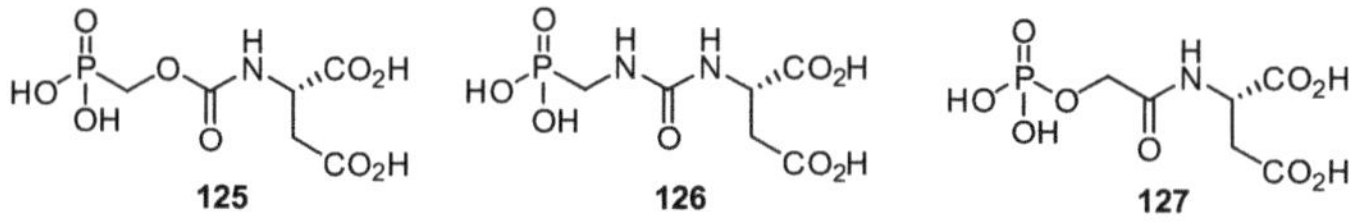

Fig. 18 Phosphonic acids **125** and **126**, and phosphoric acid **127**

different directions were identified by Montchamp to improve the efficiency of PALA-related drugs. First, modifying the design of PALA-related prodrugs for masking charges, preferentially around the phosphorus atom to improve bioavailability, and, second, the development of cyclic or constrained compounds miming the enzymatic transition-state.

5.3 *Inhibitors of Tubulin Polyglutamylation*

Microtubules are major structural components of the cytoskeleton, and are formed by the oriented polymerization of α- and β-tubulin from the centrosome to the periphery of the cell [106]. They play essential functions by regulating cell shape, intracellular transport and cell motility. Post-translational modifications (PTM) of tubulin subunits modified their biophysical properties affecting their dynamic and interaction with the microtubules-associated proteins. Indeed, modified tubulins are not equally distributed along microtubules and it is now admitted that a biochemical "tubulin code" can be recognized by factors which interact with tubulin [107]. Among these post-translational modifications (PTM), the C-terminal tail of the tubulin subunit undergoes reversible changes, such as detyrosination, polyglycylation, and polyglutamylation [108, 109]. Polyglutamylation consists of α or γ side chains branching on the C-terminal of glutamic acid within the C-terminal domain of both α and β tubulin. Deregulation of PTM affects neuronal development [110, 111]. It is also implied in hereditary spastic paraglesias [112], mitosis [113], pancreatic cancer cells [114] and neurodegeneration [115]. Enzymes which catalyze tubulin glutamate ligation for initial α- or γ-elongation are members of the "tubulin-tyrosine ligase-like" (TTLL) family (Scheme 26) [116].

Therefore, the development of small molecule inhibitors of TTLL enzymes has been perceived as a pertinent therapeutic response, disrupting tubulin PTM levels. To select specific inhibitors, inhibitory potentials were evaluated against the different TTLL E-ligases. Tanner and Roll-Mecak targeted three different potential TTLL inhibitors **128–130** [117]. These structures were merely mimicking the tetrahedral transition states of every intermediate (Fig. 19).

The preparations of inhibitors **128** and **130** were performed applying a divergent synthetic pathway (Scheme 27). First, the free radical hydrophosphination of vinyl glycine **131** [118], followed by the addition to an electro-deficient olefin **132** [119] and the methylation with trimethylsilyldiazomethane afforded phosphinate ester **133**.

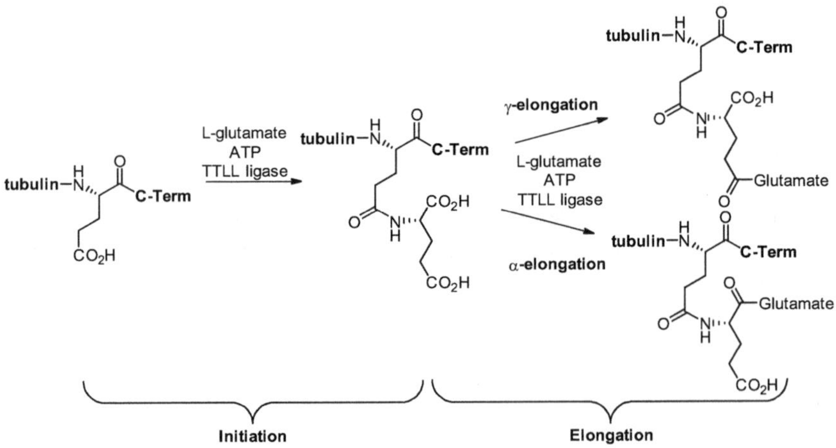

Scheme 26 Tubulin polyglutamylation catalyzed by the TTLL enzymes

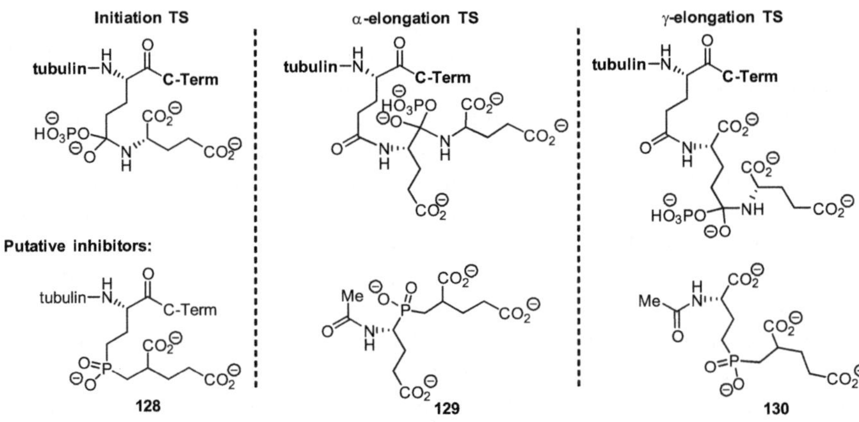

Fig. 19 Putative tetrahedral TS of tubulin polyglutamylation and their analogues

Thereafter, the cleavage of the benzyl group followed by acetylation and amide coupling or esterification gave amide **134** or methyl ester **135**. Deprotection of the methyl ester and acidification gave phosphinic acids **128** and **130**, each as a mixture of two diastereomers.

The first steps for the synthesis of phosphinic acid **129** were similar to those of **128** and **130**. Starting from *H*-phosphinic acid **136**, methyl phosphinate **137** was obtained in 88% yield (Scheme 28) [120]. An oxidative cleavage of the phenyl group using $RuCl_3$ and $NaIO_4$ produced carboxylic acid **138**, which was directly transformed by trimethylsilyldiazomethane into tetramethyl ester **139**. The Boc protecting group was removed with TFA and the free amine was acetylated, affording phosphinate **140**, which, after basic hydrolysis and acidification, furnished phosphinic acid **129** as a mixture of four stereoisomers.

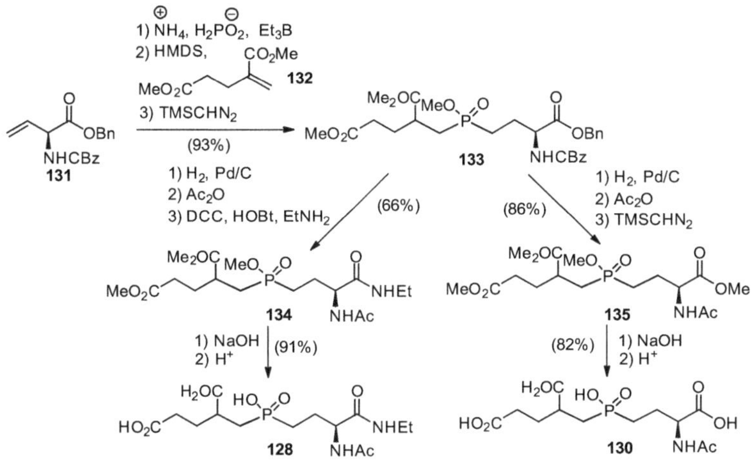

Scheme 27 Preparation of TTLL inhibitors **128** and **130**

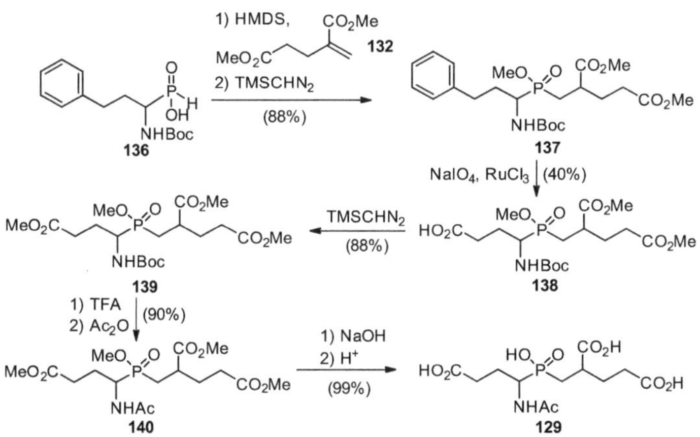

Scheme 28 Preparation of TTLL inhibitor **129**

The response of the recombinant mouse variant of the TTLL7 enzyme to the products **128–130** was evaluated using a synthetic *N*-acetylated βIVb-tubulin C-terminal peptide in the presence of glutamate, ATP and Mg^{2+}. The most powerful inhibitory effect was noticed for compound **129** which displayed an IC_{50} value of 150 μM [117]. In the case of compounds **128** and **130**, inhibitions were observed only in the millimolar range of concentrations. It was suggested that stronger binding of inhibitor **129** reflects the ability of TTLL7 to catalyze the α-elongation step for polyglutamination. Thereafter, all three compounds **128–130** were tested against the related enzyme, tubulin-tyrosine ligase (TTL). None of these compounds were able to inhibit the reaction catalyzed by TTL, indicating a selectivity for glutamylase TTLL7. The authors suggested that the modest inhibition of

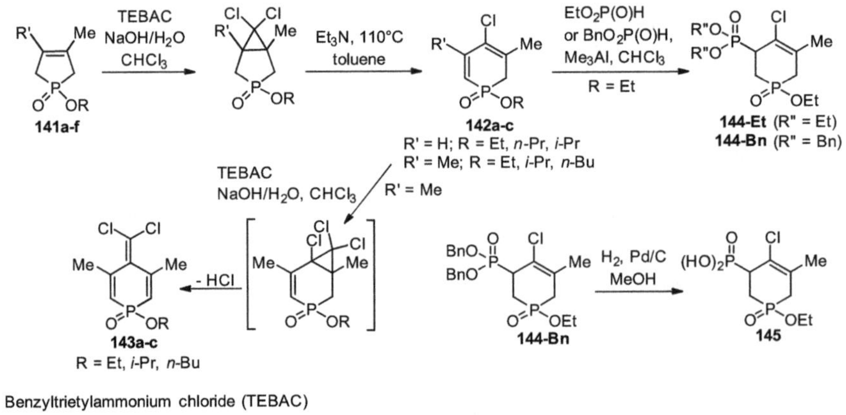

Scheme 29 Preparation of 1,2- and 1,4-dihydrophosphinine oxides **142** and **143**, and 1,2,3,6-tetrahydro analogues **144** and **145**

compound **129** could be caused by its small size and its insufficient global negative charge compared to natural tubulin-polypeptide substrates.

5.4 Antiproliferative Phosphinates

Although resistant mechanisms are encountered, organophosphorus alkylating agents such as cyclophosphamide, ifosfamide and thiopa are intensively used in clinic to treat cancer [121]. Numerous other organophosphorus molecules, such as phosphinic and phosphonic acid derivatives, aminophosphonates, bisphosphonic acid derivatives, phosphonocarboxylate–platinum complexes and aminophosphonate–platinum complexes were studied and displayed anticancer properties [121]. Keeping this background in mind, Hudson and Keglevich reported the evaluation of six-membered ring *P*-heterocycles [122], such as 1,2- and 1,4-dihydrophosphinine oxides **142**, **143** [123, 124] and 1,2,3,6-tetrahydrophosphosphinines **144**, **145** as anticancer agents (Scheme 29) [125]. 1,4-Dihydrophosphinines **142**, **143** were obtained by ring expansion of 2,5-dihydro-1*H*-phosphole oxides **141a–f** using dichlorocarbene cycloaddition followed by an intramolecular rearrangement. 1,2,3,6-Tetrahydrophosphosphinines **144a**, **b** were produced by Michael addition of diethyl phosphite or dibenzyl phosphite. Cleavage of the benzyl group by hydrogen from **144b** afforded phosphonic acid **145**.

Antiproliferative activities of these structures have been evaluated in vitro against the NCI 60-cell line panel of human malignant cells [125]. The best total growth inhibition (TGI) activities on the 60-cell line were obtained for the phosphinate series **142**, **143** with values of 38.1 μM for **142a** and 29.2 μM for **143c**. Tetrahydrophosphinines **144-Et**, **144-Bn** and **145** were less active. In parallel,

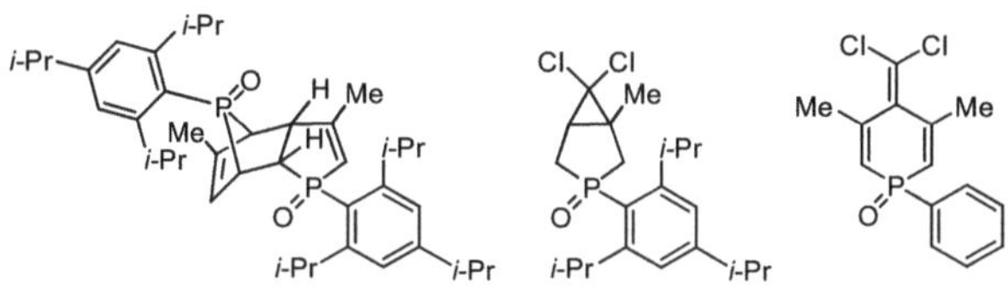

Fig. 20 Structures of phosphine oxides **146–148** evaluated for antiproliferative activity

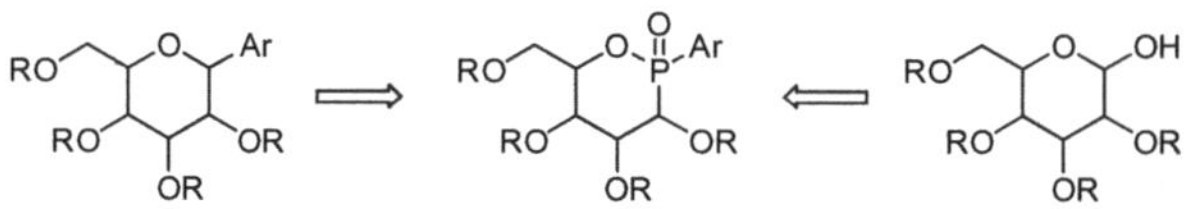

Fig. 21 Phostine a combination of C-arylglycosides and glycopyranosides

among phosphine oxides **146–148** tested, structure **146** presented the best antiproliferative activity (TGI = 6.63 μM) (Fig. 20).

5.5 *Phosphinates as Sugar Mimetics*

Gliomas are the most frequent primary central nervous system (CNS) tumours. They account for over 40% of CNS tumours and 78% of CNS malignancies in adults. The median survival is only 12–15 months for patients with glioblastomas, 2–5 years for patients with anaplastic gliomas and 4–10 years for patients with low-grade gliomas. The best treatment, which consists of surgical resection followed by chemotherapy with Temozolomide (TMZ) and radiotherapy, gives a median survival of 15 months as compared to 55 days with exeresis alone. To develop a new antiglioma therapy, Bakalara et al. presented, in 2012, a new family of compounds, phostines, targeting CNS cancers without affecting normal astrocytes [126, 127]. Phostines, which present a 1,2-oxaphosphinane heterocyclic core, were designed as a combination of glycopyranosides and *C*-arylglycosides (Fig. 21).

The phosphinolactone group of phostines can be regarded as a surrogate of the lactol. Their synthesis first involved the nucleophilic Pudovik addition of *H*-phosphinate anions **149** to the open-chain form of arabinofuranose **150**. The second step consisted of a subsequent prototropy and a 6-exo-tet cyclization, affording six-membered ring phosphinolactones in 21–67% yield (Scheme 30). This procedure offered an easy tuning of the aryl substituent directly bound to the phosphorus atom. Four enantiopure diastereomers were formed, arising from the

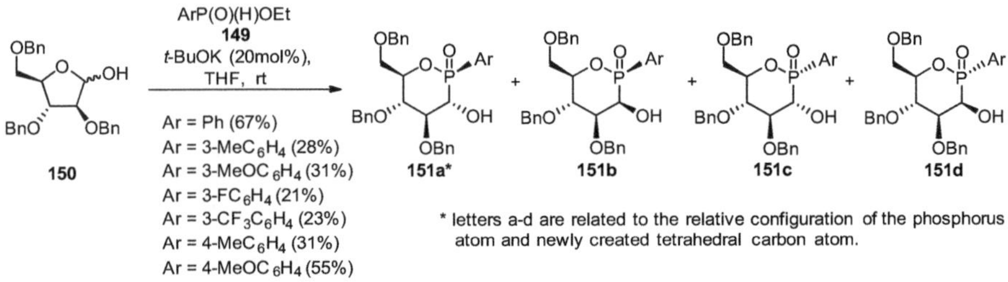

Scheme 30 Synthetic route of oxazaphosphinanes **151a–d**

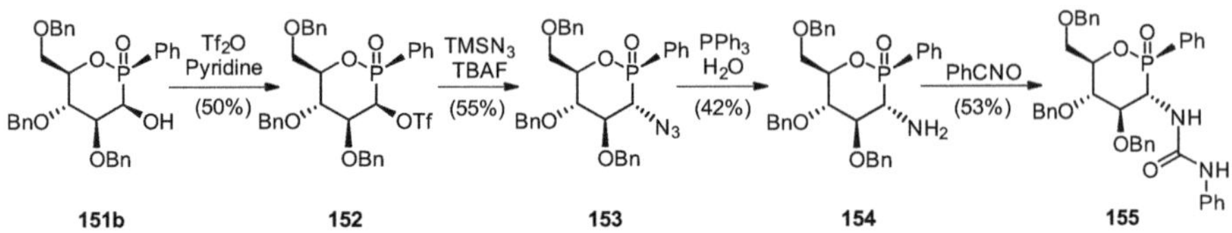

Scheme 31 Synthetic route to aminophosphino-α-D-glucose derivatives

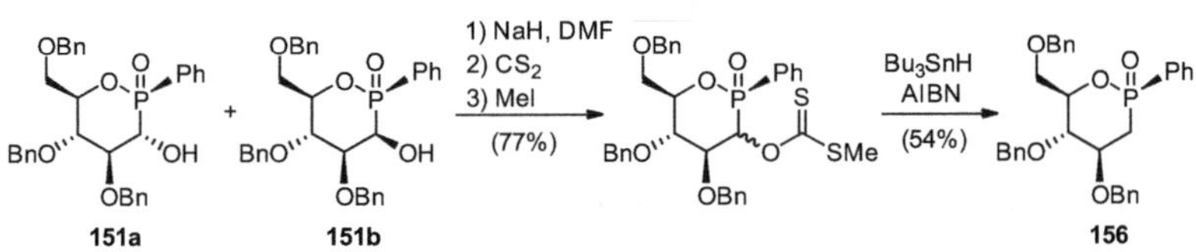

Scheme 32 Synthetic route to deoxyphosphinolactone **156**

creation of two chiral centres with weak diastereoselectivities, but they were separated by chromatography.

In a second set of reactions, different chemical modifications were introduced at the carbon adjacent to the phosphorus to synthesize amino and urea derivatives of phosphino-α-D-glucose. The transformation of the alcohol function into a good leaving group led to triflate **152** in 50% yield. The substitution by TMS-N_3 gave azido derivative **153** with an inversion of configuration. The Staudinger reaction furnished the glucose-like amino-1,2-oxaphosphinane **154**. Reaction with phenylisocyanate gave the corresponding urea **155** in 53% yield (Scheme 31).

To estimate whether the presence of the free hydroxyl group affects the IC_{50}, the corresponding deoxyphosphinolactone **156** was synthesized, starting from the mixture of diastereoisomers **151a** and **151b** and using Barton–McCombie deoxygenation (Scheme 32). As expected, the methylene derivative **156** was isolated as a single enantiomer.

Other modifications were introduced by the formation of the corresponding ester **157** and ether **158** using acylation or Williamson reactions (Scheme 33).

The direct synthesis of amino-oxaphosphinanes **160** and **161** was performed starting from alkylamino-arabinofuranose **159**, directly obtained by the treatment of

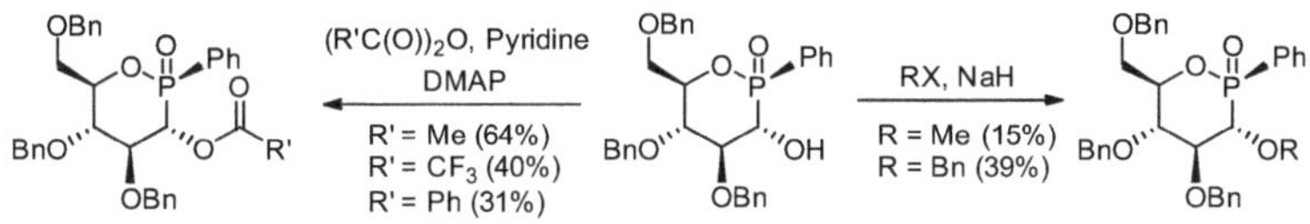

Scheme 33 Synthesis of ester **157** and ether **158** phosphinolactone derivatives

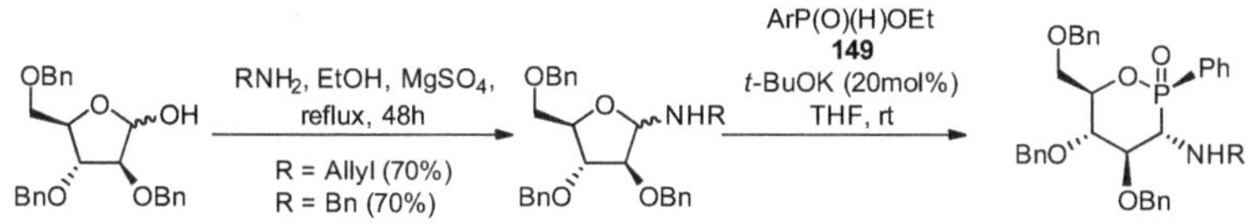

Scheme 34 Synthesis of amino oxaphosphinanes **160a** and **161a**

2,3,5-tri-*O*-benzyl-D-arabinofuranose **150** with an excess of benzylamine or allylamine (Scheme 34). Only one stereoisomer, aminophosphino-α-D-glucose-like, was obtained for each reaction (R = Allyl **160a** or R = Bn **161a**).

Twenty-six compounds have been screened for their antiproliferative activity against a large panel of NCI cancer cell lines. The first data indicated that this new family of compounds targeted CNS cancer cell lines without affecting normal astrocytes. Furthermore, compound **151a** has better in vitro antiproliferative activity on glioma than TMZ and no toxicity at 100 μM.

In 2014, Bakalara et al. described new phostines, D-glycero-D-talo- and D-glycero-D-galactopyranose analogues, which were tested for their inhibition on the invasion and migration on both GBM primary cultures (Gli7 and Gli4) and GBM cancer cell lines (C6, SNB75) [128].

Their syntheses were achieved by the reaction of 2,3,5,6-di-*O*-isopropylidene-α-D-mannofuranose **162** with various ethyl alkyl- or methyl phenylphosphinates **163** in the presence of catalytic amounts of potassium *tert*-butoxide. Only three pure diastereomers **164–166** were obtained by chromatography. Modifications from β-D-glycero-D-talo-heptopyranose **164d** were introduced by the formation of the corresponding ester **167d**, **168d** and ether **169d–172d** derivatives using acylation or Williamson reactions (Scheme 35).

Amino- and acetamido-D-glycero-D-galactopyranose analogues were obtained starting from 1-amino-2,3,5,6-di-*O*-isopropylidene-α-D-mannofuranose **173**, directly obtained by the treatment of 2,3,5,6-di-*O*-isopropylidene-α-D-mannofuranose **162** with an excess of aryl- or benzylamine. Only one stereoisomer, aminophosphino-α-D-glycero-D-galactopyranoses **174–177**, was obtained in each case in moderate yield (Scheme 36). *N*-Acetyl derivative **178** was obtained after cleavage of benzyl protecting group, followed by acetylation of the free amine.

One phostine **176** (R = 1-naphth) inhibited invasion and migration on fibronectin, vitronectin and laminin on both GBM primary cultures (Gli7 and Gli4) and

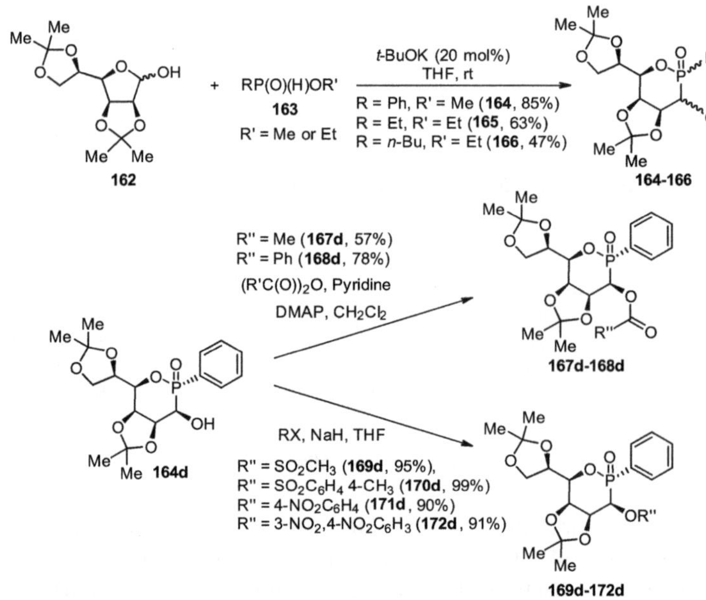

Scheme 35 Synthetic routes of D-glycero-D-talo-heptopyranose analogues

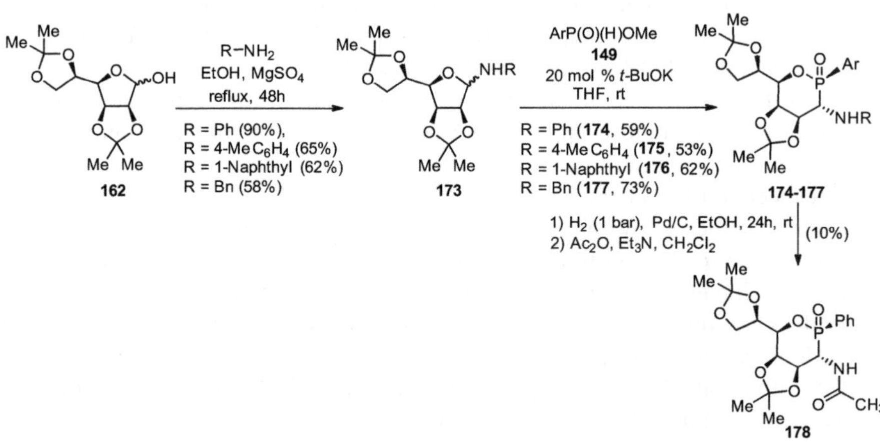

Scheme 36 Synthesis of amino and acetamido-α-D-glycero-D-galactopyranose analogues

GBM cancer cell lines (C6, SNB75). K_i values for Gli7 and Gli4 migration inhibition on fibronectin were of 16 and 31 nM, respectively. These activities were associated with an antiproliferative effect on Gli4 ($EC_{50} = 5.22$ μM) and Gli7 ($EC_{50} = 2.33$ μM).

The heptopyranose mimetic **176**, devoid of toxicity on astrocyte and cortical neuron cultures (at concentrations below 100 μM), opens new therapeutic perspectives to treat glioblastoma.

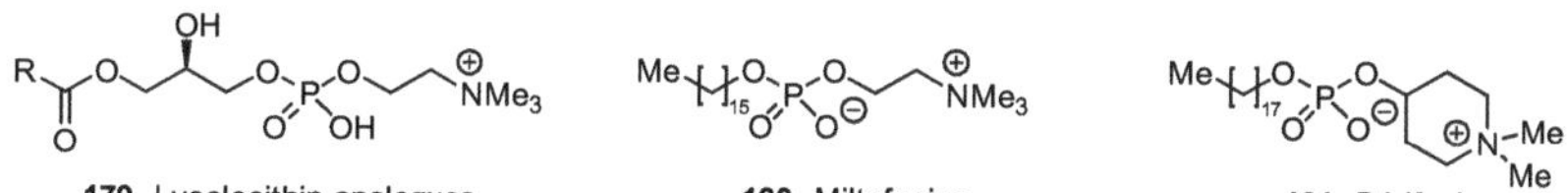

179, Lysolecithin analogues — 180, Miltefosine — 181, Périfosine

Fig. 22 Structure of lysolecithin analogues **179**, miltefosine **180** and perifosine **181**

Phosphinate miltefosine analogues, **182** — Phosphinate perifosine analogue, **183**

n= 15: **182a** R = NHMe; **182b** R = NH_3; **182c** NMe_3
n = 17: **182d** R = NMe_3

Fig. 23 Phosphinate analogues of miltefosine **182a–c** and perifosine **183**

5.6 Phospholipases Inhibitors

Analogues of alkyl phospholipids, such as stable analogues of lysophosphatidylcholine **179**, synthesized in the late 1960s for potent immune modulators, have also been investigated in ascites tumours. In 1978 Tarnowski et al. showed that lysolecithin analogues **179** induced tumour inhibition against Ehrlich in mice [129]. In the late 1980s, Eibl and Unger identified miltefosine **180** as the minimal structure to observe an anti-tumour activity for synthetic analogues of phospholipids [130]. In the 1990s, Stekar et al. synthesized perifosine **181**, a second generation of phospholipid analogues with potent anti-tumour activity [131, 132] (Fig. 22).

In 2011 and 2013, Regan et al. described the synthesis of phosphinate analogues of miltefosine **180** and perifosine **181** [133, 134]. The replacement of the phosphate group in miltefosine **180** and perifosine **181** by a phosphinate should make these analogues **182** and **183** less prone to degradation by phospholipid-metabolizing enzymes such as phospholipases C and D. The presence of two stable P–C bonds in analogues **182** and **183**, along with one negative charge at the oxygen, makes them suitable bioisosteres (Fig. 23).

The synthesis of phosphinate precursor **186** was accomplished using free radical addition of sodium phosphinate **184** to the appropriate terminal alkenes **185** in high yields ($n = 15$ or 17:91%) (Scheme 37). However, subsequent hydrophosphinylation of *N*,*N*-dimethylallylamine with hexadecylphosphinic acid **186** under the same conditions only afforded 3% yield of the desired adduct **187**. Then the phosphinate salt was dissolved in ethyl acetate and treated with concentrated hydrochloric acid, giving the zwitterionic form **182a** in 97.5% yield. A more efficient synthesis of phosphinate analogues of miltefosine was obtained by combining the first high-yielding radical reaction with a silyl phosphonite mediated Michael addition to acrylonitrile, followed by acidic hydrolysis, producing **188** in high yield (>90%). Esterification of **188** with trimethyl orthoformate and hydrogenation at atmospheric pressure in the presence of Raney nickel afforded amines

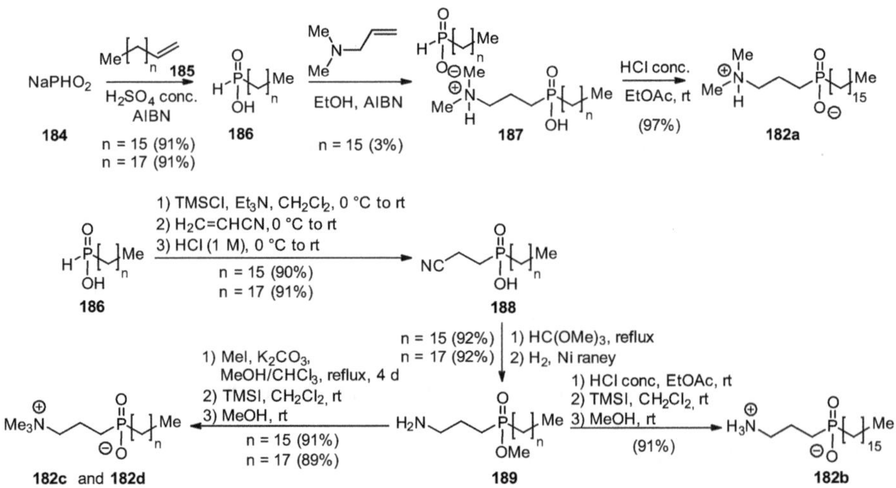

Scheme 37 Synthesis of phosphinate analogues of miltefosine **182**

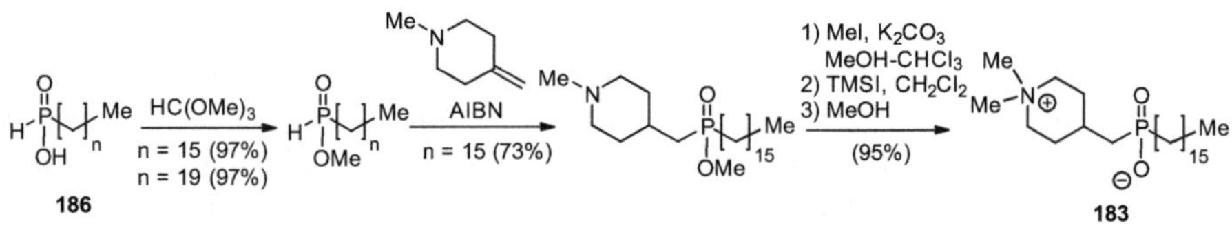

Scheme 38 Synthesis of phosphinate analogues of perifosine **183**

189 in 92%yields. An excess of methyl iodide in the presence of anhydrous potassium carbonate led to the quaternization of the primary amine. Then deprotection of methyl phosphinate esters was achieved using iodotrimethylsilane followed by methanolysis to afford ammonium phosphinate **182b–d** inner salts in high yields.

Phosphinate perifosine analogue **183** was synthesized using the same methodology described previously in good overall yield (Scheme 38).

All these described compounds are actually under biological investigation in order to establish a structure–activity relationship.

6 Phosphinic Acids as Antibiotics or for Parasitic Diseases

6.1 *Phosphonophosphinic Acid: Quinolone Prodrugs*

Bisphosphonates (BPs) are compounds belonging to an important therapeutic class for bone disease treatment (Paget's disease, metastatic and osteolytic bone disease and osteoporosis). Since the beginning of their clinical use in the late 1960s, the

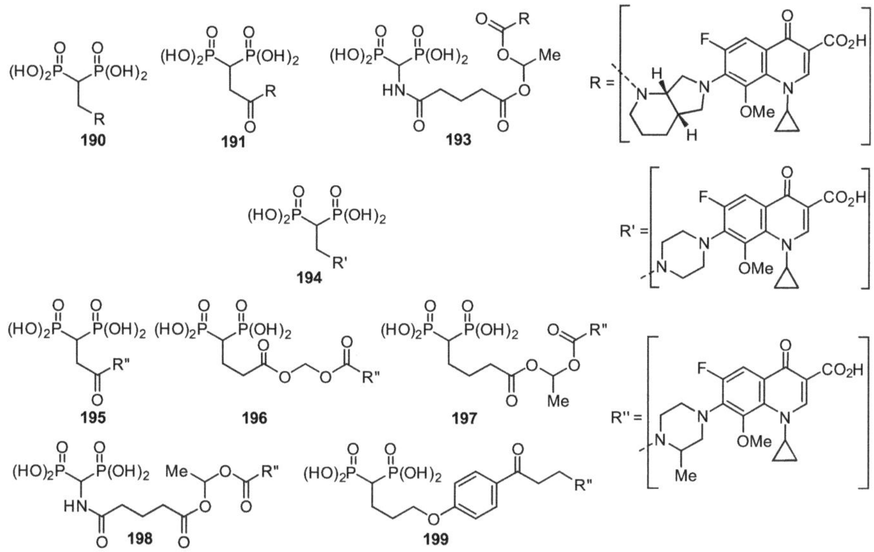

Fig. 24 Bisphosphonic acids **190–199** tested as antibacterial agents

bisphosphonates, and particularly their chemistry, has been the subject of continuous development, demonstrated by the regular appearance of publications [135, 136]. In addition, because of their potential to accumulate in the hydroxyapatite bone matrix, bisphosphonates appeared of interest as useful agents for the vectorization of drugs towards bones. In this context, Far and collaborators were interested in linking a fluoroquinolone antibiotic to bisphophosphonics acids **190–199** (Fig. 24) or phosphomethylphosphinic acids **200** and **201a–c** [137]. The basic idea was to provide a cure for osteomyelitis, a notoriously problematic bacterial infection located in bones.

The first phosphonomethylphosphinic acid prodrug **200** has been prepared according to a seven-step sequence beginning with dibromopropiophenone **202** (Scheme 39). After hydrolysis of the bromine atom and transformation to the α-hydroxyketone intermediate, 4-(4-bromophenyl)-5-methyl-1,3-dioxol-2-one **203** was obtained by cyclization with phosgene. A palladium-catalyzed coupling reaction between **203** and phosphonomethyl-*H*-phosphinate **204** [138], followed by an allylic bromination, led to carbonate **205**. A direct nucleophilic substitution with gatifloxacin gave the protected prodrug **206**, which finally produced the free acid **200** after total deprotection.

Prodrugs **201a–c** were synthesized from palladium-catalyzed phosphinylation of 4,4′-dimethyldihydrocoumarins **207** [139] with phosphonomethyl-*H*-phosphinate **204** (Scheme 40). Cleavage of the lactone function and reaction with benzyl bromide gave the ester **208**. Different substituents were introduced at the phenol function, affording esters **209a–c**. Cleavage of the benzyl group produced acids

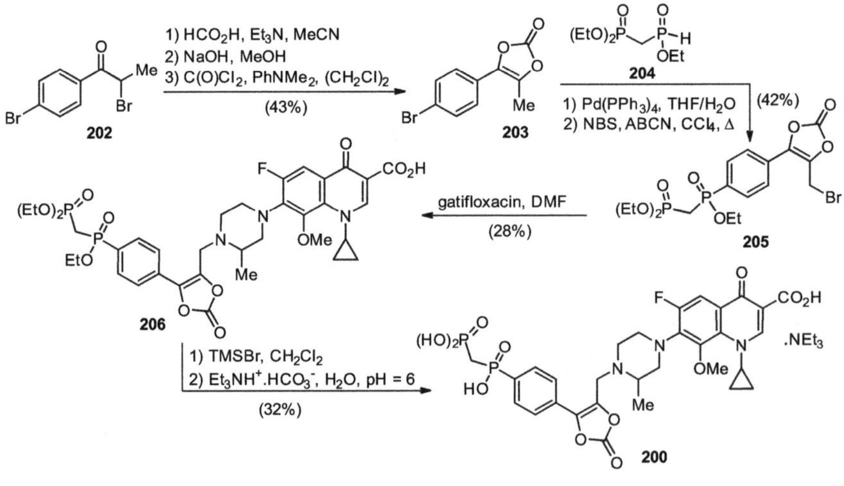

Scheme 39 Synthesis of phosphonomethylphosphinic acid prodrug of gatifloxacin **200**

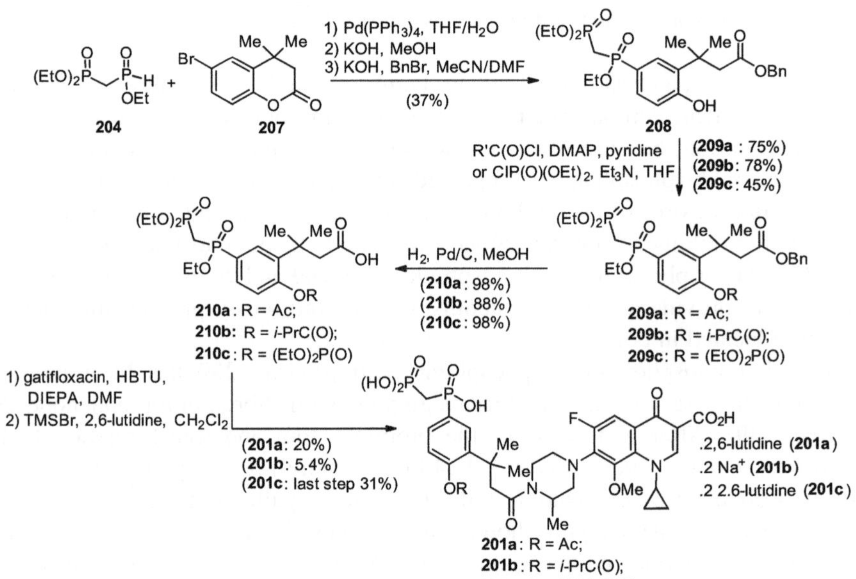

Scheme 40 Preparation of prodrugs **201a–c**

210a–c, which coupled with gatifloxacin in the presence of HBTU to yield prodrugs **201a–c** after deprotection.

In vitro binding bone powder tests showed that most of the bisphosphonates were taken up with an efficiency greater than 80%. To a lesser extent, phosphonyl-phosphinic acids **200** and **201a, b** exhibit a binding capacity comprised of 35–76%.

However, the results obtained in vitro for their antibacterial activities against *Staphylococcus aureus* ATCC13709 were less convincing. Compared to gatifloxacin (minimum growth inhibition concentration = 0.06 μg/mL), compounds **200**, **199** and **196** activity was lower (0.12, 0.12 and 0.5 μg/mL, respectively). According to the authors, the reduced observed activity of prodrugs would only be caused by the release of the fluoroquinolone drug during the course of the assay.

Because of their capacity to release their parent antibiotic drug and of their high bond bone affinity, bisphosphonic acids derivatives **196** and **199** were selected for in vivo evaluation against *S. aureus* in rat model of osteomyelitis. Interestingly, the results revealed that, in contrast to gatifloxacin, both compounds **196** and **199** were able to prevent the infection booting. These drugs are therefore likely to be further investigated for their capacity to treat osteomyelitis.

6.2 Strained Cyclic Phosphinic Acids, Inhibitors of Glutamate Racemase (GR)

The rise in antibiotic resistance of pathogenic organisms stimulated the discovery of new targets and de facto the development of new antibacterial agents. L-Amino acids prevail in nature, but their enantiomers, D-amino acids, may be used as specific metabolites, regulators or as key components by living organisms. For instance, D-glutamate is an essential amino acid for elaboration of peptidoglucane layer of bacterial cell wall, protecting them against osmotic lysis [140–142]. Glutamate racemase (GR) is the enzyme controlling the stereoconversion of L-glutamate into D-enantiomer. In this perspective, Pal and Bearne reported cyclic phosphinic acid **211**, analogues of natural glutamate substrate, as modest and partial noncompetitive inhibitors of glutamate racemase from *Fusobacterium nucleatum* (FnGR) [143].

Phosphinane **211** was prepared through a six-step sequence starting from hypophosphorous acid (Scheme 41) [143]. After transformation into silyl phosphonite by treatment with trismethoxy(propyl)silane, a Michael addition

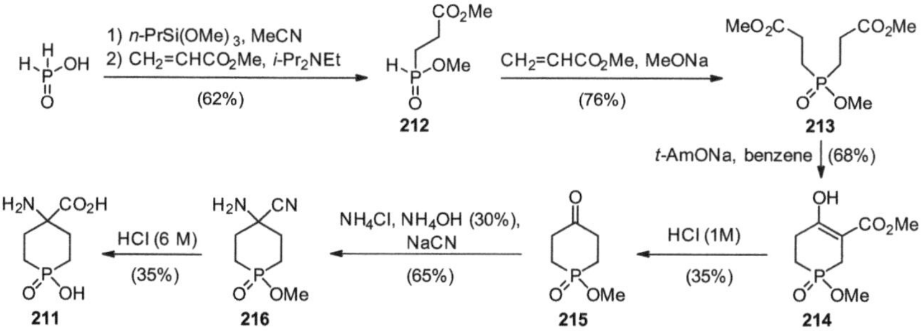

Scheme 41 Synthesis of phosphinane **211** as inhibitor of glutamate racemase

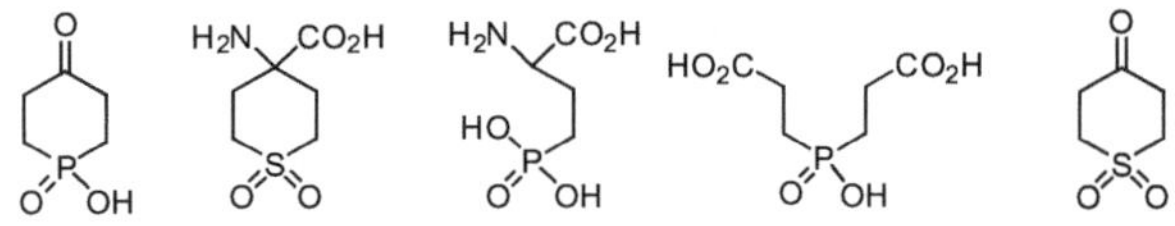

Fig. 25 Glutamate analogues **217–221** tested as glutamate racemase inhibitors

afforded *H*-phosphinic acid **212** in 62% yield. 4-Oxophosphinane **215** was synthesized from **212** by applying the strategy already described by Wróblewski and Verkade and involving a Michael addition followed by a Dieckmann cyclization [144]. Employing Strecker reaction from **215**, 4-amino-4-cyanophosphinane **216** was obtained as a diastereomeric mixture (1:2) and was deprotected to the final structure **211** by acidic hydrolysis.

The inhibitory properties of phosphinane **211** were evaluated against FnGR [143]. With a K_i value of 24.1 ± 3.8 mM, compound **211** was considered as a modest inhibitor of FnGR. Moreover, even when used at the highest foreseeable inhibitory concentration of 37 mM, a complete inhibition was not reached. Furthermore, with IC_{50} ranging from 47.7 to 215 mM (Fig. 25), the racemic mixture of **211** was the best inhibitor among others glutamate substrate analogues tested, such as compound **217** (produced by acidic hydrolysis of **215**), cyclic sulfone **218** (produced by deprotective reaction of commercial available *N*-Boc derivative), acyclic phosphonic acid derivative **219** (commercially available) and symmetric phosphinic acid **220** (produced by acidic hydrolysis of **213**). The only one giving a slightly better response was the commercially available cyclic γ-ketosulfone **221**, showing an $IC_{50} = 21.2 \pm 1.5$ mM. Nevertheless, the inhibition kinetics of **211** highlighted an affinity twice as weak ($K_i = 3.1$ mM) as the natural substrate, but about six times better than **221** ($K_i = 18.4$ mM).

6.3 *Cyclobutyl-1,3-Thiazolyl Phosphinic Acids*

In 2011, Koparir et al. reported aminophosphinic acids **223a–l** containing cyclobutane and 1,3-thiazole rings with antibacterial properties (Scheme 42) [145]. Their preparation consisted of a condensation, mediated by hydrochloric acid, of aryl aldehyde derivatives with hypophosphorous acid and 2-amino-thiazoles **222a–c** incorporating a substituted cyclobutyl group in position 4. The precursors **222a–c** were prepared according to the method given previously by the Koparir group [146–149].

Compound **223f** exhibited a significant activity against *Staphylococcus aureus*, and **223g** revealed the best product tested against *Mycobacterium fortuitum* with a minimum inhibitory concentration (MIC) value of 32 μg/mL.

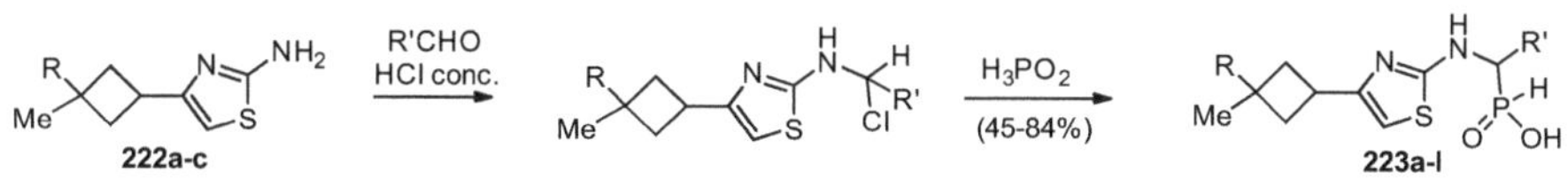

223a: R = mesityl, R' = 2-hydroxynaphthalen-1-yl
223b: R = mesityl, R' = 4-nitrophenyl
223c: R = mesityl, R' = 4-methoxyphenyl
223d: R = mesityl, R' = 4-chlorophenyl
223e: R = phenyl, R' = 2-hydroxynaphthalen-1-yl
223f: R = phenyl, R' = 4-nitrophenyl
223g: R = phenyl, R' = 4-methoxyphenyl
223h: R = phenyl, R' = 4-chlorophenyl
223i: R = tetralino, R' = 2-hydroxynaphthalen-1-yl
223j: R = tetralino, R' = 4-nitrophenyl
223k: R = tetralino, R' = 4-methoxyphenyl
223l: R = tetralino, R' = 4-chlorophenyl

Scheme 42 Synthesis of aminophosphinic acids **223a–l**

6.4 Antimalarial Phosphinates: Inhibitors of Dihydrofolate Synthase

Dihydropteroate synthase (DHFS, EC 6.3.2.12) is an enzyme which catalyzes the ATP-dependent conversion of 7,8-dihydrodropteroate (DHP) into 7,8-dihydrofolate (DHF) by insertion of a single L-glutamate moiety (Fig. 26) [150]. This enzymatic step precedes the folylpoly-γ-glutamyl synthase (FPGS) which catalyzes the ATP-dependent addition of glutamate moieties to folate. Unlike human cells, microorganisms such as *Plasmodium falciparum*, the causative agent of malaria, express this enzyme as well as the 7,8-dihydropteroate synthase (PfDHPS) [151], which plays an important role for the de novo synthesis of reduced folylmono- and folylpoly-γ-glutamates. Moreover, *Plasmodium falciparum* has a unique feature since DHFS and FPGS are present in a single bifunctional protein (PfDHFS-FPGS). If DHPS and dihydrofolate reductase (DHFR) inhibitors have been extensively investigated, little attention was given to the design of PfDHFS-FPGS inhibitors. For this reason, Coward and Yang planned to prepare several arylphosphinic acid derivatives **224**, as possible PfDHFS-FPGS inhibitors, by mimicking, after introduction of a phosphate group on **224**, the tetrahedral intermediate formed during enzymatic amide bond formation (Fig. 26) [152].

For instance, arylphosphinic acid **224a** was prepared by applying a five-step sequence (Scheme 43) [152]. The first reaction was a palladium-catalyzed phosphinylation of benzyl 4-iodophenylcarbamate **225a** with anilinium hypophosphite in the presence of palladium(II) acetate and a bidentate phosphine ligand. Methyl arylphosphinate **226a** was transformed into phosphonite formed in situ by reaction with *N*,*O*-(bistrimethylsilyl)acetamide (BSA) and was treated with dimethyl 2-methyleneglutarate **227**, affording **228a**. Hydrogenolysis of the CBZ protecting group with hydrogen and palladium on charcoal afforded almost quantitatively the deprotected methyl 4-methylaminophenylphosphinate derivatives **229a**. Alkylation reaction of **229a** with an excess of 6-(bromomethyl)pteridines **230a** in dimethylacetamide gave, after 14 days at room temperature, a mixture of ester **231a-OMe** and phosphinic acid **231a-OH** in 52% yield. The mixture **231a-OMe/231a-OH** was separable by semipreparative HPLC and **231a-OMe** was isolated in low yield. Compound **224a** was obtained from **231a** after basic

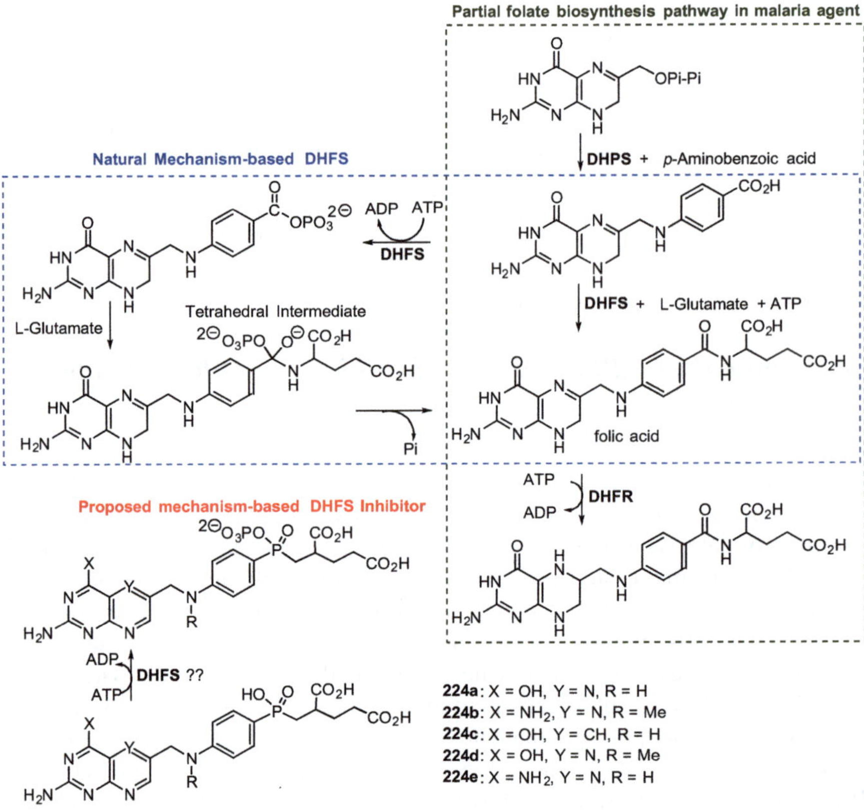

Fig. 26 Partial folate biosynthesis pathway in malaria agent (*Plasmodium falciparum*), natural mechanism-based DHFS, and proposed mechanism-based DHFS inhibitor

hydrolysis in methanol, followed by acidification with hydrochloric acid. Compounds **224b**, **224d** and **224e** were prepared using a similar strategy adapting reactants, reagents and conditions.

Phosphinic acid **224c** was accessible from **229a**, proceeding by a reductive amination instead of a direct alkylation and producing precursor **231c**. Subsequent deprotections gave the fully deprotected phosphinic acid **224c** (Scheme 44).

Biological evaluation of compounds **224a–e** revealed that they were modest inhibitors of *Escherichia coli* glutamate DHFS-FPGS-catalyzed ligation reaction [152]. Further study highlighted that the reduced 7,8-dihydroderivative **232** (Scheme 45) [153] was accessible by reduction with dithionite [154, 155] of phosphinic acid **224a** and inhibited *Plasmodium falciparum* DHFS activity with an IC_{50} of 0.41 μM. This compound **232** was also five times more potent for the inhibition of PfDHFS ($K_i = 140$ nM) compared to PfFPGS ($K_i = 630$ nM). In contrast to these results, reduced 7,8-dihydroderivative **234** (Scheme 45), obtained from **233**, was 19 times more active on PfFPGS ($K_i = 91$ nM) than on PfDHFS

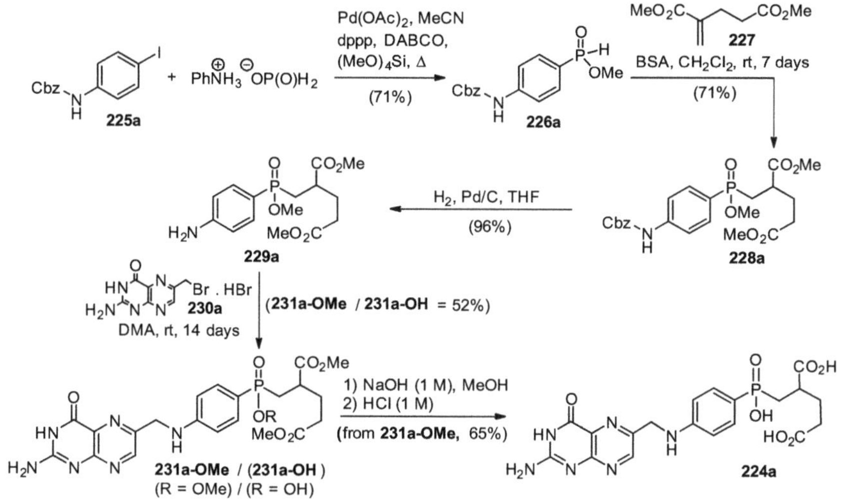

Scheme 43 Synthesis of arylphosphinic acid **224a**

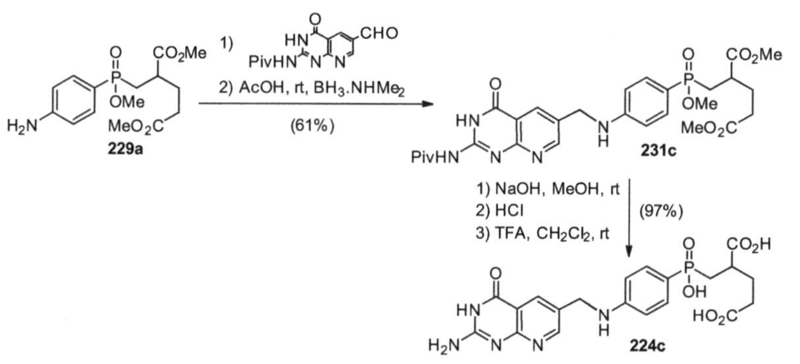

Scheme 44 Synthesis of phosphinic acid **224c**

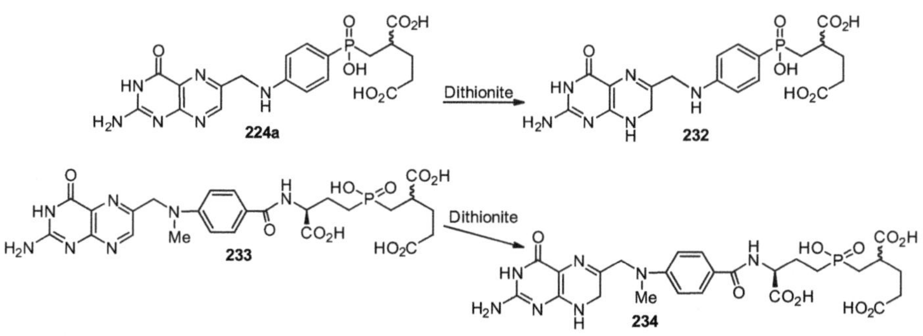

Scheme 45 Preparation of reduced compounds **232** and **234**

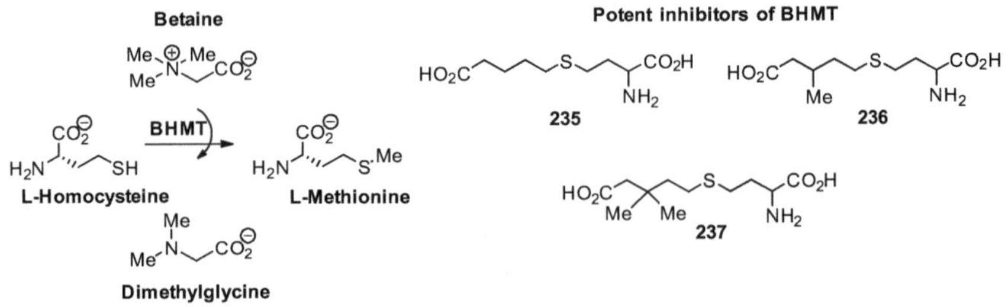

Fig. 27 Enzymatic synthesis of L-methionine – Potent inhibitors **235–237** of BHMT

($K_i = 1{,}690$ nM). Nevertheless, preliminary data indicated that these reduced compounds were ineffective at inhibiting parasite growth in culture.

7 Phosphinic Acids for Metabolic Disorders

7.1 *Inhibitors of Betaine-Homocysteine Methyltransferase*

Betaine-homocysteine methyltransferase (BHMT; EC 2.1.1.5) belongs to the Pfam02574 family of zinc- and thio/selenol-dependent methyltransferase [156]. It catalyzes the transfer of a methyl group from betaine to homocysteine, producing dimethylglycine and methionine (Fig. 27). Because methionine is a source of *S*-adenosylmethionine (SAM), which is considered as a major methyl donor in the cell, BHMT plays a central role in the methylation cycle [157]. The BHMT pathway is particularly active in the liver and the kidney cortex, which are the main organs which store large amounts of betaine. The transient inhibition of BHMT enzyme by SAM regulates cellular osmolytic equilibrium by modulating betaine concentration. Thus, BHMT is considered as a potential target to restore osmotic balance during unwanted diuresis.

In the past, Jiráček et al. described several selective and potent inhibitors of BHMT, belonging to phosphinic pseudopeptides [158] and *S*-alkyl derivatives of homocysteine [159, 160]. Among the compounds tested, three of them **235–237** incorporating an *S*-alkyl homocysteine were detected as potent inhibitors (Fig. 27).

Continuing their rational design, Jiráček and al. described α-aminophosphinic acids **239**, **240**, structurally close to compounds **235** and **237** [161]. Each structure was obtained in 56–64% yield, by mixing ethyl 5-bromovalerate or ethyl 5-bromo-3,3-dimethylpentanoate with phosphinomethionine **238** in the presence of sodium and ammonia (Scheme 46). The *H*-phosphinic acid intermediate **238** was readily accessible by a sequence of three steps from acrolein [162–164].

Both *H*-phosphinic acids **239** and **240**, used at 20 μM, exhibited in vitro almost a complete inhibition of the human recombinant BHMT in a buffer containing 0.25 μM of betaine and 0.1 mM of D,L-homocysteine. The efficiency of both products **239** and **240** was very similar to that of the parent carboxylic acid **235**,

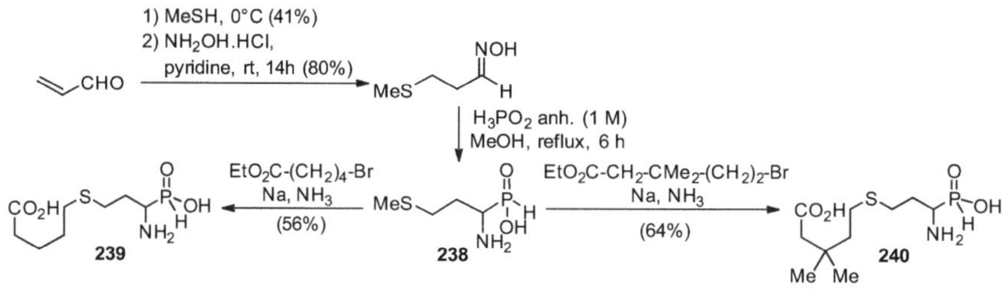

Scheme 46 Synthetic pathway for preparation of BHMT inhibitors **239** and **240**

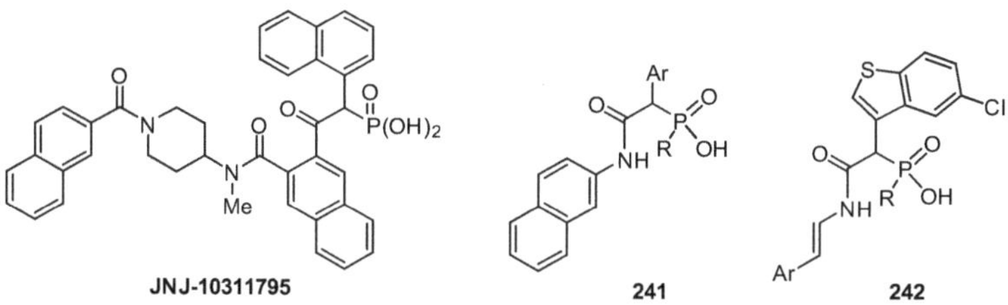

Fig. 28 Chymase inhibitor JNJ 10311975, and targeted compounds **241** and **242**

which gave under the same conditions a complete inhibition. Nevertheless, the parent molecule **235** is a better inhibitor with an IC_{50} of 0.093 μM, at least one order of magnitude lower than those of the *H*-phosphinic acid analogues **239** and **240** (IC_{50} = 5.0 and 1.1 μM, respectively). Unlike *H*-phosphinic acids **239**, phosphonic acid analogue is clearly less active, displaying a BHMT inhibition of 58%, and an IC_{50} of 219 μM, which is about 40-fold higher.

7.2 *Inhibitors of Human Mast Cell Chymase*

One important factor in the establishment of immune-mediated inflammatory diseases is the imbalance which exists between protease and antiprotease. As a matter of fact, the recruitment in the injured tissue of neutrophiles and mast cells leads to protease secretions such as Cathepsin G (EC 3.4.21.20) and Chymase (EC 3.4.21.39). The role of these chymotrypsin-like proteases is to degrade the extracellular matrix to facilitate leukocytes migration and tissue remodelling [165–169]. However, tissue injury is accentuated when the protease activity is no longer under the control of antiprotease [170, 171]. Therefore, a therapeutic indication for pulmonary inflammatory diseases, such as asthma and chronic obstructive pulmonary disease (COPD), could be solved by administration of an exogenous inhibitor of serine proteases, bringing back a benefit balance. With this objective, Greco and Maryanoff proposed a series of β-carboxamido-phosphinic acids **241** and **242** as potent chymase inhibitors (Fig. 28) [172]. These structures were based on

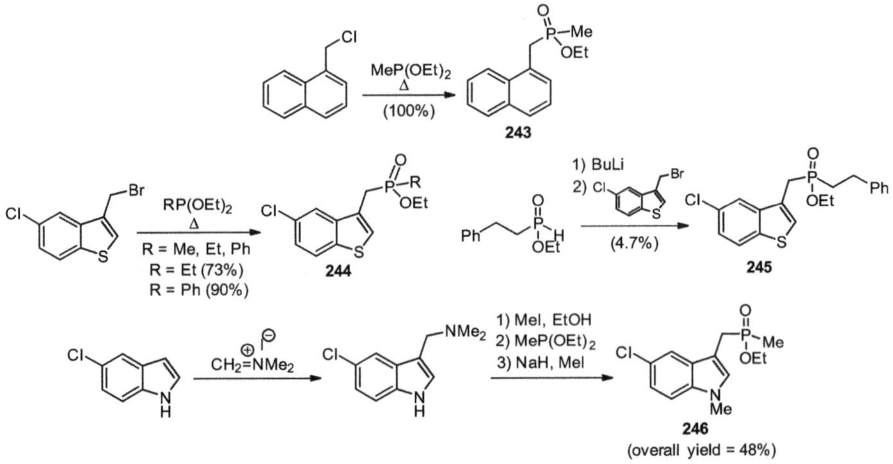

Scheme 47 Preparation of alkylphosphinate precursors **243–246**

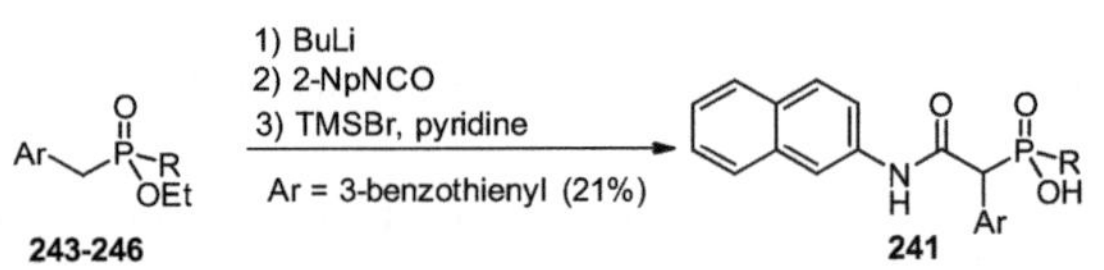

Scheme 48 Synthesis of carbamoylphosphinic acids **241**

modifications of the molecule **JNJ 10311795**, already described by the same authors as a potent dual inhibitor of human Cathepsin G ($K_i = 38$ nM) and human chymase ($K_i = 2.3$ nM) [173].

Preparation of the first compound series **241** required preparation of phosphinate intermediates **243–246** [172] (Scheme 47). An Arbusov reaction between 1-chloronaphthalene and diethyl methylphosphonite gave phosphinate **243**. Similarly, reaction of 3-(bromomethyl)-5-chlorobenzothiophene with methyl, ethyl or phenylphosphonite formed phosphinates **244**. Another strategy was applied for preparation of phosphinate **245** using ethyl phenethyl-*H*-phosphinate as starting material. The latter was deprotonated by *n*-butyllithium, and the anion was trapped by 3-bromo-5-chlorobenzothiophene, affording **245**. The indolyl derivative **246** was prepared by a four-step sequence. The first reaction consisted of the S_EAr reaction of 5-chloro-indole with Eschenmoser's salt. The resulting dimethylamino product was quaternized by methyl iodide. After substitution of the ammonium group by a phosphinate according to an Arbusov reaction, and methylation of the indole ring, compound **246** was produced in 48% overall yield.

All targeted molecules **241** were accessible by treating intermediates **243–246** with butyllithium, trapping the carbanion with 2-naphthylisocyanate and, finally, deprotecting the phosphinate ester with TMSBr (Scheme 48) [172].

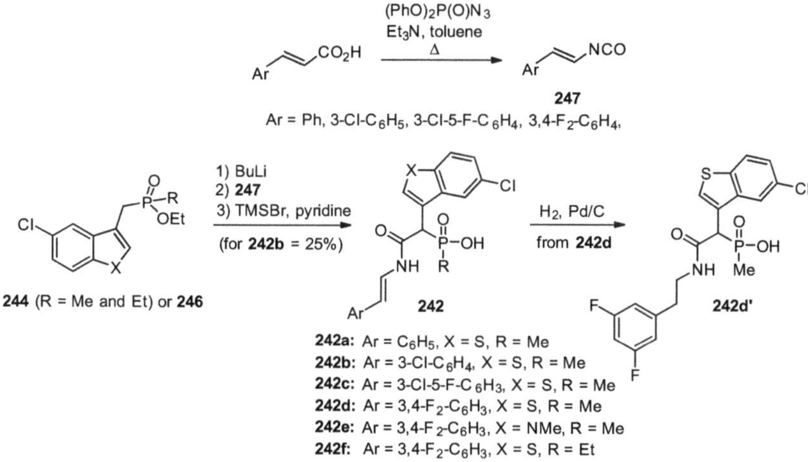

Scheme 49 Preparation of phosphinic acids **242a–f** and **242d′**

The second series has been prepared from isocyanates **247**, which were readily produced by Curtius rearrangement of cinnamic acid with diphenylphosphoryl azide in the presence of triethylamine (Scheme 49). Isocyanates **247** were engaged with lithium derivatives **244** or **246**. After deprotective reaction of phosphinate ester, phosphinic acids **242a–f** were obtained [172]. One example of saturated structure **242d′** was synthesized by reduction of the alkene [172].

In vitro tests performed on compounds **241** and **242** highlighted that the methyl group linked to the phosphorus atom gave the best inhibition response against the human chymase. In the subclass of compounds **241**, the combination indole (intermediate **246**) with the methyl group linked to the phosphorus atom gave the strongest inhibition for this series ($IC_{50} = 10$ nM) [172]. This value is equivalent to the phosphonic acid derivative ($IC_{50} = 13$ nM). Nevertheless, (*E*)-styryl derivative **242b** ($IC_{50} = 3.5$ nM) was revealed as the most efficient phosphinic acid tested, being even slightly better than the reference structure **JNJ-10311795** ($IC_{50} = 4.5$ nM). Moreover, the *E*-styryl group in comparison to the naphthyl group is essential for getting favourable pharmacokinetic properties, as demonstrated in rats with **242c**. A per os administration compound **242c** was shown to be pharmacologically active in a hamster inflammation model [172]. A more recent study reported that compound **242d** (**JNJ-18054478**) and its 3,5-dichloro analogue (**JNJ-28838017**) showed a remarkable species selectivity for human and macaque chymases compared to dog, sheep, guinea pig and hamster [174]. Both compounds were indeed about 39 to more than 1,000-fold less potent against non-human chymase species.

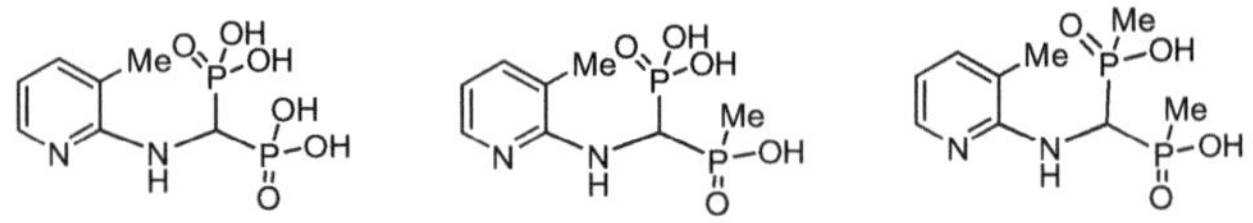

Parent bisphosphonic acid 248 249

Fig. 29 Inhibitors of FFPS

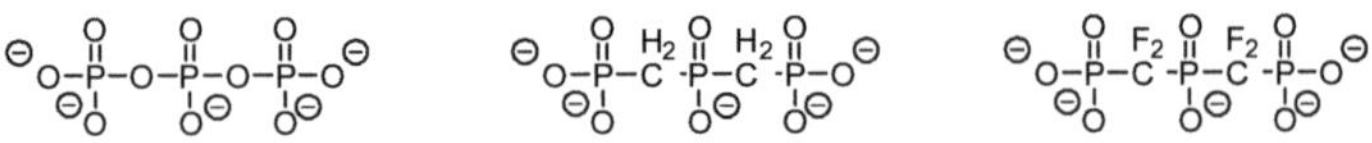

250 251 252

Fig. 30 Methylene **251** and difluoromethylene **252** analogues of triphosphate **250**

7.3 Phosphinate Inhibitor of Farnesylpyrophosphate Synthetase

It has been established that the principal mode of action of the nitrogen-containing bisphosphonate drugs (N-BPs) is the mevalonate pathway inhibition of isoprenoid lipids. The target enzyme is farnesylpyrophosphate synthase (FPPS) [175, 176], which catalyzes the sequential condensation reaction of dimethylallyl pyrophosphate with two units of 3-isopentenyl pyrophosphate to form farnesyl pyrophosphate. Dunford et al. have recently concluded that inhibition of FPPS by N-BPs (such as methyl-phosphinate analogues **248**, **249**) is consistent with a time-dependent isomerization of the human enzyme-inhibitor complex (Fig. 29) [177]. FPPS inhibitory effect of phosphono-phosphinic acid **248** was shown to be higher than the parent bisphosphinic acid. Surprisingly, bisphosphinic acid **249** was shown to be inactive. Moreover, it appeared during this study that in vivo potency of N-BPs was correlated on one hand to their capacity to inhibit FPPS enzyme, and on another hand to their ability to maintain the enzyme under its isomerization form, strengthening the inhibition global potency.

7.4 Phosphinic Acids as Pyrophosphate or Triphosphate Analogues

The concept of hydrolytically resistant phosphorus-based compounds emerged early in the 1960s. In this context, different non-hydrolysable ATP analogues or modified nucleotide triphosphates where the oxygen atoms of the triphosphate unit **250** were replaced by methylene groups **251** or difluoromethylene groups **252** were developed (Fig. 30).

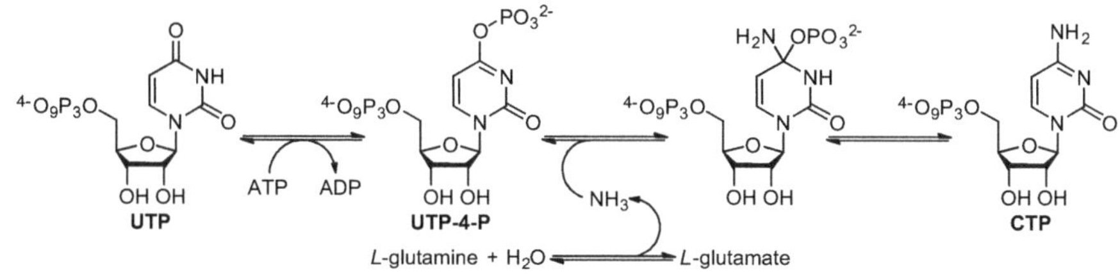

Scheme 50 Transformation of UTP mediated by cytidine 5′-triphosphate synthase

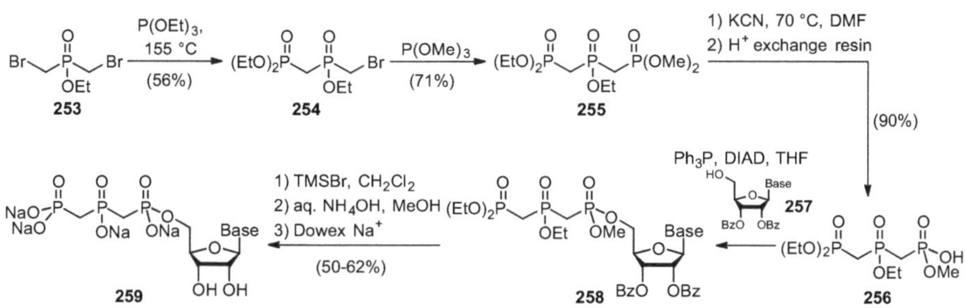

Scheme 51 Synthesis of bismethylene triphosphate analogue **259** of nucleosides

7.4.1 Bismethylene Triphosphate Analogues

Cytidine 5′-triphosphate synthase (CTPS, EC 6.3.4.2) catalyzes the formation of cytidine 5′-triphosphate (CTP) from uridine 5′-triphosphate (UTP), the last committed steps in pyrimidine nucleotide biosynthesis (Scheme 50) [178, 179]. The reaction proceeds by an ATP-dependent phosphorylation and requires ammonia. CTPS is a relevant target in some forms of cancers (i.e. non lymphocytic leukaemia) [180].

Synthesis of the modified triphosphate subunit was accomplished starting from ethyl bis(bromomethyl)phosphinate **253** (Scheme 51). An Arbusov reaction with triethyl phosphite and trimethyl phosphite, respectively, produced bis (phosphonomethyl)phosphinate **255** in 40% overall yield. A selective monodemethylation by cyanide and acidification led to monophosphonic acid **256**. Bismethylene triphosphate (BMT) **258** analogues of nucleosides were obtained by coupling of the 5′-free OH nucleosides **257** with phosphonate **256** under Mitsunobu conditions [181, 182]. TMSBr mediated deprotection followed by treatment with aqueous ammonia and cation exchange afforded bismethylene analogues **259** of nucleoside triphosphate in yields up to 62%.

The high IC_{50} value observed for **259a** means that BMT is not a good mimic of the triphosphate group for UTP analogue (Fig. 31) [183]. By contrast, CTPS inhibition by **259b** appeared at a lower value with an IC_{50} of 0.42 mM. When compared to **259b**, the lack of binding discrimination for **259c** or **259d** (0.13 and 0.17 mM, respectively), the closest analogues of UTP-4-P suggests that the BMT

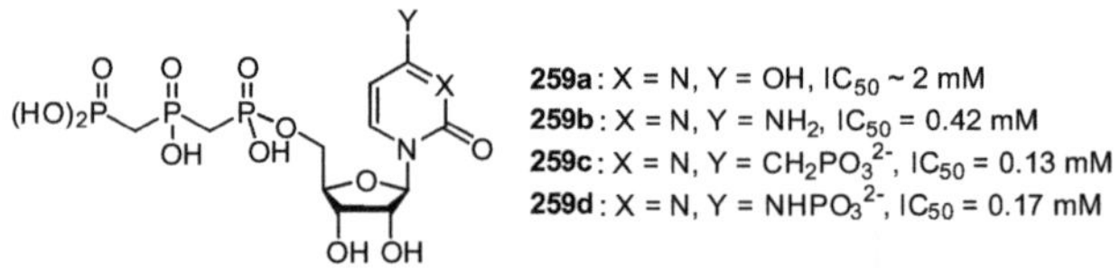

Fig. 31 Activities of BMT analogues **259a–d**

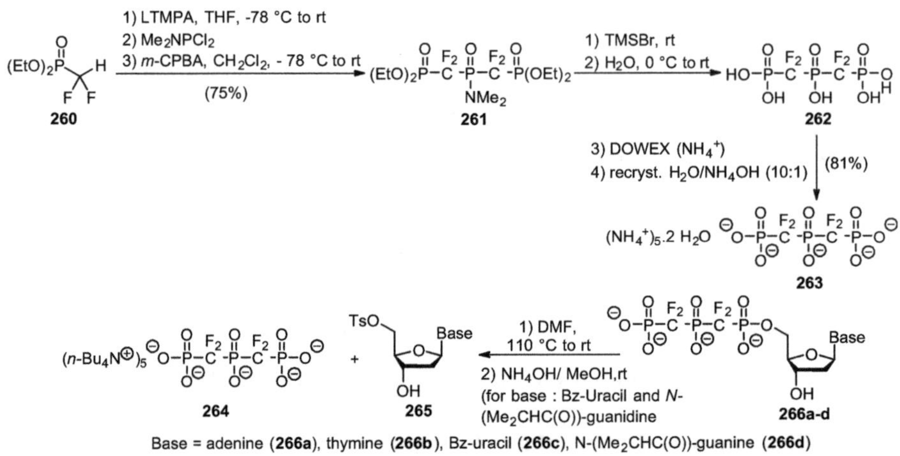

Scheme 52 Preparation of hydrolytically stable triphosphate analogues **266** of nucleosides

group probably binds differently to CTPS and induces this poor difference of selectivity.

7.4.2 Fluorinated Deoxynucleotide Analogues

The combination of fluorine and phosphorus heteroatoms led to relevant bioactive molecules [184]. In this context, the phosphate function can be advantageously replaced by an α,α′-difluoromethylphosphonate or phosphinate, which were generally considered as better bioisosteres than the methylene derivatives [185–193]. The difluoromethylene pattern (CF_2) is sterically and electronically a better mimic of the oxygen atom when compared to the methylene group. The electron-withdrawing properties of fluorine influences pK_a, bringing them within the range of phosphate. Among the other desirable properties, the possibility of fluorine hydrogen bonding, an increased redox stability and a better lipophilicity compared to the non-fluorinated parent compound can be expected.

Tetrafluorophosphinobisphosphonates **266a–d** were prepared starting from difluoromethylphosphonate **260** (Scheme 52) [194]. Once deprotonated by LTMPA and treated with diethylaminodichlorophosphine, the resulting aminophosphine intermediate was oxidized by *m*-CPBA, affording phosphinamide

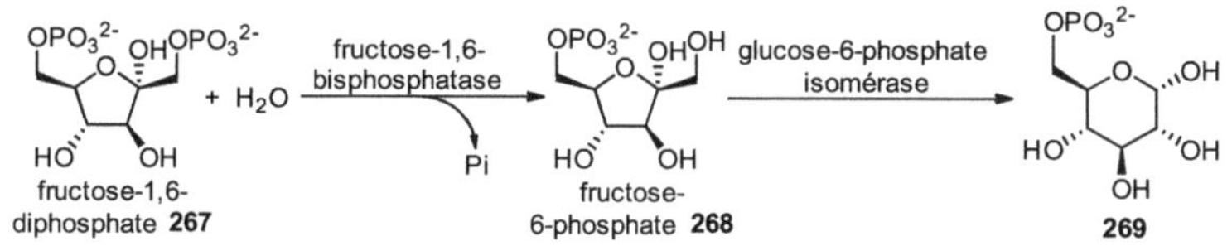

Scheme 53 Reaction catalyzed by the fructose-1,6-bisphosphatase

261 in 75% yield. Phosphonate deprotection with TMSBr followed by hydrolysis gave bis(phosphonodifluoromethylene)phosphinic acid **262** in almost quantitative yield. After conversion into ammonium salt by passage through DOWEX ion-exchange resin, the resulting product was recrystallized in aqueous ammonia, affording **263** in 61% yield. Phosphorylation at the 5′-position was accomplished by reaction with nucleoside. Using Blackburn's method, the nucleophilic substitution of 5′-tosylate nucleoside **265** by the difluoromethylene triphosphate analogue **264** gave fluoronucleosides **266** in 10–91% yields.

Preliminary biological results showed that thymine analogue **266b** was stable to enzymatic hydrolysis. In addition, a co-crystallization of **266b** with DNA polymerase (DNA Pol β) showed that the substitution of oxygen atoms by difluoromethylene group prevented the dissociation of the triphosphate unit. The binding constant of **266b** obtained by competing with desoxythymidine triphosphate was found to be similar to that of the natural substrate ($K_i = 1.4$ mM).

7.5 *Fructose-1,6-Bisphosphatase Inhibitors*

Fructose-1,6-bisphosphatase (FBPase, EC 3.1.3.11) is a key enzyme of gluconeogenesis (GNG) which catalyzes the hydrolysis of fructose 1,6-bisphosphate **267** to fructose 6-phosphate **268**, a precursor of glucose 6-phosphate **269** (Scheme 53) [195]. This enzyme is reported as the rate-limiting enzyme of the GNG pathway. Therefore, inhibition of FBPase is considered as a promising treatment for type 2 diabetes. A structure activity relationship revealed that adenosine monophosphate (AMP) mimics can compete in the active site of the enzyme and, more specifically, that the phosphate binding site might be accessed from the 8-position of the purine base [196].

Dang et al. synthesized various adenosine monophosphate mimics (Scheme 54) [197]. Stille coupling of allyl-tributylstannane with 8-bromo-9-phenethyladenine **270** gave 8-allyladenine **271** in only 15% yield. Then radical mediated hydrophosphinylation of alkene by reaction of sodium hypophosphite in acidic or neutral condition afforded the branched *H*-phosphinate **272** or the linear *H*-phosphinate **273** in, respectively, 2.1% and 4.9% yields. Ozonolysis of **271** followed by reduction of the aldehyde gave alcohol **274**. Using the procedure developed by Froehler, *O*-adenyl-*H*-phosphonate **275** was synthesized in 34% yield [198]. For comparison, the phosphate analogue **276** was obtained from Johns' procedure by the reaction of

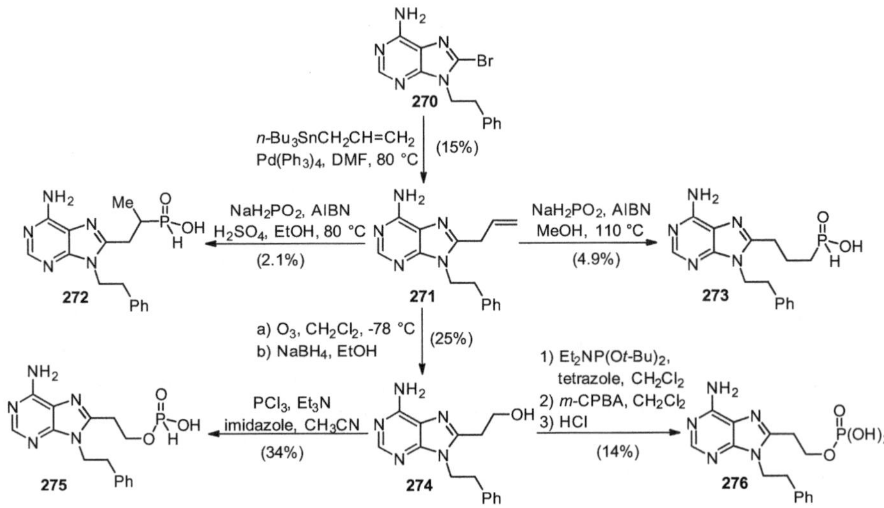

Scheme 54 Synthesis of adenine based inhibitors **272**, **273**, **275** and **276**

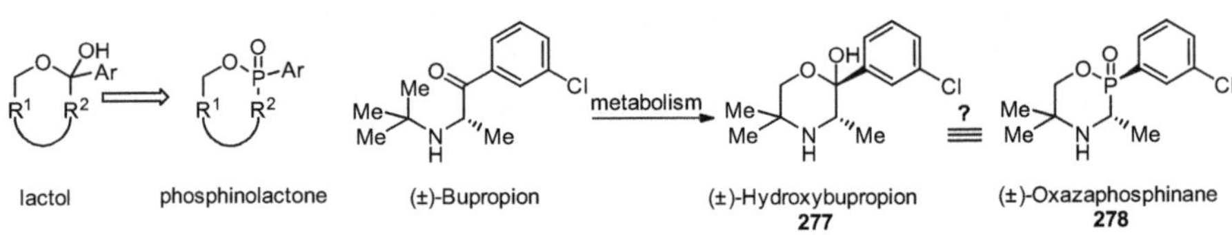

Fig. 32 Analogy between lactol and phosphinolactone group

tert-butylphosphoramidite with alcohol **274** and was followed by an oxidation of the trivalent phosphorus intermediate and a hydrolysis [199].

These compounds were tested in vitro for inhibition of human liver FBPase in order to find a suitable phosphate mimic. Phosphate **276** has an IC_{50} of 40 μM, lower than AMP ($IC_{50} = 1$ μM). Unfortunately, both *H*-phosphinate derivatives **272** and **273** were completely inactive. *H*-Phosphonate **275** exhibited the same lack of activity. By contrast, the phosphonate analogue of **273** (P–H replaced by a P–OH) has an IC_{50} of 100 μM. This result denoted that the presence of two negative charges at this position has a deep impact on the biological activity.

7.6 *Phosphinolactones as Anti-Depressants*

In 2010, Volle et al. described the first in vivo activity of phosphinolactone function by making bioisostere of the lactol group constituting the bioactive molecule, hydroxybupropion **277** (Fig. 32). Bupropion marketed in 1989 by GSK in the USA is an atypical drug which has been successfully used both as antidepressant

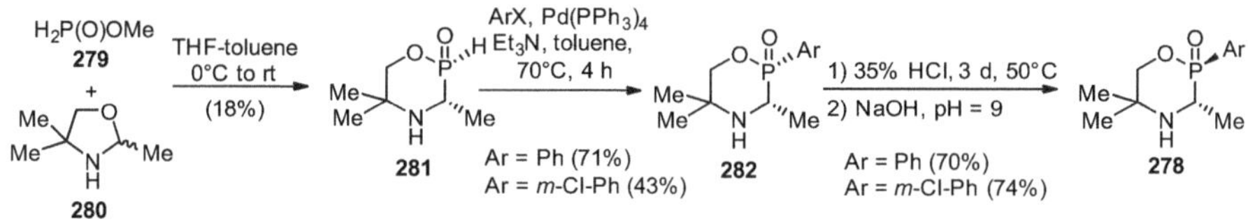

Scheme 55 Synthesis of oxazaphosphinanes

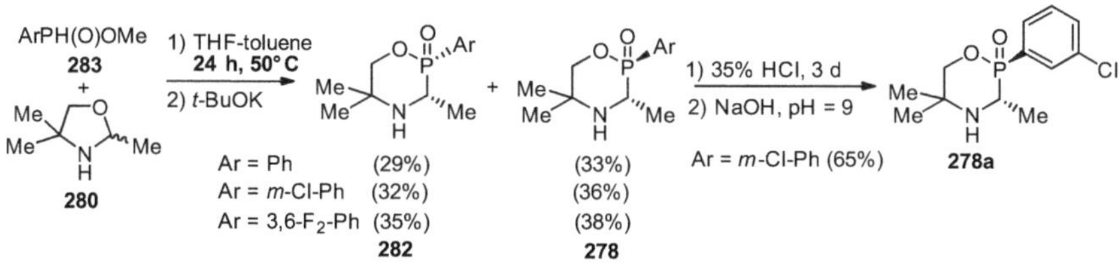

Scheme 56 Synthetic route to oxazaphosphinanes **278**

(Wellbutrin) and as a smoking-cessation aid (Zyban). It has been proposed that Bupropion activity would come from its metabolite, (2*S*,3*S*)-hydroxybupropion (Radafaxine). In order to evaluate the potential of phosphinolactone as lactol bioisostere, various (±)-oxazaphosphinanes **278** were synthesized and evaluated in vivo in a forced swimming test in mice [200].

The first synthetic route consisted of a condensation of methyl hypophosphite **279** with 2,2,4-trimethyl-1,3-oxazolidine **280** affording *H*-oxazaphosphinane **281** in 18% yield (Scheme 55). Then, after purification by chromatography, P–H function was arylated through a pallado-catalyzed reaction to give in good yields, oxazaphosphinanes **282**. The last step was a non-classical and selective epimerization at the phosphorus atom in concentrated HCl at 50 °C followed by neutralization by NaOH until pH 9 to afford (±)-oxazaphosphinanes **278** in moderate overall yield.

The most efficient synthetic route was a condensation of methyl arylphosphinates **283** with oxazolidine **280**. Precursors **283** were prepared by esterification of corresponding phosphinic acid, using a procedure described by Afarinkia (Scheme 56) [201]. When *H*-phosphinic acids were not commercially available, they were obtained by pallado-catalyzed arylation of anilinium hypophosphite [202]. The condensation gave a mixture of two diastereomers (*syn*/*anti*). The last step was selective epimerization at the phosphorus atom in concentrated HCl.

(±)-Oxazaphosphinanes **278**, bupropion and (±)-hydroxybupropion **277** were screened for their biological activity in the forced swimming test in mice, a model chosen to delineate antidepressant-like activity in rodents. The first outstanding result is the ability of 1,4,2-oxazaphosphinanes **278** to diffuse through the blood–brain barrier in mice to reproduce the original activity observed with Bupropion. Furthermore, the dose–response study showed that compounds **278c** (Ar = 2,4-F_2-C_6H_4) and

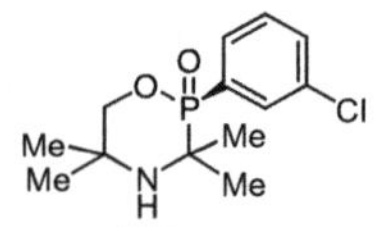

284

Fig. 33 Tetramethyl-[1, 4, 2]-oxazaphosphinane **284**

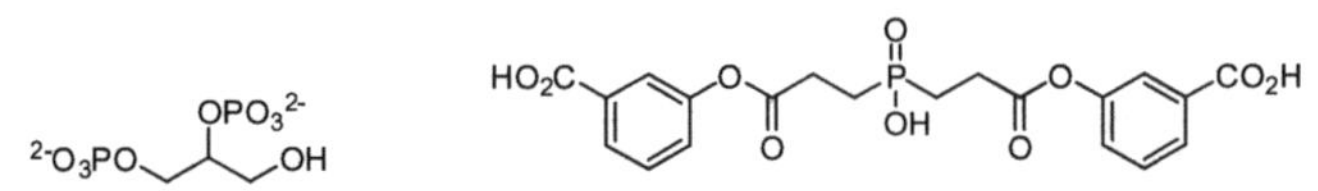

DPG **285** *m*-BCCEP **286**

Fig. 34 Natural 2,3-diphosphoglycerate **285** and *m*-BCCEP **286** cross linking reagent

284 (Fig. 33) had a significant effect at 10 mg/kg, and that **284** could be considered as twice as potent as (±)-hydroxybupropion **277**. Thus, Volle et al. showed that the phosphinolactone group might be considered as a biostere of lactol group and used as an unprecedented scaffold for the elaboration of new drug candidates.

7.7 Phosphinic Acids as Cross-Linking Reagent

Replacing human blood by substitutes was investigated early in the 1900s but, while many approaches were developed, it still remains an unmet requirement [203–205]. There are several attempts reporting the direct use of haemoglobin (Hb), the oxygen transporter protein, instead of red blood cells (RBC) [206]. The lack of immune response is probably the main advantage of this approach but it is impeded by three major problems: the short retention time of haemoglobin, a low oxygen release caused by high affinity, and its high kidney toxicity [207].

Because of the loss of the 2,3-diphosphoglycerate (DPG) linker **285** when Hb is removed from RBC, the tetrameric assembly splits into the two αβ dimer units inducing short lifetime and toxicity (Fig. 34). DPG **285** is negatively charged at physiological pH and acts as an electrostatic glue by interaction with the positively charged amino acid residues on the β-subunit. Cross-linking reagents such as bis [2-(3-carboxyphenoxy)carbonylethyl]phosphinic acid, *m*-BCCEP **286** were developed to maintain the initial behaviour of haemoglobin by tight binding of the two αβ dimer units [208].

m-BCCEP **286**, was synthesized from the double Michael addition of bis (trimethylsilylphosphonite **287** generated in situ to acrylate ester **288** (Scheme 57). After deprotection of the terminal carboxylic acids using TFA, *m*-BCCEP **286** was isolated in 73% overall yield. The trisodium salt **286-Na** was obtained by elution on an ion exchange resin (AG 50 W-X8 sodium form) and isolated in 77% yield.

The effect of *m*-BCCEP on the modified Hb showed that the oxygen affinity (P_{50}) was considerably lowered by the linkage ($P_{50} = 25.8$ Torr) and remained

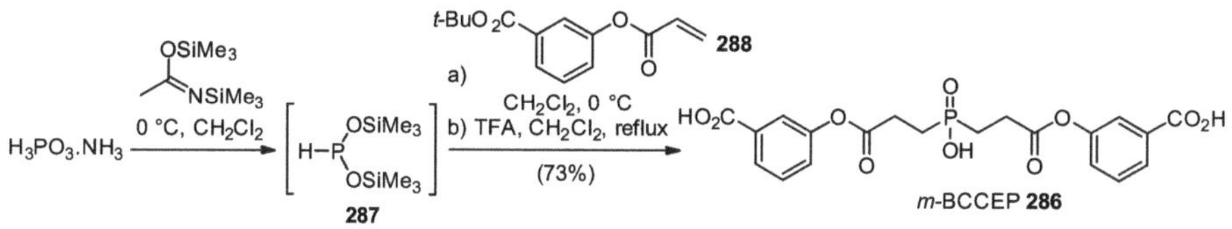

Scheme 57 Preparation of cross-linking reagent, *m*-BCCEP **286**

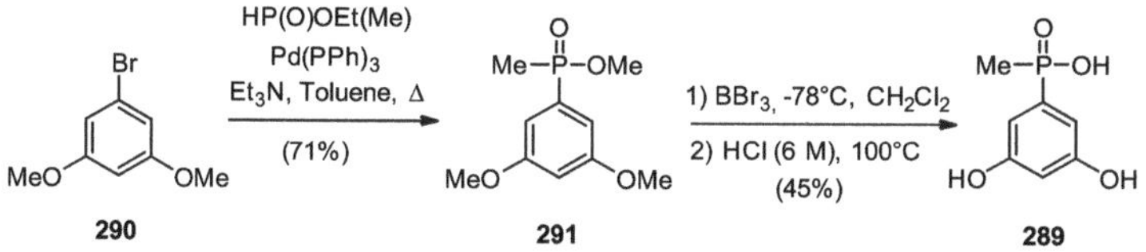

Scheme 58 Synthesis of (3,5-dihydroxyphenyl)(methyl)phosphinic acid **289**

comparable to the affinity of the whole blood ($P_{50} = 27$ Torr). The cross-linked Hb also retained some of the oxygen binding cooperativity between the two αβ dimers. The reactive sites of the cross-linked Hb were in the β-cleft (β1 and β2 subunits).

7.8 *Lactate Receptor (GPR81) Agonists with Antilipolytic Effects*

Stimulation of G-protein-coupled receptors in adipose tissue leads to a decrease in the lipolysis of triglycerides by hormone-sensitive lipase, which results in reduced transport of fatty acids to the liver and a decrease in hepatic triglycerides [209, 210]. Following the characterization of lactate receptor (GBR81), in vitro GTPγS binding assays were performed by Dvorak et al. on several substituted 3-hydroxybenzoic acids which showed an anchor point for the carboxylic acid moiety of the endogenous ligands [211]. Phosphinic acid **289** was prepared from a palladium catalyzed cross-coupling reaction between ethyl methylphosphinate and bromo-3,5-dimethoxybenzene **290**, giving phosphinate ester **291** in 60% yield (Scheme 58). The methyl ethers were then cleaved with BBr_3 followed by hydrolysis of the phosphinate group, affording **289** in 73% yield.

The phosphinic acid **289** analogue of hydroxybenzoic acid did not show better agonist selectivity on GPR81 ($EC_{50} > 1{,}000$ μM) than the lead, 3-chloro-5-hydroxybenzoic acid ($EC_{50} = 16$ μM). Lead in vivo effects on lipolysis gave significant reductions in free fatty acids at all doses tested, and further investigations of GPR81 agonists are ongoing.

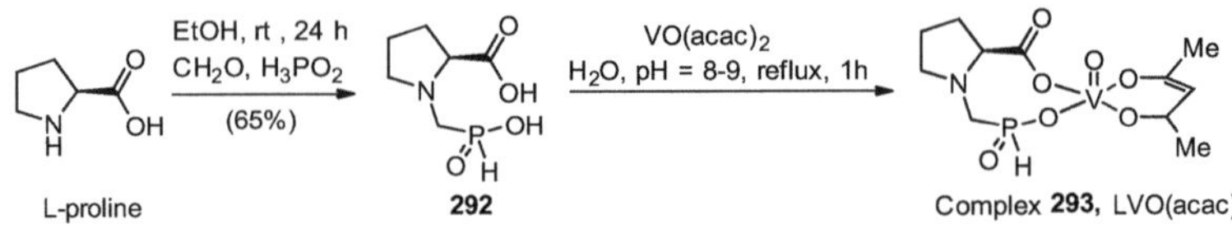

Scheme 59 Preparation of LVO(acac) **293** from L-proline

7.9 *Inhibitors of α-Amylase and α-Glucosidase*

In 1985, Heyliger et al. first demonstrated a serendipitous discovery that oral administration of 0.8 mg/mL of sodium orthovanadate in drinking water to streptozotocin-induced diabetic rats resulted in normoglycemia [212]. It has been shown that vanadium salt can reduce dependence on exogenous insulin (type 1 diabetes), and might replace other oral hypoglycemic agents for type 2 diabetes. To date, the use of vanadium compound for therapeutic treatment of diabetic patients has failed. This is likely because of inadequate biological responses. In addition, gastrointestinal distress is observed at weaker doses than in animal models [213]. For this reason, research is oriented to increase the bioavailability of vanadium(IV) complexes. Even though bis(ethylmaltolato)oxovanadium(IV) (BEOV) has been successfully evaluated in a phase 2 clinical trial, a probable kidney side-effect prevented further clinical applications. Nevertheless, vanadium complexes are still considered as relevant for future treatment of human diabetes. In this perspective, Kaboudin et al. designed a proline-based aminophosphinic acid as ligand **292** for development of an effective and safe antidiabetic agent by vanadyl complex formation **293** (LVO(acac)) [214]. The targeted complex **293** incorporating phosphinic acid moiety ($–PO_2–CH_2$) was designed potentially to inhibit α-amylase and α-glycosidase.

The preparation of complex **293** has been implemented in two steps (Scheme 59). First, Mannich-type reaction of L-proline with formaldehyde and hypophosphorous acid in aqueous solution produced ligand **292**. After addition of $VO(acac)_2$ in basic aqueous solution containing ligand **292** with a molar ratio of 1:1, complex **293** was obtained by precipitation.

Tests of compound **293** through yeast α-glycosidase and pancreatic α-amylase highlighted a significant inhibitory activity on both enzymes. This inhibitory effect was similar to that observed for the commercial Acarbose. To control the glycaemic index, this novel complex **293** might be a preferred alternative to powerful synthetic α-glycosidase inhibitors such as Voglibose, which may cause hepatic disorders at high doses.

8 Other Uses of Phosphinic Acids and Derivatives

8.1 Phosphinic Acids for MRI or Radioligand Delivery

Magnetic resonance imagery emerged during the last 30 years as a technique of choice in medicinal diagnostics, offering the opportunity to produce high resolution images of any part of the whole human body. If physical principles of MRI rely on the properties of water in the body environment, the lack of contrast or resolution prompted the development of contrast enhancing agents (CA). Among them, the T_1-CAs are complexes of paramagnetic cations and they induce a positive contrast enhancement. Gadolinium has a high magnetic moment and a long electronic relaxation time compared to other lanthanides. These characteristics make it a very attractive paramagnetic metal for this application. The main disadvantage is associated with the "heavy metal character" when the cation is free of ligands. Body retention of the free ionic gadolinium(III) was found in various tissues, including liver, lymph nodes and bones, and it interferes with different calcium-dependent pathways. In kidney, nephrotoxicity (renal toxicity) is the most prominent pathology induced by contrast agents based on Gd^{III} and is generally associated with the high retention of the complexes in the glomerular zone [215].

When coordinated, Gd^{III} accepts octadentate ligands with a bicapped trigonal prism geometry. The ninth site is occupied by water. The different families of radiopharmaceutical chelators have been intensively reviewed during the past few years [216–218]. Most of the ligands developed are based on an azaalkyl skeleton completed by charged organic functions (i.e. carboxylate, phosphonate and phosphinate) or neutral siderophores (*N*-hydroxypyridone). Prototypical chelators are represented by the 1,4,7,10-tetrazacyclododecane core (**294–296**), close analogues of carboxylic derivative DOTA **297**, 1,4,7-triazacyclononane **298**, or linear patterns such as DTTAPR **299** and H5-XT **300** (Fig. 35).

Gadolinium complexes of modified DOTA conjugated to cyclodextrin scaffolds were screened as MRI contrast agents for higher magnetic fields. H_5do3ap^{PrA} **295a**

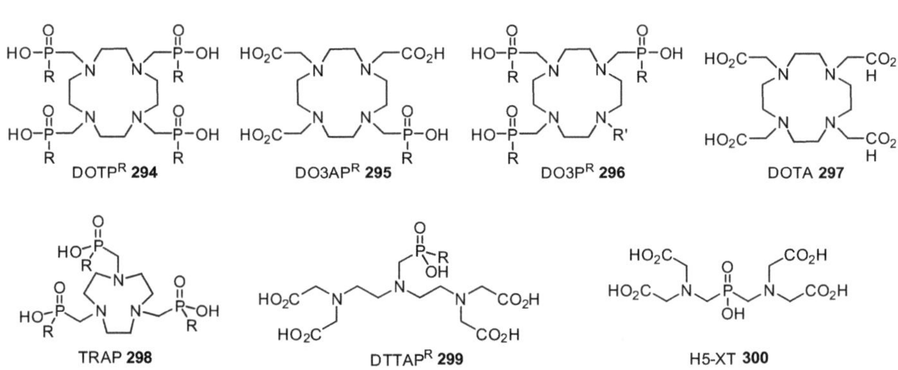

Fig. 35 Representative phosphinate cores **294–300** as complexing units

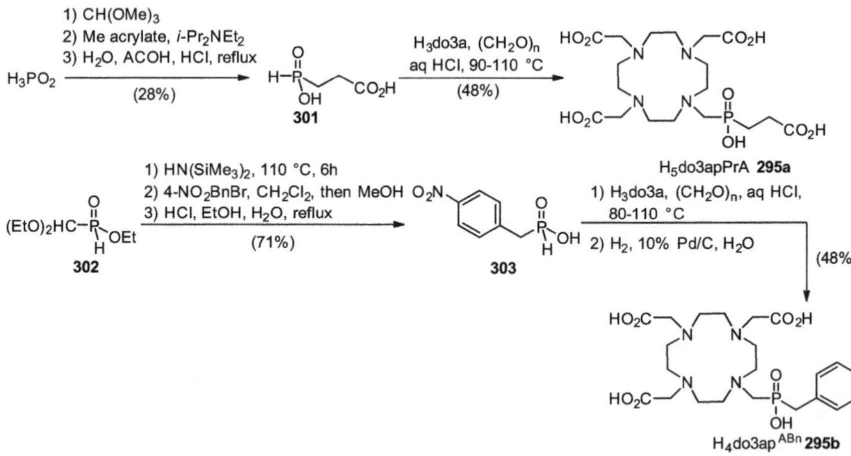

Scheme 60 Synthesis of phosphinates **295a, b**

was synthesized from hypophosphorous acid (Scheme 60) [219]. Methyl hypophosphite was formed in situ and was reacted directly with methyl acrylate, affording, after hydrolysis, phosphinic acid **301** in 28% yield. The formation of the chelating target **295a** was accomplished by three-component Kabachnik–Fields reaction of **301** with 1,4,7-triscarboxymethyl-1,4,7,10-tetrazacyclododecane (H_3do3a) and an excess of paraformaldehyde in 48% yield. From *H*-protected phosphinate **302**, a silyl Arbuzov reaction followed by the hydrolysis of the ester function led to nitrobenzylphosphinic acid **303** obtained in 86% yield [220]. Then the reaction with H_3do3a under Kabachnik–Fields conditions gave, after reduction of the nitro group, H_4do3ap^{ABn} **295b** in 48% yield.

Complexation and biodistribution of ^{111}In and ^{90}Y complexes with DOTA analogues **295a** and **295b** were studied [221]. Both ligands bind the radiometals in the same way as DOTA **297**. Pharmacokinetic and distribution in rats with radiolabelled ligands showed a rapid radioactivity decrease in blood and organs, excepted for kidneys, highlighting a fast elimination of the complexes by urinary excretion. The decrease of radioactivity in kidneys is correlated to a comparable increase in bladder and then in urine. Phosphinates **295a** and **295b** were not accumulated in calcified tissues, also confirming the high binding properties of such ligands and their rapid clearance.

In another direction, ligand **295b** was modified by introduction of per amino 6- or 7-α,β-cyclodextrins (Scheme 61). Cyclodextrins are known to be biocompatible and the increased number of gadolinium ions could result in multiplication of relaxation enhancement. In this context, DO3P^{NCS} **295c** was obtained from the reaction of the amine **295b** with thiophosgene [222]; then its reaction with the appropriate α- or β-cyclodextrin afforded the metal free conjugates **295d** in yields ranging from 66% to 78% [223]. The lanthanide complexes were obtained by mixing conjugates **295d** with a large excess. Low molecular weight impurities or reagents are removed by ultrafiltration, affording the pure conjugates **Gd-295d**.

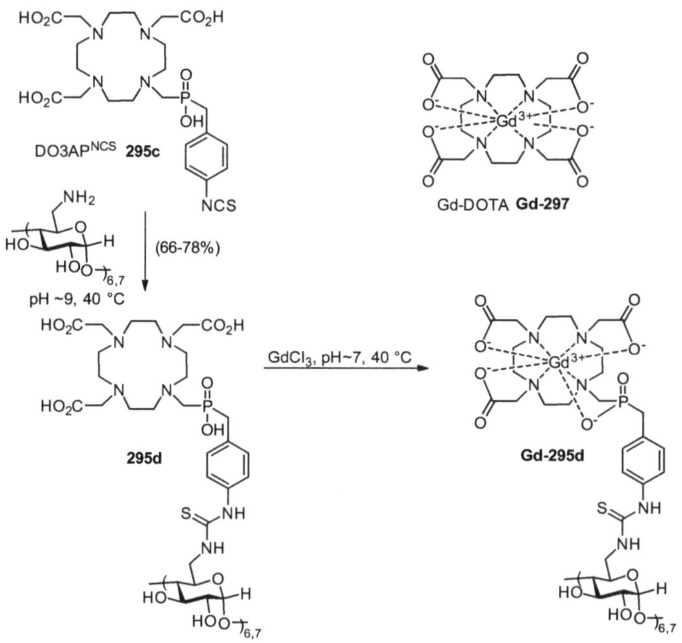

Scheme 61 Gadolinium complexes Gd-**297** – cyclodextrin conjugates Gd-**295d** synthesis

The physico-chemical properties of complexes from **295d** show that they endow a fast water exchange rate close to the optimum. The high relaxivity at magnetic fields over 3 T and temperature (37 °C) (for $(GdL)_7$-β-CD **Gd-295d**, $r_1 = 12.3\ s^{-1}\cdot mM^{-1}$) makes them more compatible with the high field systems compared to the simple $[Gd(DOTA)H_2O]^-$ complex **Gd-297** ($r_1 = 3.9\ s^{-1}\cdot mM^{-1}$). Conjugates **295d** are biocompatible and no toxicity was detected. The contrast agents prepared were used for imaging pancreatic islets or stem cells [224]. The MRI probe based on a cyclodextrin scaffold was internalized into cells and resulted in sufficient labelling for MRI visualization.

Dendrimers are multibranched macromolecules with nanometer-scale dimensions. They are considered to be ideal candidates for biological applications. Such macromolecules are monodisperse or closely monodisperse and they give reliable drug or gene deliveries compared to other polymers. Dendrimers are then used for the formation of labelled-ligand bioconjugates [225].

A bifunctional ligand was attached to different generation of ethylenediamine cored polyamidoamine dendrimer (PAMAM). Then G1, G2 or G4 PAMAM dendrimer (8, 16 or 64 terminal primary amines) were mixed in water with an equimolecular amount of DO_3AP^{NCS} **295c** per number of primary amine (Scheme 62). Gd(III) and Gd(III)/Y(III) complexes were prepared by mixing 2 equiv. of metal cation per thiourea function and removing the excess by washing with disodium EDTA [226, 227]. The reaction was extended to the fourth

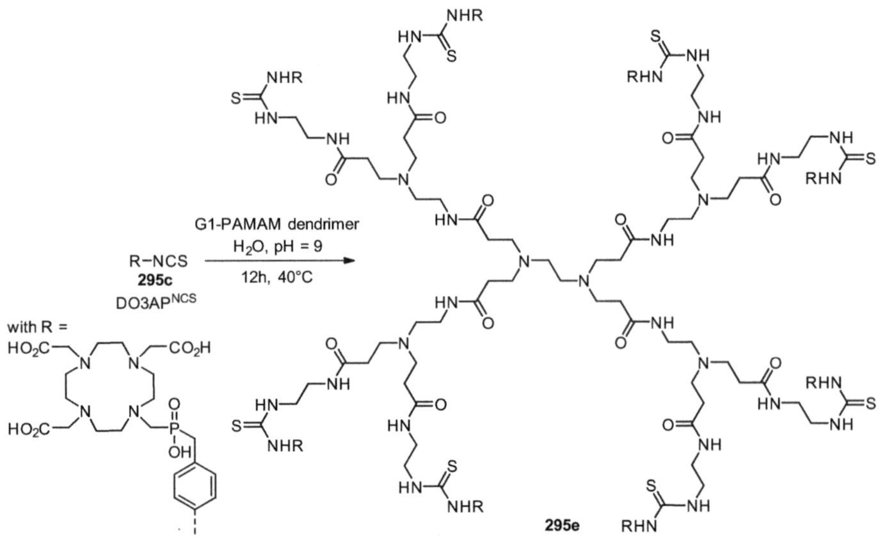

Scheme 62 Gadolinium complexes – cyclodextrin conjugates **295e** synthesis

Table 2 Physico-chemical parameters of **Gd-295** and **G4** compounds

Grafted dendrimer	r_1 ($s^{-1}.mM^{-1}$)	τ_M (ns)
$[Gd(H_2O)(do3a\text{-}P^{ABn})]^-$ Gd-295b	4.2/5.7	16
$G1\text{-}[Gd(H_2O)(do3a\text{-}P^{BnN\{SC\}})]_8^{x-}$ Gd-295e	10.1/14.8	48
$G4\text{-}[Gd(H_2O)(do3a\text{-}P^{BnN\{SC\}})]_{59}^{x-}$	18.6/25.8	68

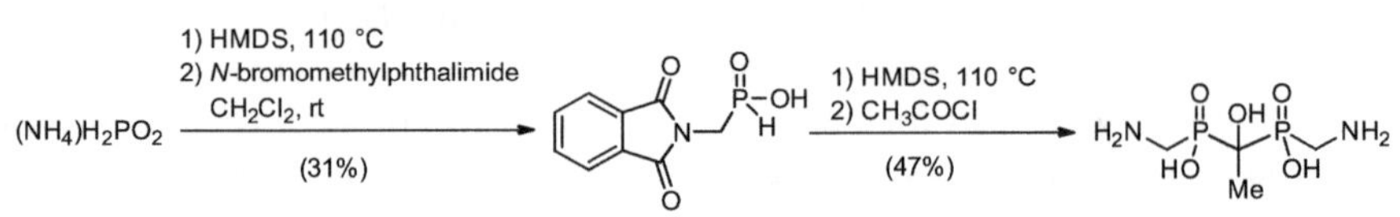

Scheme 63 Synthesis of hydroxyethyl-bis(aminomethylphosphinic) diacid **305**

generation PAMAM. These two modified dendrimers were obtained with high loading (100% for 295e and 92% for the**G4**).

The relaxometric measurement was determined at 10.1 and 18.6 $s^{-1}{\cdot}mM^{-1}$ for dendrimer **Gd-295e** and **G4**, respectively (Table 2). Measurement of the relaxivity depended on the pH and was increased by a factor of 1.3 when the core amino group is protonated (14.8 and 25.8). The water residence life time was also increased from 16 ns for **Gd-295b** to 68 ns for **G4** as the molecular size of the conjugate increased. It also appeared that the conjugation did not affect the water binding site. Their properties make them suitable for MRI.

By analogy with bisphosphonates, bisphosphinic acids were recently synthesized for their complexing abilities (Scheme 63) [228]. The synthesis was

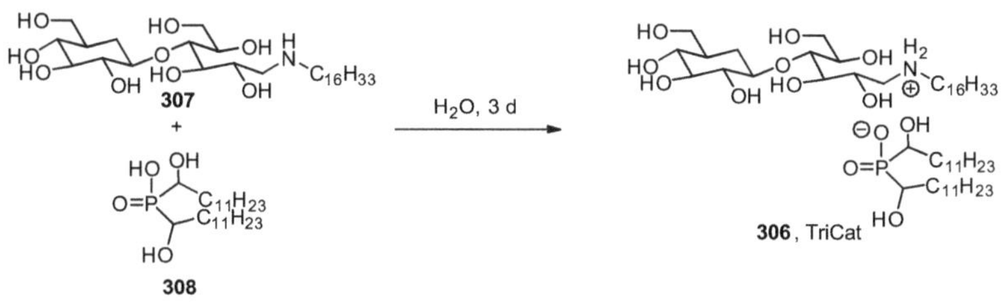

Scheme 64 Preparation of TriCat **306**

accomplished in two steps, the first involving a silyl-Arbuzov reaction of bromomethylphthalimide with bis(trimethylsilyl)phosphonite obtained in situ from ammonium hypophosphite. Bis(phosphinate) **305** was formed by the usual reaction of *H*-phosphinate **304** with HMDS followed by P(III) intermediate trapping with acetyl chloride in 47% yield.

8.2 Drug Delivery Surfactants

Liposomes are highly useful components in the pharmaceutical field, particularly for drug delivery [229–232]. Among liposomes, cationic vesicles constituted by the association of ionic surfactants of opposite charges, have drawn attention for their efficiency to modulate and control substances delivery towards a target. To understand the mechanisms of drug delivery, Blanzat et al. studied the action of tricatenar cationic surfactant, 1-*N*-hexydecylammonium-1-deoxyactitol-bis(*R*-hydroxydodecylphosphinate) **306**, so-called "TriCat" [233].

"TriCat" **306** was prepared simply by mixing *N*-hexadecylamino-1-deoxylactitol **307** and bis(hydroxydodecyl)phosphinic acid **308** in water (Scheme 64) [234]. Both precursors were prepared according to procedures already described in the literature [235–237].

The incorporation of "TriCat" inside cells using TriCat vesicles labelled with the fluorescent probe Fluocat has been proven for phagocytic and non-phagocytic cells. Thereafter, using various endocytosis inhibitors such as cold temperature or sugar solutions, mechanisms of intrusion into phagocytic and non-phagocytic cells have been studied, and the ability of TriCat **306** to interact with phospholipid membrane has been evaluated. The results highlighted that, whatever the cell type, using an appropriate methodology, cationic vesicles were incorporated in cells through an active and/or passive endocytosis process. A better understanding of the universal behaviour of cationic vesicles is required before considering them as an alternative encapsulation system for therapeutic applications.

9 Phosphinic Acids in Agrochemistry

9.1 *Herbicidal Properties of [α-(Substituted Phenoxyacetoxy)(Substituted Phenyl)Methyl or Alkyl] (Methyl)Phosphinates*

Transformation of pyruvate to acetyl-CoA is linked to glycolysis and to the Krebs cycle and oxidative phosphorylation. The pyruvate dehydrogenase complex (PDHc) catalyzes the oxidative decarboxylation of pyruvate and acetylation of coenzyme A (CoA) to acetyl-CoA. This enzyme has been viewed as a potential herbicide target [238]. In the perspective of structural optimization of a potent new herbicide, He et al. synthesized a series of structures **307** [239–241], which were a priori designed to inhibit the PDHc.

The preparation of these compounds started by a Pudovik reaction of methyl-*H*-phosphinate **308** with 2-pyridinecarboxaldehyde or benzaldehyde derivatives to produce α-hydroxy-phosphinates **309**. Further coupling of **309** with substituted phenoxyacetic chloride **310** in presence of triethylamine gave the final compounds **307** (Scheme 65).

The preliminary herbicidal tests showed that **307a–d** (Ar = 2-pyridyl, Ar′ = 4-Cl-C_6H_4, 3-CF_3-C_6H_4, 4-Cl-5-Me-C_6H_3, and 2,4-Cl-C_6H_4) harboured excellent inhibitory effects (inhibition percentage ≥90%) on the root of barnyard grass and rape at a dose of 10 ppm [239]. Compounds **307a**, **307c** and **307d** also displayed excellent inhibitory activities against stalk of rape at the same dose. Others structures were tested to 900 g/ha in pre- and post-emergence through six different plants, derivatives **307e** (Ar = phenyl, Ar′ = 4-F-C_6H_4), **307f** (Ar = 2-Cl-C_6H_4, Ar′ = 4-F-C_6H_4) and **307g** (Ar = 2,4-Cl_2-C_6H_3, Ar′ = 4-F-C_6H_4), exhibiting an excellent control in post-emergence against all plants tested (70≤ inhibition rate ≤100%) [240]. For instance, **307e** was efficient in post emergence against *Echinochloa crusgalli*, *Digitaria sanguinalis* and *Brassica juncea* with 90%, 85% and 100% of growth inhibitory rate, respectively. Another study led to 150 g/ha in pre- and post-emergence against *Echinochloa crusgalli*, *Digitaria sanguinalis* and *Brassica napus*, highlighting that the most efficient compound tested, exhibiting 95–100% of relative inhibition of growth, was structure **307h** (Ar = 4-Cl-C_6H_4, Ar′ = 4-CF_3-C_6H_4). These results underlined the beneficial effect produced by the trifluoromethyl group in this strategic position for enhancement of herbicidal activity [241].

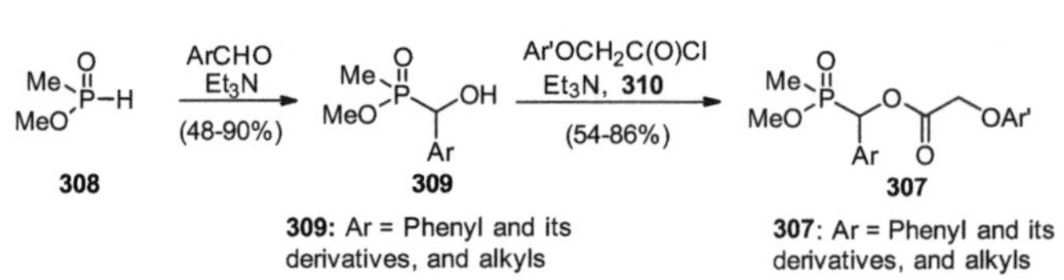

Scheme 65 Preparation of herbicide candidates **307**

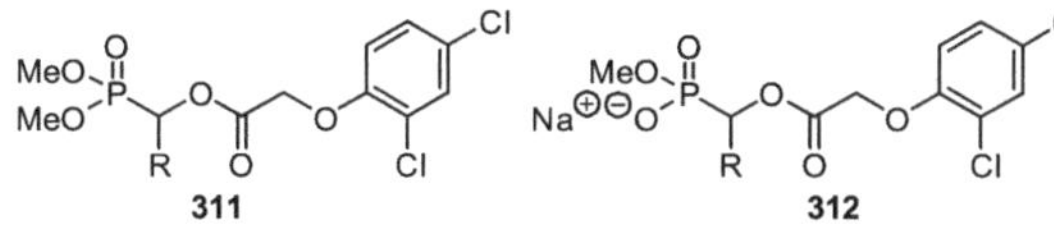

Fig. 36 Phosphonates and phosphonic acid monosodium salts **311** and **312**

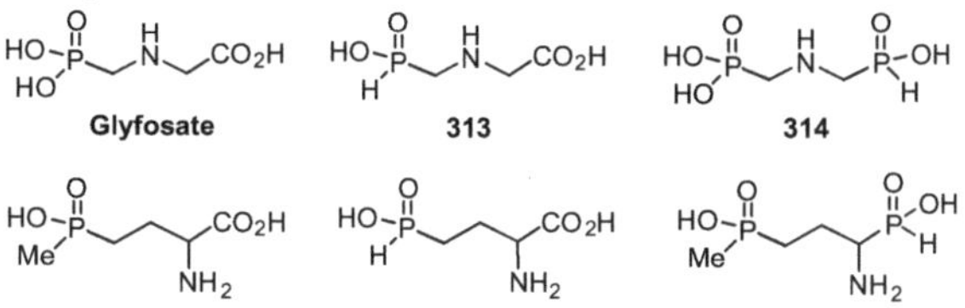

Fig. 37 Glyphosate and glufosinate, and their analogues **313**, **314**, **315** and **316**

Nevertheless, unlike phosphonate series **311** [242] and phosphonic acid monosodium salts series **312** [243], no inhibitory activity in vitro against PDHc from plants has been reported to confirm that PDHc was the biological target (Fig. 36).

9.2 Structural Modification of Commercial Herbicides Glyphosate and Glufosinate

In 2008, Hall published an overview of the role of phosphorus compounds as weed control agents [244]. The author mentioned that both phosphinate analogues **313**, **314** of glyphosate and modified glufosinate analogues **315**, **316** were tested for their herbicidal activities (Fig. 37). The structural modifications of parent molecules led to a dramatic loss of activity, except for compound **316**, which showed a good activity against Setaria to a dose of 4 kg/ha.

Conclusion and Future Outcomes

To date, two phosphinates have reached the market as a commercial drug or agrochemical. In 1991, Bristol-Myers Squibb launched Fosinopril, an angiotensin converting enzyme inhibitor, which is used for the treatment of hypertension and chronic heart failure. In agroscience, Bayer (formerly Hoescht in 1984) launched glufosinate for commercial use as non-selective contact herbicide with some systemic action. Since this period, the number of molecules incorporating a phosphinate core increased dramatically on several and original biological targets (GABA, antivirals, anticancers. . .). Based upon the diversity and richness of structural motifs listed in this chapter, we

(continued)

expect great success and applications for phosphinic acids or derivatives in the near future.

In recent years it has become increasingly apparent that more highly reduced phosphorus compounds (i.e. phosphonic and phosphinic acids) also play prominent roles in life science [245–247]. The mechanisms which have driven the chemical evolution of phosphate-based biochemistry are still under discussion [248]. Alternative prebiotic mechanisms for polyphosphate formation using reduced phosphorus species and particularly *H*-phosphinates ($H_2PO_2^-$) are emerging as probable explanations.

References

1. Butusov M, Jernelöv A (2013) Phosphorus: an element that could have been called Lucifer. Springer, New York
2. Frank AW (2013) Chemistry of plant phosphorus compounds. Elsevier, Amsterdam
3. Peck SC, Gao J, Van der Donk WA (2012) Methods Enzymol 516:101
4. Metcalf WW, Van der Donk WA (2009) Annu Rev Biochem 78:65
5. Nair SK, Van der Donk WA (2011) Arch Biochem Biophys 505:13
6. Bougioukou DJ, Mukherjee S, Van der Donk WA (2013) Proc Natl Acad Sci U S A 110:10952
7. Meanwell NA (2011) J Med Chem 54:2529
8. Elliott TS, Slowey A, Ye Y, Conway SJ (2012) Med Chem Commun 3:735
9. Mucha A, Kafarski P, Berlicki L (2012) J Med Chem 54:5955
10. Crofts P, Kosolapoff GM (1953) J Am Chem Soc 75:5738
11. Jencks WP, Regenstein J (1976) In: Fasman GD (ed) Handbook of biochemistry and molecular biology. Physical and Chemical Data, p 305
12. Haake P, Hurst G (1966) J Am Chem Soc 88:2544
13. Kalgutkar AS, Daniels JS (2010) In: Smith DA (ed) RSC drug discovery series N 1. Royal Society of Chemistry, p 99
14. Ballatore C, Huryn DM, Smith AB III (2013) ChemMedChem 8:835
15. Schramm VL (2011) Annu Rev Biochem 80:703
16. Hall RG, Riebli P (1999) Synlett 10:1633
17. Hall RG, Riebli P (2002) Phosphorus Sulfur Silicon 177:1557
18. Mehellou Y, de Clercq E (2010) J Med Chem 53:521
19. Alexandre FR, Amador A, Bot S, Caillet C, Convard T, Jakubik J, Musiu C, Poddesu B, Vargiu L, Liuzzi M, Roland A, Seifer M, Standing D, Storer R, Dousson CB (2011) J Med Chem 54:392
20. Cheek MA, Sharaf ML, Dobrikov MI, Shaw BR (2013) Antiviral Res 98:144
21. Shaw BR, Madison J, Sood A, Spielvogel BF (1993). Chapter 11 Oligonuleoside boranophosphate (borane phosphonate). In: Methods in Molecular Biology (vol 20): protocols for oligonucleotides and analogues. Sudhir Agrawal Humana Press, p 225
22. Barral K, Priet S, de Michelis C, Sire J, Neyts J, Balzarini J, Canard B, Alvarez K (2010) Eur J Med Chem 45:849
23. Sheldon J, Barreiro P, Soriano V (2007) Expert Opin Investig Drugs 16:1171
24. Fried MW, Hadziyannis SJ (2004) Semin Liver Dis 24:47
25. Lin C, Amberg SM, Chambers TJ, Rice CM (1993) J Virol 67:2327
26. Kolykhalov AA, Mihalik K, Feinstone SM, Rice CM (2000) J Virol 74:2046
27. Bode JG, Brenndorfer ED, Haussinger D (2007) Arch Biochem Biophys 462:254

28. Lamarre D, Anderson PC, Bailey M, Beaulieu P, Bolger G, Bonneau P, Bos M, Cameron DR, Cartier M, Cordingley MG, Faucher AM, Goudreau N, Kawai SH, Kukolj G, Lagace L, LaPlante SR, Narjes H, Poupart MA, Rancourt J, Sentjens RE, St George R, Simoneau B, Weldon SM, Yong CL, Llinas-Brunet M (2003) Nature 426:186
29. Reesink HW, Zeuzem S, Weegink CJ, Forestier N, van Vliet A, de Rooij J, McNair L, Purdy S, Kauffman R, Alam J, Jansen PLM (2006) Gastroenterology 131:997
30. Sarrazin C, Rouzier R, Wagner F, Forestier N, Larrey D, Gupta SK, Hussain M, Shah A, Cutler D, Zhang J, Zeuzem S (2007) Gastroenterology 132:1270
31. Jacobson IM, McHutchison JG, Dushieko G, Di Bisceglie AM, Reddy KR, Bzowej NH, Marcellin P, Muir AJ, Ferenci P, Flisiak R, George J, Rizzetto M, Shouval D, Sola R, Terg RA, Yoshida EM, Adda N, Bengtsson L, Sankoh AJ, Kieffer TL, George S, Kauffman RS, Zeuzem S (2011) N Engl J Med 364:2405
32. Clarke MO, Chen X, Cho A, Delaney WE IV, Doerffler E, Fardis M, Ji M, Mertzman M, Pakdaman R, Pyun H-J, Rowe T, Yang CY, Sheng XC, Kim CU (2011) Biorg Med Chem Lett 21:3568
33. Sheng XC, Casarez A, Cai R, Clarke MO, Chen X, Cho A, Delaney WE IV, Doerffler E, Ji M, Mertzman M, Pakdaman R, Pyun H-J, Rowe T, Wu Q, Xu J, Kim CU (2012) Bioorg Med Chem Lett 22:1394
34. Pompei M, Di Francesco ME, Koch U, Liverton NJ, Summa V (2009) Bioorg Med Chem Lett 19:2574
35. McCauley JA (2008) Abstract of papers, 235th ACS National Meeting, New Orleans
36. Casarez A, Chaudhary K, Cho A, Clarke M, Doerffler E, Fardis M, Kim CU, Pyun H-J, Sheng XC, Wang J (2008) WO patent 005565 A2
37. Pyun H-J, Chaudhary K, Somoza JR, Sheng XC, Kim CU (2009) Tetrahedron Lett 50:3833
38. Sheng XC, Pyun H-J, Chaudhary K, Wang J, Doerffler E, Fleury M, McMurtrie D, Chen X, Delaney WE IV, Kim CU (2009) Bioorg Med Chem Lett 19:3453
39. Liverton NJ (2008) HCV NS3 protease inhibitors, Patent WO 2008/051477
40. Das K (2012) J Med Chem 55:6263
41. Thorlund K, Awad T, Boivin G, Thabane L (2011) BMC Infest Dis 11:134
42. Meier C, Habel L, Laux W, De Clercq E, Balzarini J (1995) Nucleos Nucleot 14:759
43. Neyts J, De Clercq E (1997) Antimicrob Agents Chemother 41:2754
44. Kralikova S, Budesinsky M, Masojidkova M, Rozenberg I (2000) Nucleos Nucleot 19:1159
45. Snoeck R, Holy A, Dewolf-Peeters C, Van Den Oord J, De Clercq E, Andrei G (2002) Antimicrob Agents Chemother 46:3356
46. Szymanska A, Szymczak M, Boryski J, Stawinski J, Kraszewski A, Collu G, Sanna G, Giliberti G, Loddo R, La Collo P (2006) Bioorg Med Chem 14:1924
47. Korshin EE, Pozdeev OK (2013) Tetrahedron 69:11109
48. Mukhametov FS, Khorshin EE, Nekhoroshkov VM, Efremov YY (1989) Russ J Gen Chem 59:1309
49. Foster AC, Kemp JA (2006) Curr Opin Pharmacol 6:7
50. Bloom FE, Iversen LL (1971) Nature 229:628
51. Niswender CM, Conn PJ (2010) Annu Rev Pharmacol Toxicol 50:295
52. Javitt DC (1987) J Clin Psychiatry 9:12
53. Duty S (2010) J Pharmacol 161:271
54. Selvam C, Goudet C, Oueslati N, Pin J-P, Acher F (2007) J Med Chem 50:4656
55. Acher FC. Selvam C, Pin J-P, Bertrand HO (2010) Diastereoisomers of hypophosphorous acid derivatives, Patent WO 2010/106526
56. Beurrier C, Lopez S, Révy D, Selvam C, Goudet C, Lhérondel M, Gubellini P, Kerkerie-LeGoff L, Acher FC, Pin J-P, Amalric M (2010) FASEB J 23:3619
57. Selvam C, Oueslati N, Lemasson IA, Brabet I, Rigault D, Courtiol T, Cesarini S, Triballeau N, Bertrand H-O, Goudet C, Pin J-P, Acher FC (2010) J Med Chem 503:2797
58. Wieronska JM, Stachowicz K, Acher F, Lech T, Pilc A (2012) Psychopharmacology (Berl) 220:481

59. Goudet C, Vilar B, Courtiol T, Deltheil T, Bessiron T, Brabet I, Oueslati N, Rigault D, Bertrand H-O, McLean H, Daniel H, Amalric M, Acher FC, Pin J-P (2012) FASEB J 26:1682
60. Acher F, Busserolles J, Cesarini S, Commare B, Eschalier A, Goudet C, Lemasson I, Pin J-P, Rigault D Hypophosphorous acid derivatives having antihyperalgic activity and biological applications thereof, Patent WO 2012/156931 A1
61. Paredes RG, Agmo A (1992) Neurosci Biobehav Rev 16:145
62. O'Connell AW, Fox GB, Kjoller C, Gallagher HC, Murphy KJ, Kelly J, Regan CM (2001) Eur J Pharmacol 424:37
63. Zheng W, Xie W, Zhang J, Strong JA, Wang L, Yu L, Xu M, Lu L (2003) J Biol Chem 278:48321
64. Sernagor E, Young C, Eglen SJ (2003) J Neurosci 23:7621
65. Wisor JP, DeLorey TM, Homanics GE, Edgar DM (2002) Brain Res 955:221
66. Zhang Q, Lehmann A, Rigda R, Den J, Holloway RH (2002) Gut 50:19
67. Alstermark C, Amin K, Dinn SR, Elebring T, Fjellström O, Fitzpatrick K, Geiss WB, Gottfries J, Guzzo PR, Harding JP, Holmén A, Kothare M, Lehmann A, Mattsson JP, Nilsson K, Sundén G, Swanson M, Von Unge S, Woo AM, Wyle MJ, Zheng X (2008) J Med Chem 51:4315
68. Somekh L, Shanzer A (1982) J Am Chem Soc 104:5836
69. Locock KES, Yamamoto I, Tran P, Hanrahan JR, Chebib M, Johnston GAR, Allan RD (2013) J Med Chem 56:5626
70. Kumar RJ, Chebib M, Hibbs DE, Kim H-L, Johnston GAR, Salam NK, Hanrahan JR (2008) J Med Chem 51:3825
71. Daly AM, Gilheany DG (2003) Tetrahedron Asymmetry 14:127
72. Hansen HI, Kehler JA (1999) Synthesis 1925
73. Gavande G, Yamamoto I, Salm NK, Ai TH, Burden PM, Johnston GAR, Hanrahan JR, Chebib M (2011) Med Chem Lett 2:11
74. Li X, Cao JH, Li Y, Rondard P, Zhang Y, Yi P, Liu JF, Nan FJ (2008) J Med Chem 51:3057
75. Belley M, Sullivan R, Evans J, O'Neill G, Ng GYK (1999) Bioorg Med Chem 7:2697
76. Li ZG, Mintzer E, Bittman R (2006) J Org Chem 71:1718
77. Hatanaka Y, Hashimoto M, Kurihara H, Nakayama H, Kanaoka Y (1994) J Org Chem 59:383
78. Kolb HC, Sharpless KB (2003) Drug Discov Today 8:1128
79. Masharina A, Reymond L, Maurel D, Umezawa K, Johnsson K (2012) J Am Chem Soc 134:19026
80. Gavande M, Kim HL, Doddareddy MR, Johnston GAR, Chebib M, Hanrahan JR (2013) Med Chem Lett 4:402
81. Shane B (1989) Vitam Horm 45:263
82. McBurney MW, Whitmore GF (1974) Cell 2:173
83. Taylor RT, Hanna ML (1975) Arch Biochem Biophys 171:507
84. McGuire JJ, Coward JK (2003) Drug Future 28:967
85. Coward JK, McGuire JJ (2008) In: Litwack G (ed) Vitamins hormones (folic acid and folates). Elsevier, p 347
86. Tsukamoto T, Haile WH, McGuire JJ, Coward JK (1998) Arch Biochem Biophys 355:109
87. Valiaeva N, Bartley D, Konno T, Coward JK (2001) J Org Chem 66:5146
88. McGuire JJ, Haile WH, Valiaeva N, Bartley D, Guo J, Coward JK (2003) Biochem Pharmacol 65:315
89. Bartley DM, Coward JK (2005) J Org Chem 70:6757
90. McGuire JJ, Bartley DM, Tomsho JW, Haile WH, Coward JK (2009) Arch Biochem Biophys 488:140
91. Rosowsky A, Freisheim JH, Bader H, Forsch RA, Susten SS, Cucchi CA, Frei E (1985) J Med Chem 28:660
92. Kensler TW, Cooney DA (1981) Adv Pharmacol Chemother 18:273
93. Christopherson RI, Lyons SD (1990) Med Res Rev 10:505
94. Collins KD, Stark GA (1971) J Biol Chem 246:6599

95. Grem JL, King SA, O'Dwyer PJ, Leyland-Jones B (1988) Cancer Res 48:4441
96. Fleming RA, Capizzi RL, Muss HB, Smith S, Fernandes DJ, Homesley H, Loggie BW, Case LD, Morris R, Russell GB, Richards F (1996) Clin Cancer Res 2:1107
97. Chan TCK, Young B, King ME, Taetle R, Howell SB (1985) Cancer Treat Rep 69:425
98. Coudray L, Kantrowitz ER, Montchamp J-L (2009) Bioorg Med Chem Lett 19:900
99. Coudray L, Pennebaker AF, Montchamp J-L (2009) Bioorg Med Chem 17:7680
100. Bravo-Altamirano K, Montchamp J-L (2006) Org Lett 8:4169
101. Bravo-Altamirano K, Montchamp JL (2008) Org Synth 85:96
102. Cristau H-J, Herve A, Virieux D (2004) Tetrahedron 60:877
103. Coudray L, Bravo-Altamirano K, Montchamp J-L (2008) Org Lett 10:1123
104. Bravo-Altamirano K, Abrunhosa-Thomas I, Montchamp JL (2008) J Org Chem 73:2292
105. Furata T, Torigai H, Osawa T, Iwamura M (1993) J Chem Soc Perkin Trans 1:3139
106. Wade RH (2009) Mol Biotechnol 43:177
107. Verhey KJ, Gaertig J (2007) Cell Cycle 17:2152
108. Garnham CP, Roll-Mecak A (2012) Cytoskeleton 69:442
109. Janke C, Rogowski K, Wloga D, Regnard C, Kajava AV, Strub J-M, Temaruk N, van Dijk J, Boucher D, Dorsselaer AV, Suryavanshi S, Gaertig J, Edde B (2005) Science 308:1758
110. Ikegami K, Mukai M, Tsuchida J, Heier RL, MacGregor GR, Setou M (2006) J Biol Chem 281:30707
111. Ikegami K, Heier RL, Taruishi M, Takagi H, Mukai M, Shimma S, Tairo S, Hatanaka K, Morone N, Yao I, Campbell PK, Yuasa S, Janke C, MacGregor GR, Setou M (2007) Proc Natl Acad Sci U S A 104:3213
112. Roll-Mecak A, McNally FJ (2010) Curr Opin Cell Biol 22:96
113. Zhang D, Rogers GC, Buster DW, Sharp DJ (2007) J Cell Biol 177:231
114. Kashiwaya K, Nakagawa H, Hosokawa M, Mochizuki Y, Ueda K, Piao L, Chung S, Hamamoto R, Eguchi H, Ohigashi H, Ishikawa O, Janke C, Shinomura Y, Nakamura Y (2010) Cancer Res 70:4024
115. Rogowski K, van Dijk J, Magiera MM, Bosc C, Deloulme J-C, Bosson A, Peris L, Gold ND, Lacroix B, Grau MB, Bec N, Larroque C, Desagher S, Holzer M, Andrieux A, Moutin M-J, Janke C (2010) Cell 143:564
116. Janke C, Bulinski JC (2011) Nat Rev Mol Cell Biol 12:773
117. Liu Y, Garnham CP, Roll-Mecak A, Tanner ME (2013) Bioorg Med Chem Lett 23:4408
118. Liu S, Ben RN (2005) Org Lett 7:2385
119. Feng Y, Coward JK (2006) J Med Chem 49:770
120. Georgiadis D, Matziari M, Vassiliou S, Dive V, Yiotakis A (1999) Tetrahedron 55:14635
121. Colvin MD (2000) Cancer medicine, 5th edition. Chapter 48 alkylating agents and platinum antitumor compounds
122. Hudson HR, Keglevich G (2008) Phosphorus Sulfur Silicon Relat Elem 183:2256
123. Keglevich G, Brlik J, Janke F, Toke L (1990) Heteroatom Chem 1:419
124. Keglevich G, Toke L, Kovacs A, Toth G, Ujszaszy K (1993) Heteroatom Chem 4:61
125. Keglevich G, Sipos M, Szieberth D, Nyulaszi L, Imre T, Ludanyi K, Toke L (2004) Tetrahedron 60:6619
126. Clarion L, Jacquard C, Sainte-Catherine O, Loiseau S, Filippini D, Hirlemann MH, Volle JN, Virieux D, Lecouvey M, Pirat JL, Bakalara N (2012) J Med Chem 55:2196
127. Pirat JL, Virieux D, Clarion L, Volle JN, Bakalara N, Mersel M, Montbrun J, Cristau HJ (2009) New phosphorus containing heterocyclic compounds, sugar analogs, and compositions having anticancer activity containing the same. Patent WO 2009/004096 A1
128. Clarion L, Jacquard C, Sainte-Catherine O, Decoux M, Loiseau S, Rolland M, Lecouvey M, Hugnot JP, Volle JN, Virieux D, Pirat JL, Bakalara N (2014) J Med Chem 57:8293
129. Tarnowski GS, Moutain IM, Stock CC, Weltzien HU, Wesphal O (1978) Cancer Res 38:339
130. Eibl H, Unger C (1990) Cancer Treat Rev 17:233
131. Stekar J, Hilgard P, Voegeli R, Maurer HR, Engel J, Kutscher B, Nöbner G, Schumacher W (1993) Cancer Chemother Pharmacol 32:437
132. Hilgard P, Klenner T, Stekar J, Nössner G, Kutscher B, Engel J (1997) Eur J Cancer 33:442
133. Markoulides MS, Regan AC (2011) Tetrahedron Lett 52:2954

134. Markoulides MS, Regan AC (2013) Org Biomol Chem 11:119
135. Abdou WM, Shaddy AA (2009) ARKIVOC 9:143
136. Russell RGG (2011) Bone 49:2
137. Houghton TJ, Tanaka KSE, Kang T, Dietrich E, Lafontaine Y, Delorme D, Ferreira SS, Viens F, Arhin FF, Sarmiento I, Lehoux D, Fadhil I, Laquerre K, Liu J, Ostiguy V, Poirier H, Moeck G, Parr TR, Far AR (2008) J Med Chem 51:6955
138. Luke GP, Shakespeare WCA (2002) Synth Commun 32:2951
139. Chandraratna RA (1995) US Pat 5399561
140. Schleifer KH, Kandler O (1972) Bacteriol Rev 36:407
141. Höltje J-V (1998) Microbiol Mol Biol Rev 62:181
142. van Heijenoort J (2001) Nat Prod Rep 18:503
143. Pal M, Bearne SL (2014) Bioorg Med Chem Lett 24:1432
144. Wróblewski AE, Verkade JG (1996) J Am Chem Soc 118:10168
145. Koparir P, Karaarslan M, Orek C, Koparir M (2011) Phosphorus Sulfur Silicon Relat Elem 186:2368
146. Ahmedzade M, Cukurovali A, Koparir M (2003) J Chem Soc Pak 25:51
147. Koparir M, Cansiz A, Ahmedzade M, Cetin A (2004) Heteroatom Chem 15:26
148. Koparir M, Cansiz A, Ahmedzade M (2004) Russ J Org Chem 40:1813
149. Koparir M, Cansiz A, Ahmedzade M (2006) Chem Heterocycl Comp 3:424
150. Nzila A, Ward SA, Marsh K, Sims PFG, Hyde JE (2005) Trends Parasitol 21:334
151. Nzila A, Ward SA, Marsh K, Sims PFG, Hyde JE (2005) Trends Parasitol 21:292
152. Yang Y, Coward JK (2007) J Org Chem 72:5748
153. Wang P, Wang Q, Yang Y, Coward JK, Nzila A, Sims PFG, Hyde JE (2010) Mol Biochem Parasitol 172:41
154. Pastore EJ, Friedkin M, Jardetzky O (1963) J Am Chem Soc 85:3058
155. Hillcoat BL, Blakley RL (1964) Biochem Biophys Res Commun 15:303
156. Evans JC, Huddler DP, Jiracek J, Castro C, Millian NS, Garrow TA, Ludwig ML (2002) Structure 10:1159
157. Obeid R (2013) Nutrients 5:3481
158. Collinsova M, Castro C, Garow TA, Yiotakis A, Dive V, Jiracek J (2003) Chem Biol 10:113
159. Jiracek J, Collinsova M, Rosenberg I, Budesinsky M, Protivinska E, Netusilova H, Garrow TA (2006) J Med Chem 49:3982
160. Vanek V, Budesinsky M, Kabeleova P, Sanda M, Kozisek M, Hanclova I, Mladkova J, Brznda J, Rosenberg I, Koutmos M, Garrow TA, Jiracek J (2009) J Med Chem 52:3652
161. Picha J, Vanek V, Budesinsky M, Mladkova J, Garrow TA, Jiracek J (2013) Eur J Med Chem 65:256
162. Catch JR, Cook AM, Graham AR, Heilbron IJ (1947) J Chem Soc 1609
163. Zhukov YN, Khomutov AR, Osipova TI, Khomutov RM (1999) Russ Chem Bull 48:1348
164. Liboska R, Picha J, Hanclova I, Budesinsky M, Sanda M, Jiracek J (2008) Tetrahedron Lett 49:5629
165. Caughey GH (1994) Am J Respir Crit Care Med 150:S138
166. Bank U, Ansorge S (2001) J Leukocyte Biol 69:197
167. Owen CA, Campbell EJ (1999) J Leukocyte Biol 65:137
168. Hart PH (2001) Immunol Cell Biol 79:149
169. Caughey GH (1995) Mast cell proteases in immunology and biology. Marcel-Decker, New-York, p 305
170. Stockley RA (1999) Am J Respir Crit Care Med 160:S49
171. Simon SR (1993) Agents Actions Suppl 42:27
172. Greco MN, Hawkins MJ, Powell ET, Almond HR, De Garavilla L, Hall J, Minor LK, Wang Y, Corcoran TW, Di Cera E, Cantwell AM, Savvides SN, Damiano BP, Maryanoff BE (2007) J Med Chem 50:1727
173. de Garavilla L, Greco MN, Sukumar N, Chen Z-W, Pineda AO, Mathews FS, Di Cera E, Giardino EC, Wells GI, Haertlein BJ, Kauffman JA, Corcoran TW, Derian CK, Eckardt AJ, Damiano BP, Andrade-Gordon P, Maryanoff BE (2005) J Biol Chem 280:18001

174. Kervinen J, Crysler C, Bayoumy S, Abad MC, Spurlino J, Deckman I, Greco MN, Maryanoff BE, de Garavilla L (2010) Biochem Pharmacol 80:1033
175. Dunford JE, Thompson K, Coxon FP, Luckman SP, Hahn FM, Poulter CD, Ebetino FH, Rogers MJ (2001) J Pharmacol Exp Ther 296:235
176. Bergstrom JD, Bostedor RG, Masarachia PJ, Reszka AA, Rodan G (2000) Arch Biochem Biophys 373:231
177. Dunford JE, Kwaasi AA, Rogers MJ, Barnett BL, Ebetino FH, Russell RGG, Oppermann U, Kavanagh KL (2008) J Med Chem 51:2187
178. Lieberman I (1956) J Biol Chem 222:765
179. Long CW, Levitzki A, Koshland DE (1970) J Biol Chem 245:80–87
180. Verschuur AC, van Gennip AH, Leen R, Muller EJ, Elzinga L, Voûte PA, van Kuilenburg ABP (2000) Eur J Cancer 36:627
181. Taylor SD, Mirzaei F, Bearne SL (2006) Org Lett 8:4243
182. Taylor SD, Mirzaei F, Bearne SL (2006) J Org Chem 73:1403
183. Taylor SD, Lunn FA, Bearne SL (2008) ChemMedChem 3:1853
184. Romenko VD, Kukhar VP (2006) Chem Rev 106:3868
185. Blackburn GM, England DA, Kolkmann FJ (1981) Chem Commun 930
186. Blackburn GM, Kent DE, Koklman FJ (1984) J Chem Soc Perkin Trans 1:1119
187. Davisson VJ, Woodside AB, Neal TR, Stremler KE, Muehlbacher M, Poulter CD (1986) J Org Chem 51:4768
188. Chambers RD, O'Hagan D, Lamont RB, Jaina SC (1990) Chem Commun 1053
189. Halazy S, Ehrhard A, Danzin C (1991) J Am Chem Soc 113:315
190. Thatcher GR, Campbell AS (1993) J Org Chem 58:2272
191. Burke TR, Smyth MS, Otaka A, Nomizu M, Roller PP, Wolf G, Case R, Shoelson SE (1994) Biochem 33:6490
192. Berkowitz DB, Bose M (2001) J Fluorine Chem 112:13
193. Blackburn GM, Turkmen H (2005) Org Biomol Chem 3:225–226
194. Surya Prakash GK, Zibinsky M, Upton TG, Kashemirov BA, McKenna CE, Oertell K, Goodman MF, Batra VK, Pedersen LC, Beard WA, Shock DD, Wilson SH, Olah GA (2010) Proc Natl Acad Sci U S A 107:15693
195. McCormack JG, Westergaard N, Kristiansen M, Brand CL, Lau J (2001) Curr Pharm Des 7:1451
196. Erion MD, van Poelje VD, Dang Q, Kasibhatla SR, Potter SC, Reddy MR, Reddy KR, Jiang T, Lipscomb WN (2005) Proc Natl Acad Sci U S A 102:7970
197. Dand Q, Brown BS, Liu Y, Rydzewski RM, Robinson ED, van Poelje PD, Reddy MR, Erion MD (2009) J Med Chem 52:2880
198. Froehler BC, Ng PG, Matteuci MD (1986) Nucleic Acids Res 14:5399
199. Perish JW, Johns RB (1988) Synthesis 142
200. Volle JN, Filippini D, Krawczy B, Kaloyanov N, Van der Lee A, Maurice T, Pirat JL, Virieux D (2010) Org Biomol Chem 8:1438
201. Afarinkia K, Yu HW (2003) Tetrahedron Lett 44:781
202. Montchamp JL, Duromnd YR (2001) J Am Chem Soc 123:510
203. Napolitano LM (2009) Crit Care Clin 25:279
204. Squires JE (2002) Science 295:1002
205. Spahn DR (2000) Adv Drug Deliv Rev 40:143
206. Winslow RM (2007) Semin Hematol 44:51
207. Everse J, Hsia N (1997) Free Radic Biol Med 22:1075
208. Cai H, Roach TA, Dabek M, Somerville KS, Acharya S, Hosmane RS (2010) Bioconjug Chem 21:1494
209. Tunaru S, Kero J, Schaub A, Wufka C, Blaukat A, Pfeffer K, Offermanns S (2003) Nat Med 9:352
210. Wise A, Foord SM, Fraser NJ, Barnes AA, Elshourbagy N, Eilert M, Ignar DM, Murdock PR, Steplewski K, Green A, Brown AJ, Dowell SJ, Szekeres PG, Hassall DG, Marshall FH, Wilson S, Pike NB (2003) J Biol Chem 278:9869

211. Dvorak CA, Liu C, Shelton J, Kuei C, Sutton SW, Lovenberg TW, Carruthers NI (2012) Med Chem Lett 3:637
212. Heyliger CE, Tahiliani AG, McNeill JH (1985) Science 227:1474
213. Thomson KH, Orvig C (2000) J Chem Soc Dalton Trans 2885
214. Kaboudin B, Moradi K, Safaei E, Dehghan H, Salehi P (2012) Phosphorus Sulfur Silicon Relat Elem 187:1521
215. Perazella MA (2009) Curr Drug Saf 3:67
216. Hermann P, Kubicek V, Lukes I (2008) Dalton Trans 3027
217. Ramogida CF, Orvig C (2013) Chem Commun 49:4720
218. Price EW, Orvig C (2014) Chem Soc Rev 43:260
219. Forsterova M, Svobodova I, Lubal P, Tarborsky, Kotek J, Hermann P, Likes I (2007) Dalton Trans 535
220. Rudovsky J, Botta M, Hermann P, Lukes I, Maienro V, Aime S (2005) Org Biomol Chem 112
221. Forsterova M, Petrik M, Laznickova A, Laznicek M, Hermann P, Likes I, Melichar F (2009) App Rad Isotop 67:21
222. Rorudovsky J, Botta M, Hermann P, Koridze A, Aime S (2006) Dalton Trans 2323
223. Kotkova Z, Helm L, Kotek J, Hermann P, Lukes I (2012) Dalton Trans 13509
224. Kotkova Z, Kotek J, Jirak D, Jendelova P, Herynek V, Berkova Z, Hermann P, Lukes I (2010) Chem Eur J 16:10094
225. Mintzer MA, Grinstaff MW (2011) Chem Soc Rev 40:173
226. Rudovsky J, Botta M, Hermann P, Hardcastle KI, Lukes I, Aime S (2006) Bioconjug Chem 17:975
227. Polasek M, Hermann P, Peters JA, Geraldes CFGC, Lukes I (2009) Bioconjug Chem 20:2142
228. David T, Prochazkova S, Havlickova J, Kotek J, Kubicek V, Hermann P, Lukes I (2013) Dalton Trans 42:2414
229. Pinto-Alphandary H, Andremont A, Couvreur P (2000) J Antimicrob Agents 13:155
230. Moses MA, Brem H, Langer R (2003) Cancer Cell 4:337
231. Allen TM, Cullis PR (2004) Science 303:1818
232. Faraji AH, Wipf P (2009) Biorg Med Chem 17:2950
233. Boudier A, Castagnos P, Soussan E, Beaune G, Belkhelfa H, Ménager C, Cabuil V, Haddioui L, Roques C, Rico-Lattes I, Blanzat M (2011) Int J Pharm 403:230
234. Soussan E, Mille C, Blanzat M, Bordat P, Rico-Lattes I (2008) Langmuir 24:2326
235. Blanzat M, Perez E, Rico-Lattes I, Promé D, Promé JC, Lattes A (1999) Langmuir 15:6163
236. Blanzat M, Perez E, Rico-Lattes I, Lattes A (1999) New J Chem 23:1063
237. Brun A, Etemad-Moghadam G (2002) Synthesis 10:1385
238. Baillie AC, Wright K, Wright BJ, Earnshaw CG (1988) Pest Biochem Physiol 30:103
239. Wang T, He HW (2008) Phosphorus Sulfur Silicon Relat Elem 183:1884
240. Deng X, Liao G, Long Q, Gao Y, Peng H, He H (2013) Phosphorus Sulfur Silicon Relat Elem 188:663
241. Li M, Peng GY, He H (2013) J Pest Sci 38:78
242. He H-W, Yuan J-L, Peng H, Chen T, Shen P, Wan S-Q, Li Y, Tan H-L, He Y-H, He J-B, Li Y (2011) J Agric Food Chem 59:4801
243. He H-W, Peng H, Wang T, Wang C, Yuan J-L, Chen T, He J, Tan X (2013) J Agric Food Chem 61:2479
244. Hall R (2008) Phosphorus Sulfur Silicon Relat Elem 183:258
245. Chang W-C, Mansoorabadi SO, Liu H-W (2013) J Am Chem Soc 135:8153
246. Lee J-H, Bae B, Kuemin M, Circello BT, Metcalf WW, Nair SK, van der DonK WA (2010) Proc Natl Acad Sci USA 107:17557
247. Evans BS, Zhao C, Gao J, Evans CM, Ju K-S, Doroghazi JR, van der Donk WA, Kelleher NL, Metcalf WW (2013) ACS Chem Biol 8:908
248. Bryant DE, Marriott KER, Macgregor SA, Kilner C, Pasek MA, Kee TP (2010) Chem Commun 46:3726

Top Curr Chem (2015) 360: 115–160
DOI: 10.1007/128_2014_561

Published online: 13 November 2014

Prodrugs of Phosphonates and Phosphates: Crossing the Membrane Barrier

Andrew J. Wiemer and David F. Wiemer

Abstract A substantial portion of metabolism involves transformation of phosphate esters, including pathways leading to nucleotides and oligonucleotides, carbohydrates, isoprenoids and steroids, and phosphorylated proteins. Because the natural substrates bear one or more negative charges, drugs that target these enzymes generally must be charged as well, but small charged molecules can have difficulty traversing the cell membrane by means other than endocytosis. The resulting dichotomy has stimulated a great deal of effort to develop effective prodrugs, compounds that carry little or no charge to enable them to transit biological membranes, but able to release the parent drug once inside the target cell. This chapter presents recent studies on advances in prodrug forms, along with representative examples of their application to marketed and developmental drugs.

Keywords Bisphosphonate · Isoprenoid · Nucleotide · Phosphate · Phosphonate · Prodrugs

Contents

A.J. Wiemer
Department of Pharmaceutical Sciences, University of Connecticut, Storrs, CT 06269, USA

D.F. Wiemer (✉)
Departments of Chemistry and Pharmacology, University of Iowa, Iowa City, IA 52242-1294, USA
e-mail: david-wiemer@uiowa.edu

Abbreviations

AraC	Arabinofuranosylcytidine
AZT	Azidothymidine
CMV	Cytomegalovirus
d4TMP	2′,3′-Didehydro-3′-dideoxy-thymidine-5′-monophosphate
DOXP	1-Deoxy-D-xylulose 5-phosphate
EBV	Epstein–Barr virus
GCPR	G protein coupled receptor
GemC	Gemcitabine
HBV	Hepatitis B virus
HCV	Hepatitis C virus
HDP	Hexadecyloxypropyl
HIV	Human immunodeficiency virus
HPMPA	9-(3-Hydroxy-2-phosphonyl-methoxypropyl)adenine
HSV	Herpes-simplex virus
NA	Nucleoside analogue
PMEA	9-[2-(Phosphonomethoxy)ethyl]adenine
POC	Isopropyloxycarbonyloxymethyl
POM	Pivaloyloxymethyl
RBV	Ribavirin
SATE	*S*-Acylthioalkyl ester

1 Introduction

Metabolism employs phosphorus compounds in many ways. Phosphate esters are central to information storage, serve as the currency for energy exchange, and contribute to membrane fluidity. They are important intermediates in carbohydrate metabolism, in formation of nucleotides and their assembly into RNA and DNA, and in steroid fabrication and protein lipidation through the isoprenoid biosynthesis pathways. They hold key roles in cell signaling processes, including G-protein coupled receptors (GPCR), second messengers such as cAMP or phosphatidylinositol, and enzyme activation/deactivation through protein phosphorylation. Within the cell, the phosphate group commonly serves as a tunable leaving group. A phosphate mono- or diester is relatively stable when free in metabolic surroundings because of its negative charge at physiological pH [1], but it is readily activated upon complexation to various counterions in an enzyme's active site. Thus phosphorus compounds provide both abundant opportunities for drug design and intrinsic challenges. Chief among the challenges is the inherent conflict between a molecule which must be anionic to bind to the active site of an enzyme but first must penetrate the membrane to access that enzyme, as well as the necessity to mimic a potential leaving group with a substructure sufficiently stable to survive delivery.

A longstanding and still common strategy for preparation of more stable analogues of phosphate esters (**1**) is based upon formal replacement of the ester oxygen with carbon to afford the corresponding phosphonate (**2**) [2] (Fig. 1). Although phosphonates are known from a number of natural sources, implicating the existence of enzymes capable of C–P bond cleavage [3], it is widely recognized that phosphonates have greater metabolic stability. Replacement of the phosphate ester oxygen with a simple $-CH_2-$ group may preserve much of the size and shape of the original substrate, but it also impacts the second pK_a value, with the phosphonic acid less acidic by 0.5–1.5 pK_a units [4]. Incorporation of this carbon also eliminates the possibility of enzymatic interaction with oxygen lone pairs at this position. Both of these concerns can be addressed by addition of fluorine or oxygen substituents onto the alpha carbon, although this necessarily has an impact on steric issues and can also introduce a stereogenic center [5]. Nevertheless, numerous phosphonate analogues of biologically active phosphates have been prepared.

Phosphate monoesters are charged at physiological pH, as are the corresponding phosphonates, and those with small and/or hydrophilic substituents have difficulty in diffusion across biological membranes [6]. While this limitation may be ameliorated in the case of compounds with larger, more lipophilic substituents [7], it is always a concern that must be considered in drug design. Thus, efforts to prepare biologically active organophosphorus compounds often have been followed by studies to delineate strategies that temporarily mask any negative charges at physiological pH. Potential drugs based upon this approach may offer a number of advantages over their non-protected counterparts. In particular, addition of cell-cleavable protecting/masking groups (i.e., the prodrug approach) can: (1) increase oral bioavailability; (2) enhance cell penetration; (3) improve the specificity of tissue delivery; and (4) avoid or minimize degradation in serum via cellular sequestration. In addition, once the prodrug moieties are cleaved within the cell, the same factors which would normally limit entry of the non-protected drug into the cell can now restrict the free drug from leaving the cell. Thus, prodrugs can effectively allow the drug to achieve elevated concentrations within cells, further enhancing efficacy [8].

Study of phosphonate prodrugs may have a long history, but prodrug strategies to protect and deliver phosphate monoesters have also been of great interest.

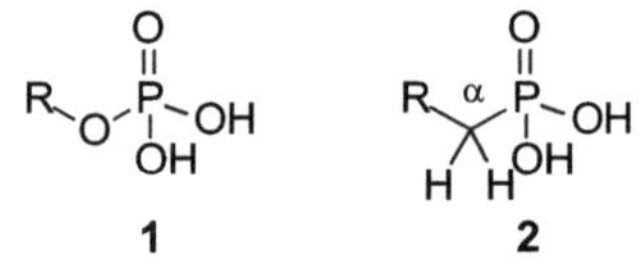

Fig. 1 General structures of the phosphate and phosphonate groups

Efficient delivery of a phosphate monoester into the cell can afford important metabolic advantages. For example, nucleoside analogues such as arabinofuranosyl cytidine (AraC) and gemcitabine (GemC) undergo activation after cell entry by conversion to the corresponding mono-, di-, and, ultimately, triphosphates [9], and use of a protected phosphate may allow intersection with natural metabolic processes at a later stage [10]. This strategy may be particularly important with nucleotide prodrugs where it can allow the agent to bypass the rate-limiting initial phosphorylation [11]. Furthermore, phosphate prodrugs may confer stability to serum phosphatases, and therefore support more effective dosing.

However, use of prodrugs is not without substantial challenges. In particular, choosing the best protecting group is difficult for several reasons. Most notably, cellular cleavage of the protecting groups can often generate products which are viewed as disadvantageous or even toxic. In addition, the protecting groups must strike a balance between allowing absorption in the intestines and allowing cleavage in the blood or target cell.

Given the tremendous importance of phosphonate and phosphate prodrugs, research in this area has grown dramatically and has been the subject of periodic reviews [12–14]. Many reviews have focused exclusively on nucleotide prodrugs [15–18] or hepatitis C (HCV) [19–21]. However, it has been more than 6 years since a comprehensive phosphorus prodrug review [22]. Therefore, we have attempted herein to compare and contrast recent investigations of phosphonate and phosphate prodrugs, and to consider the present applications and future potential of such agents. After a brief consideration of the structural factors involved, representative examples of these strategies are presented.

2 General Considerations

The majority of prodrug applications of phosphates and phosphonates seek to facilitate passive diffusion through the cell membrane by masking negative charge until the compound is within the cell. This requires attention to two primary considerations: (1) what type of derivative is employed; and (2) what process or processes is or are entrusted to remove the protecting groups once the prodrug is within the cell.

For applications in synthetic chemistry, a wide variety of protecting groups for phosphonates and phosphates is known [23]. Those employed to provide prodrugs fall into two general categories: esters (**3–6**) and amides (**7–10**). Yet even after that initial selection is made, there are a number of other concerns to address and an

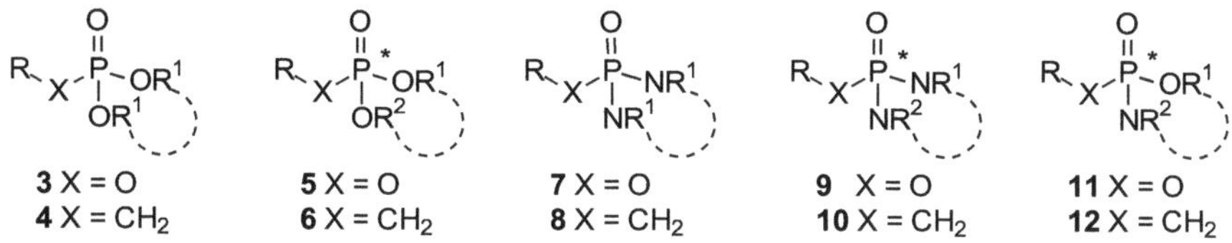

Fig. 2 General structures of phosph(on)ate prodrugs

equally large array of solutions has been developed. When two charges must be masked, will two of the same groups be employed in a diester (e.g., **3** or **4**) or will two different groups be used (e.g., **5** or **6**)? Use of two different groups necessarily results in formation of a stereogenic center at phosphorus, but may allow more flexibility in terms of the cleavage step(s). For either variation, use of two alcohols allows formation of an acyclic derivative while use of a single diol affords a cyclic derivative. Parallel considerations underlie the selection of amine-derived prodrugs (e.g. **7** or **8**), where again use of two different amines or an unsymmetrical diamine (e.g., **9** or **10**) generates a stereogenic phosphorus. Finally, mixed ester-amides (e.g., **11** or **12**) are also well known, and can offer some advantages in terms of both the cleavage step and organ or tissue targeting (Fig. 2).

Equally complex is the determination of the strategy employed for regeneration of the drug once it is within the cell. In most cases, this is assumed to involve a specific enzyme or enzymes, but these can include an esterase, an amidase, a phosphatase, or even a redox process. Furthermore, some prodrugs have been reported that rely upon thermal cleavage, and the possibility of photochemical cleavage may be available through photodynamic therapy approaches.

A dramatically different prodrug application has been to employ phosphates as anions to make a specific drug more water soluble. This strategy is particularly valuable where the drug itself is an alcohol (e.g., metronidazole, shown as the phosphate derivative **13**) [24] or a phenol (e.g., tamoxifen, shown as the nucleotide derivative **14**). In the case of metronidazole, addition of the phosphate enhances water solubility 50-fold and the prodrug is cleaved rapidly in human serum [25]. Further derivatization of metronidazole through conjugation with a nucleoside 3′-phosphate has been explored as a strategy to enhance oral bioavailability [26]. In the case of 4-hydroxytamoxifen, incorporation of the nucleotide element was desired to make this prodrug a substrate for a ribonuclease [27], a strategy designed to release a natural metabolite upon generation of the drug [26]. Variations of this approach have been employed with a number of other drugs [28–31] through functionalization of phenol substructures such as in prodrugs **15–17** [32], the nitrogen of phenyl hydantoins like derivatives **18** and **19**, and the relatively acidic carbon of β-dicarbonyl compounds [33]. While this appears to be an area of growing interest, in light of a recent review [32] it is not considered in greater depth here (Fig. 3).

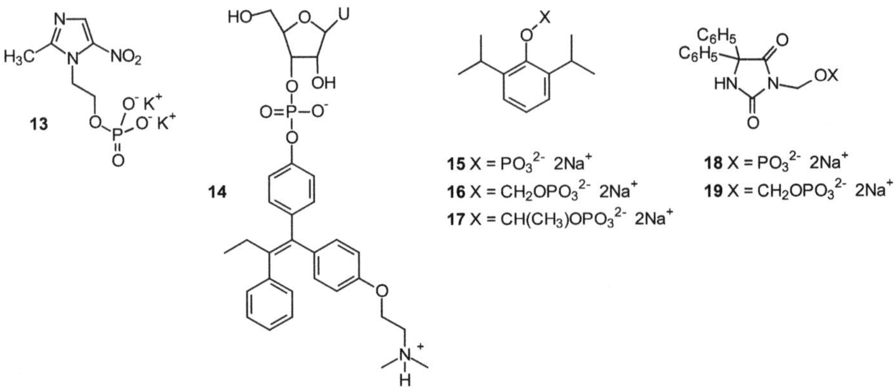

Fig. 3 Use of phosphates to enhance water solubility of drugs

Table 1 Prodrug variations based upon ester linkages

Symmetrical diesters		
	$R^1=R^2$	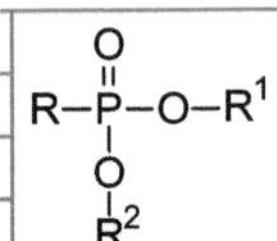
Alkyl	$-CH_3$	
Benzyl	$-CH_2C_6H_4X$	
	(X=H, OAc, OCH_3)	
Aryl	$-C_6H_5$	
Acyloxyalkyl (POM)	$-CH_2OC(O)C(CH_3)_3$	
Alkoxycarbonyloxy		
Alkyl (POC)	$-CH_2OC(O)OCH(CH_3)_2$	
S-Acylthioalkyl		
(SATE)	$-CH_2CH_2SC(O)R$	
Unsymmetrical diesters		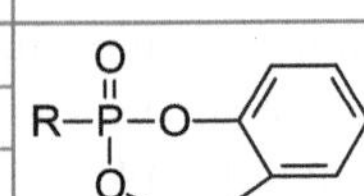
	$R^1 \neq R^2$	
cycloSal		
HepDirect		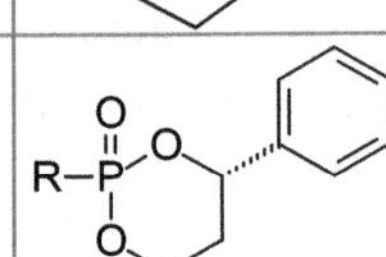
Monoesters		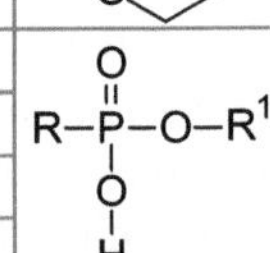
	$R^1=$	
Steroidal	Cholesteryl	
Glycerol-fatty alcohol	$-CH_2OCH_2(CH_2)_{14}CH_3$	
Internal monoesters and mixed esters		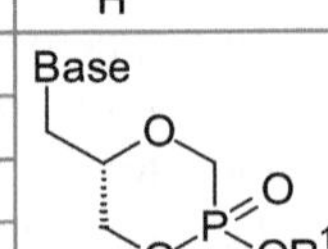
	$R^1=$	
Cidofovir and HPMPA	$-H$	
Glycerol-fatty alcohol	$-CH_2OCH_2(CH_2)_{14}CH_3$	

3 Ester Prodrugs

A great number of different alcohols can be employed to prepare phosphorus-based esters, and many examples have been reported. Some of the more useful approaches are listed in Table 1. Representative cases are described in the following sections.

3.1 *Symmetrical Diesters*

Conceptually, the simplest derivative of a phosphonic acid which would be neutral at physiological pH is a diester derived from a small alcohol such as methanol or ethanol. Dimethyl esters are readily prepared, and do not introduce stereochemical issues at phosphorus. However, despite the emergence of organophosphorus hydrolases in various bacteria [34], perhaps driven by the widespread use of phosphorus-based insecticides, simple dialkyl esters of phosphonates appear to be surprisingly stable in mammalian systems (cf. bisphosphonate examples [35–37], nucleoside phosphonate examples [38–40], and a squalene synthase inhibitor [41]). Some phosphonate dibenzyl esters have also been examined and show more promise. In one study, the parent benzyl group was converted to the free drug too slowly to be of value, but more substituted systems can have greater utility as prodrugs [42, 43]. For example, the *p*-acetoxybenzyl system can be activated by esterase-mediated hydrolysis of the acetate [38] and the *p*-methoxybenzyl esters undergo cytochrome P mediated activation [42]. While the latter process was suggested to involve cleavage of the methyl ether, it should be noted that chemical cleavage of *p*-methoxybenzyl ethers involves oxidation of the benzylic position, which could also facilitate drug regeneration within the cell [23, 44]. Whatever the mechanism of drug release, the bis[(*p*-methoxy)benzyl] prodrug showed enhanced plasma stability, which allowed intercellular drug release as well enhanced metabolism in the liver.

Aryl esters would be expected to be more chemically reactive because of the greater acidity of phenol vis-à-vis methanol. In some cases, diaryl phosphonates have proven to be useful as prodrugs [45]. This early study also confirmed that simple alkyl esters were generally incapable of functioning as prodrugs, while demonstrating that diphenyl esters performed well, releasing the free acid in vivo at high concentrations as well as exhibiting physiochemical properties that were more conducive to pharmaceutical formulation. More recently it has been suggested that the ability to tune the reactivity of diaryl esters through choice of their substituents should make discovery of useful prodrugs in this family feasible [46, 47].

Phosphonate diesters derived from more complex natural alcohols, for example glucose, might be more amenable to metabolic hydrolysis. Unfortunately, most carbohydrate-based alcohols would also add significant chemical complexity in terms of their stereogenic centers as well as potential issues of regiochemical control during their formation. Furthermore, the enzymes which metabolize carbohydrate phosphates, such as glucose-6-phosphatase, have natural substrates that are dianions or anions, and may be unlikely to hydrolyze neutral compounds. Thus, the early studies on phosphonate prodrugs often relied upon enzymatic cleavage mediated by enzymes other than phosphatases. Particularly prominent among these strategies were those that relied upon nonspecific esterases for an initial hydrolysis, followed by a chemical decomposition which liberated the desired drug [14].

Much of the early interest in phosphonate prodrugs can be traced to key studies during the quest to develop drugs against the human immunodeficiency virus (HIV), hepatitis, and other viral diseases. After De Clercq, Holy, Rosenberg, and their colleagues reported strong antiviral activity of acyclic nucleoside phosphonates such as 9-[2-(phosphonomethoxy)ethyl]adenine (PMEA) and (*S*)-9-(3-hydroxy-2-phosphonyl-methoxypropyl) adenine ((*S*)-HPMPA [48] (reviewed in [49]), the activity of these compounds stimulated great interest in the development of prodrug forms to enhance the membrane permeability of these drugs. An early focus was pivaloyloxymethyl (POM) modified phosphonates, a prodrug format which was first advanced for use with phosphate monoesters [50, 51]. However, this approach was readily adapted to phosphonates such as foscarnet esters (**20**) [52] and a phosphonate inhibitor of insulin receptor tyrosine kinase [53]. Ultimately, the POM moieties were applied to PMEA, leading (Fig. 4), after extensive studies, to the approval of adefovir dipivoxil (**21**, Hepsera) by Gilead for treatment of Hepatitis B (HBV) (reviewed by Lee and Martin [54]).

The POM group also has been applied to a variety of other phosphonates, including bisphosphonates, such as **22** [55]. This prodrug of an inositol monophosphatase inhibitor exhibits an increase of more than 2,500-fold in activity relative to the parent drug. Such a dramatic impact, together with the clinical utility of drugs such as adefovir dipivoxil, has fostered continued interest in the use of POM groups in phosphonate and phosphate prodrugs to the point where this is often the first prodrug form examined for any new phosphorus-containing drug [22, 56]. At the same time, concerns are periodically expressed about the metabolic byproducts of POM cleavage, including the toxicity [42, 43] and carcinogenicity [45] of formaldehyde as well as the burden resulting from elimination of pivalic

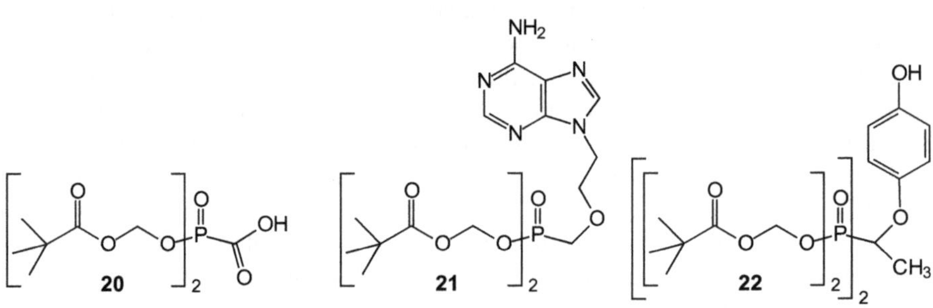

Fig. 4 Examples of pivaloyloxymethyl (POM)-modified drugs

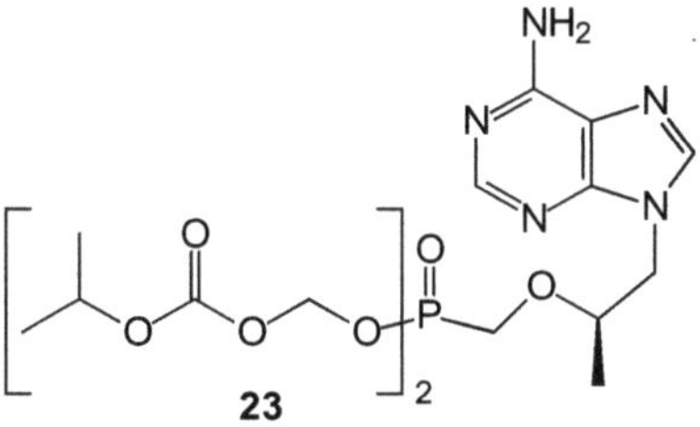

23

Fig. 5 Tenofovir disoproxil, an isopropyloxycarbonyloxymethyl(POC)-modified drug

acid as a carnitine derivative. Thus, while the POM group is a valued addition to the line-up of potential prodrug forms, research has continued into other variations.

A closely related strategy for preparation of phosphonate prodrugs involves the use of a carbonate ester such as isopropyloxycarbonyloxymethyl derivatives (POC). Esterase-mediated cleavage of this prodrug results in loss of carbonate and 2-propanol, rather than pivalic acid [57]. Therefore, in contrast to POM prodrugs, the POC prodrugs should not impact carnitine levels. An early study [58] was the basis for selecting the POC derivatives of acyclic nucleosides for development, because it showed significantly increased potency towards HIV-1 in a cell-based assay as well as a longer serum half-life than the corresponding POM compound. Ultimately, these studies have led to the clinical use of tenofovir disoproxil (**23**) for treatment of viral infections [59]. While the POC compounds may have lower levels of chemical stability than their POM analogues [60], this can be used to advantage in preparation of the parent phosphonates from POC-protected intermediates (Fig. 5) [61].

Other ester prodrugs have been based on derivatives of 2-mercapto ethanol. These prodrugs are reminiscent of the POM prodrugs in that cellular cleavage to the free drug is mediated by esterases. For example, *S*-acylthioalkyl ester (SATE) derivatives undergo enzymatic hydrolysis of the ester, thus liberating a thiol which, in turn, can attack carbon to afford episulfide and the free drug [62]. This strategy avoids formation of formaldehyde but instead yields episulfide and pivalic acid, which may interfere with the carnitine metabolism, as noted above. However, an advantage of this group is that it can be applied to both phosphonates (e.g., **24**, [63]) and phosphates (e.g., **25**, [64]) (Fig. 6).

The POM (**26**), SATE (**27**), and POC (**28**) esters all undergo cleavage through an initial ester hydrolysis, and there is some similarity in the non-drug products formed. As noted above, hydrolysis of a POM compound (**26**) releases both pivalic acid (**29**) and, after decomposition of the resulting intermediate **30**, formaldehyde (**31**) and the parent drug (**32**) (Scheme 1). Hydrolysis of a common SATE ester (**27**) also affords pivalic acid (**29**) and a thiol intermediate (**33**) which decomposes to give episulfide (**34**), and this is as problematic as formaldehyde. Hydrolysis of a POC compound (**28**) affords isopropyl alcohol (**35**) and ultimately carbon dioxide (**36**), which may be the least worrisome byproducts in this series, but issues such as the stability of the prodrug must also be considered.

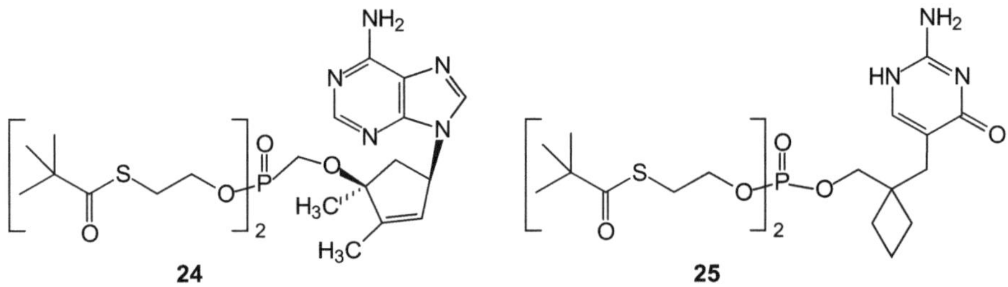

24

25

Fig. 6 *S*-Acylthioalkyl ester (SATE) modified nucleoside analogues

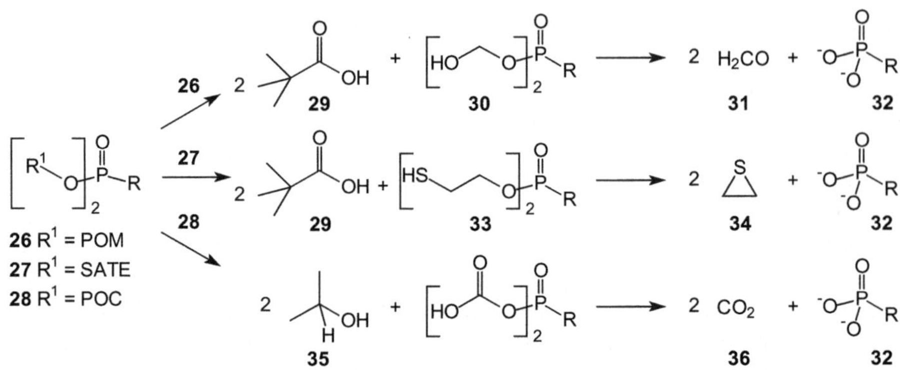

Scheme 1 The products of POM, POC, and SATE prodrug cleavage

3.2 Unsymmetrical Diesters

In contrast to symmetrical phosphonate diesters, which do not result in a stereogenic center at the phosphorus atom, unsymmetrical phosphonate diesters (or phosphate triesters) introduce asymmetry at this position. However, unsymmetrical diesters may also allow deployment of different strategies for deprotection, or even distribution of a second drug to the cell [65], while symmetrical diesters may be limited to a single strategy for unmasking the parent drug. For example, since the introduction of the cycloSal approach [66], several substituted salicyl alcohol derivatives have been employed to prepare different prodrug forms of nucleotides and carbohydrate phosphates and phosphonates [67]. Even the early studies on cycloSal, which employed derivatives designed to be removed by chemical means once within the cell, showed that this prodrug form can afford additional efficacy. For example, 100- to 600-fold increases in activity relative to their respective non-phosphorylated controls have been observed [68]. However, while comparisons of cycloSal groups in phosphate triesters with the activity of the parent phosphate ester would be more attractive, it can also be more difficult to determine if, for example, the phosphate has limited stability in the culture medium.

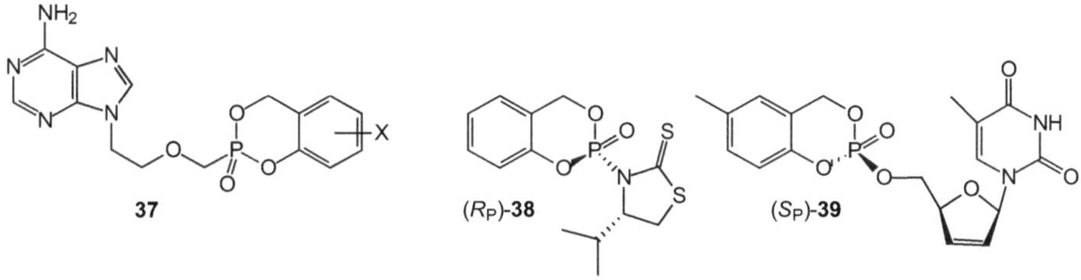

Fig. 7 CycloSal-PMEA (**37**) and some nonracemic cycloSal derivatives

Studies performed on acyclic nucleoside phosphonates prepared as cycloSal prodrugs revealed a limited impact on cell permeability and release of the prodrug **37** relative to the parent phosphonate PMEA [69]. Here the cycloSal approach imparted only one- to twofold improvement in cell activity relative to the free phosphonic acids, which was less than cell-cleavable approaches such as the bis-POM strategy. To the extent that removal of the cycloSal group is dependent upon chemical hydrolysis, with no additional increase in hydrolysis seen in the presence of plasma, once such a prodrug enters the cell it may also diffuse out rather than be unmasked, which would result in limited increases in potency. Realizing this limitation, the Meier group has prepared newer cycloSal derivatives which impart a cell cleavable component prior to the chemical hydrolysis step [70, 71]. These papers describe second and third generation cycloSal technology as well as some dramatic improvements in efficacy (Fig. 7).

Studies on cycloSal prodrugs have continued for many years, and they have been well-rewarded. For example, classical resolution of stereoisomers at phosphorus is potentially possible whenever the cycloSal prodrug is a derivative of a nucleoside or nucleoside analogue that contains stereogenic centers within itself [72]. An early observation of an 11-fold difference in the biological activity of the phosphorus stereoisomers [72] has encouraged development of methods to allow stereocontrolled synthesis. Recently it has been shown to be feasible to use valine-derived auxiliaries (e.g., **38**) to control formation of the phosphorus stereocenter via P^{V} chemistry [73]. Through controlled synthesis of the phosphorus stereocenter, and judicious placement of a methyl substituent on the aromatic ring, it was possible to observe 7- to 20-fold differences in the antiviral activity of d4TMP prodrugs (e.g., **39**) [74]. The biochemical and/or physiological basis for such significant differences in activity is/are not yet clear, but they clearly validate the effort devoted to diastereoselective syntheses and justify additional research into the cycloSal prodrugs. Applications of the cycloSal group as a leaving group in synthesis of 1,6-diglycopyranosyl-phosphates further enhances the value of this group [75, 76].

A different approach based on unsymmetrical diesters deliberately places a stereogenic center in a diol esterified to the phosphorus moiety. Although this introduces a stereogenic phosphorus atom, it can also allow tissue-targeted cleavage. Erion and colleagues have pioneered the use of this type of cyclic phosphonate prodrug, e.g., **40** (reviewed by [77]). Such compounds exhibit high stability in

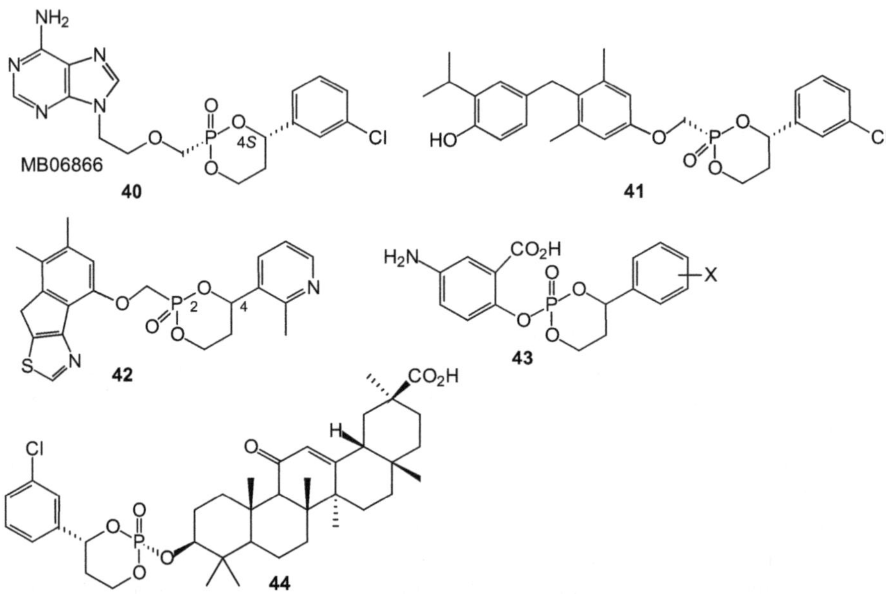

Fig. 8 The HepDirect strategy for phosph(on)ate prodrugs

plasma but undergo rapid oxidative cleavage mediated by a cytochrome P450 enzyme expressed in the liver (CYP3A4) [78] and thus are known as HepDirect prodrugs. When compared to the bis-POM prodrug of adefovir, it was found that the HepDirect approach causes higher drug accumulation in the liver and lower accumulation in the kidney and intestine in rats after oral delivery [79]. This strategy is effective with phosphonates and phosphates, given that similar distribution was seen with a HepDirect modified AraC [80]. In this study, a cyclic 1-(aryl)-1,3-propanyl prodrug of AraC was evaluated for its ability to target this compound to the liver. There was a substantial increase in mice, compared to the non-phosphorylated AraC as a control. The HepDirect technology was also applied to a 2′,3′-carbonate to make an orally available compound (39%) with high liver potency (Fig. 8) [81, 82].

The HepDirect strategy has been applied to a variety of compounds other than nucleotides or nucleotide analogues. For example, **41** was prepared as a prodrug for a thyroid hormone receptor agonist, and was shown to lower cholesterol and triglyceride levels with decreased impact on other tissues [83, 84]. A variety of compounds in this vein has been prepared, and the various stereoisomers were separable by column chromatography and HPLC on a nonracemic column. Unfortunately, the simple aryl compounds had limited aqueous solubility and functioned as inhibitors of CYP3A4. Through use of a pyridyl or substituted pyridyl ring system (e.g., **42**), inhibition of CYP3A4 was diminished, water solubility was increased, and more effective inhibition of glucose production was observed,

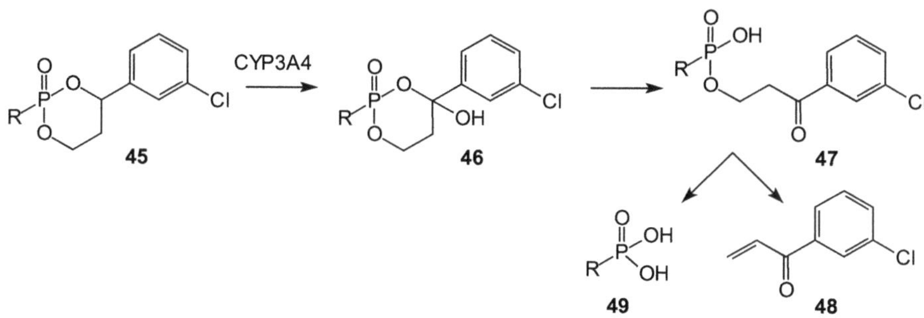

Scheme 2 The products of HepDirect prodrug cleavage

especially in the 2*R*,4*S* isomer [85]. The general strategy has also been applied to prepare prodrugs of a phenol (**43**) [86] and an alcohol (**44**) [87].

The mechanism of drug release from the prodrug **45** is believed to involve enzymatic oxidation of the benzylic position to afford the hemiketal **46** [80]. Because this oxidation takes place primarily in the liver, the prodrug effectively targets this organ. After ring opening (**47**), a β-elimination results in liberation of the drug (**49**) and an enone (**48**) derived from the masking group. It has not yet been established whether or not generation of the enone, a potential Michael acceptor, is problematic. Regardless, the authors suggest that the mechanism of HepDirect prodrug release results in gradual delivery of the original drug, therefore indirectly delaying its elimination (Scheme 2) [87].

3.3 Monoesters

While small lipophilic diesters can neutralize two negative charges in phosphates (or phosphonates) and thus facilitate passive diffusion across biological membranes, it has long been recognized that monoesters derived from larger lipophilic groups also may offer advantages. Among those advantages, monoesters avoid the complexities of phosphorus stereochemistry in the salt form (e.g., the anion of **50**). In some very early work this was exploited with cholesterol esters. For example, the cholesteryl ester of 5-fluoro-2′-deoxyuridine was prepared via a DCC coupling of the nucleotide with cholesterol [88], and subsequently tested for activity in Ehrlich ascites cells. It displayed modest activity even though it is almost insoluble in water [89]. Much more recent studies have provided more facile synthetic routes [90] to the cholesteryl esters, but further studies on the efficacy of this prodrug strategy have yet to appear (Fig. 9).

Studies with other monoesters have been both more extensive and more successful. Perhaps most notable is work with lysophospholipids (or their analogues) as phosphonate masking groups [91, 92]. Because lysophospholipids are metabolized by phospholipase A2 and readily absorbed in the GI tract, it was hypothesized

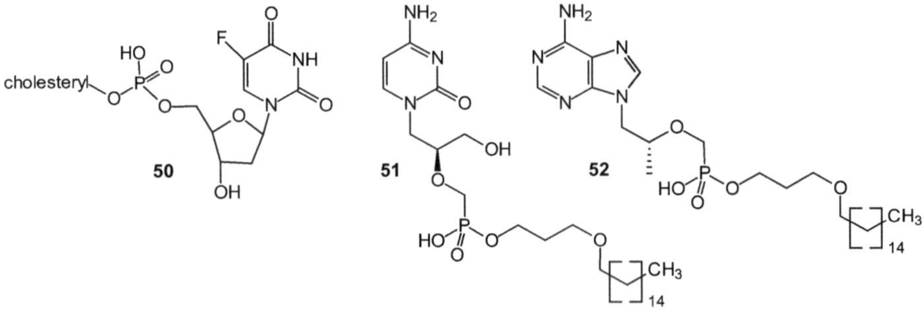

Fig. 9 Prodrugs derived from phosph(on)ate di (or mono) esters

that conjugation with nucleotide analogs would enhance the oral bioavailability of the associated drug. An early study by Hostetler demonstrated the feasibility of the phospholipid conjugation approach through preparation of acyclovir diphosphate dimyristoylglycerol, which can be viewed as an analog of the naturally occurring CDP-DAG. This compound was very effective at inhibiting replication of herpes-simplex virus (HSV) in cells deficient of thymidine kinase [93]. Likewise, several alkyl glycerol-3-phosphate derivatives of acyclovir and azidothymidine with extended alkyl chains increase intracellular levels of the monophosphate more than tenfold, with strong in vitro anti-viral activity [94]. These compounds also had an elevated and sustained oral bioavailability relative to the non-phosphorylated drug. This approach reduces the levels of the drug in the kidney and increases its levels in the liver. Differences have also been noted in the lung accumulation with varied lipid moieties [95].

Some of the strongest effects of this strategy were observed with lipid esters of the acyclic nucleoside phosphonates [91]. Across evaluation with a number of viruses, there were routinely several log unit increases in activity associated with use of the lipid esters. Of special significance was a >10,000-fold increase in in vitro potency of hexadecyloxypropyl (HDP)-(*S*)-HPMPA against HIV-1 [96], a >5,000-fold increase in potency of the HDP-HPMPA against orf virus [97], and a >2,000-fold increase in potency of HDP-CDV against Epstein–Barr virus (EBV) [98] relative to the respective free acyclic nucleoside phosphonates. These increases may be a consequence of their association with membranes, leading to increased cellular half-lives. One indication of the effectiveness of this strategy is that two lipid nucleoside conjugates are in clinical trials at present. Brincidofovir (**51**, CMX001) is an HDP-conjugated cidofovir in trials for CMV [99], while the related prodrug CMX157 (**52**) is an HDP-conjugated tenofovir in trials for HIV. This progress certainly encourages further investigation of phosphonate monoesters [100, 101].

3.4 Cyclic Monoesters and Mixed Diesters

Specific examples of acyclic nucleoside phosphonates offer unique opportunities for prodrugs based on an intramolecular cyclization. For example, cidofovir (**53**) can be converted to a cyclic monoester (cyclic cidofovir, **54**) which in turn can be further modified at the remaining acid function (reviewed by Krylov [102]). As a prodrug, the cyclic cidofovir **54** offers advantages over the parent compound in that only one ionizable function remains, and hydrolysis to the parent drug does not release any additional products. At the same time, upon ionization of **54** there is no longer a stereogenic center at phosphorus, so the only stereochemistry is that inherent in cidofovir itself. The cyclic form of cidofovir was also reported to have 10- to 40-fold lower nephrotoxicity than the parent compound [103]. However, formation of derivatives may further increase uptake and allow targeting (Fig. 10) [103].

One cidofovir derivative in this vein was based upon salicylate esters (e.g., **55**). For example, formation of the aryl ester from *n*-butyl salicylate gave a mixture of the axial and equatorial phosphorus esters, but these diastereomers were separated. The axial isomer showed greater chemical stability, and thus was assayed as an individual enantiomer. Based on the biological results, including stability in plasma and in liver and intestinal homogenates, it was hypothesized that breakdown occurs via an esterase-mediated hydrolysis of the carboxylate ester of the salicylate, followed by complete release of the drug [104]. The use of long lipid esters has also been investigated [105]. For example, the efficacy of phospholipid esters of cyclic cidofovir (e.g., **56**), cidofovir, and HPMPA has been investigated, providing valuable comparisons [106, 107].

An intriguing approach to cyclic cidofovir prodrugs has employed ester linkages to serine derivatives (**57**) [108, 109]. In addition to reduction of the phosphonate charge, this approach enables the prodrug to act as a substrate for peptide/amino acid transporters in the GI tract. However, while uptake was increased, especially with a Val-Ser derivative, it was not because of hPEPT1 transport, as these compounds bind to hPEPT1 but do not act as substrates [110]. The use of amino acid esters has also been used to prepare prodrugs of other nucleoside analogues including HPMPA, and at times improved oral availability was observed [111].

More traditional masking groups have also been used in combination with an internal ester. The acyclic nucleotide analogue based on 2,6-diaminopurine has been further modified as its POM (**58**) and HDP derivatives [112], suggesting that

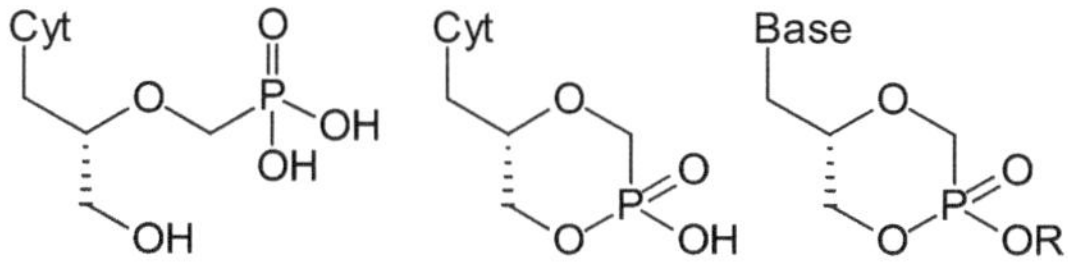

53 **54**

55 R = *n*-butylsalicylate, Base = Cyt
56 R = *n*-$C_{16}H_{33}$, Base = Cyt
57 R = Ser-Val, Base = Cyt
58 R = POM, Base = 2,6-diamino-purine

Fig. 10 Prodrugs including internal ester formation

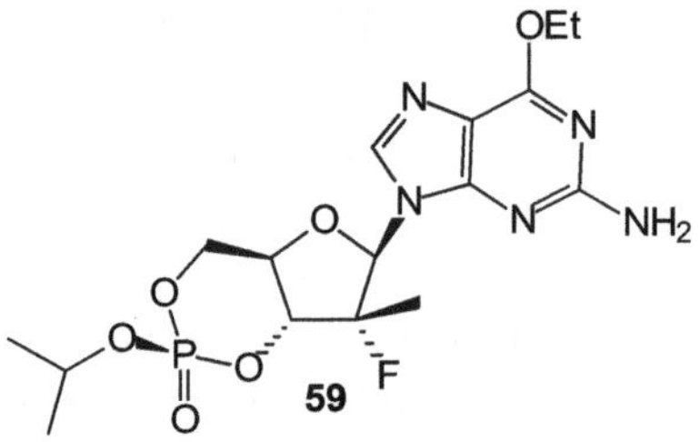
59

Fig. 11 A phosphate prodrug including internal esterification

virtually any ester (see above) or amide (see below) might be employed to obtain derivatives of the cyclic esters available from nucleotide analogues such as cidofovir.

Finally, internal esters are not limited to acyclic nucleoside phosphonates or even to phosphonates in general. For example, the cyclic nucleotide **59** has been reported to show very promising activity against the hepatitis C virus [113, 114]. In this case, the issue of phosphorus stereochemistry is ameliorated by selective formation of the desired stereoisomer, which has been accomplished on a kilogram scale, including purification by crystallization [115]. A variety of different phosphate esters of this general structure has been reported, and studies have described a mechanism of drug release which may include cytochrome P activation [114]. Again in this case, the reported studies strongly suggest that even more varied phosphate esters could be employed (Fig. 11).

4 Amidate Prodrugs

While many have used phosphorus esters to prepare prodrugs, others have developed phosphoramidate derivatives (reviewed by Mehellou [116]). As early as 1990, Devine and McGuigan had hypothesized that nucleotides containing at least one phosphoramidate linkage would exhibit cellular conversion to the free phosphate, potentially enhanced by activity of HIV protease [117]. Extensive studies in the intervening years have amply demonstrated the validity of that hypothesis with both diamidates and aryloxy phosphoramidates (reviewed in [118]), and each class offers advantages. Examples are described in the following sections (Table 2).

4.1 Symmetrical Diamidates

As described above for diesters, it can be argued that, from a conceptual standpoint, the simplest phosphordiamidates are those where symmetry avoids the issues inherent to introduction of phosphorus stereochemistry. Recently, McGuigan and colleagues have claimed just such advantages for a diamidate strategy [119,

Table 2 Some prodrug variations based upon amidate linkages

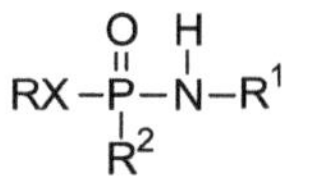

X = O phosphoramidate
X = C phosphonamidate

Symmetrical	R^1	R^2
Bisamidate	$-CHCH_3C(O)OCH(CH_3)_2$	$-NH\text{-}R^1$
Monoamidate-monoester		
Amidate/phenyl ester	$-CHCH_3C(O)OCH(CH_3)_2$	$-OC_6H_5$
Amidate/naphthyl ester	$-CHCH_3C(O)OCH(CH_3)_2$	$-OC_{10}H_7$
Amidate/alkyl ester	$-(CH_2)_4Cl$ (with NCH_3)	$-OCH_2(C_4H_3NO_3)$
Monoamidates		
Amidate	2-Pyridyl	$-OH$

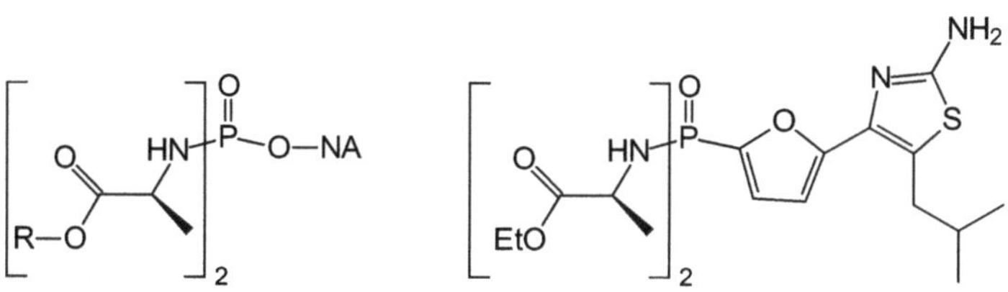

60 R = isopropyl, benzyl, cyclohexyl, etc.
NA = d4T, ddA, AZT, 3TC, RBV, etc.

61

Fig. 12 Phosphate and phosphonate prodrugs based on symmetrical diamidates

120]. When the amide is derived from an amino acid, achieving a neutral species requires esterification of the carboxyl group as well (e.g., **60**). After examination of a series of 25 diamidates for activity against HCV [120] and HIV replication, and anti-proliferative effects in several cell lines, it was found that this strategy is sufficient to afford good cellular activity with several different nucleoside analogues [119]. Preparation of symmetrical bisamidates from the symmetrical diesters also has been reported (Fig. 12) [121].

Symmetrical bisamidates have also been used to prepare prodrugs of phosphonates. For example, compound **61** was prepared as an inhibitor of fructose 1,6-bisphosphatase [122]. The phosphonodiamidate prodrug was far more potent in human hepatocytes than the free acid and, after comparison with a broad range of other prodrugs, the bisamidate was judged superior and selected for clinical trials [123]. While Phase II trials with **61** ultimately proved disappointing, studies with second generation analogues continue [124].

4.2 Amidate Esters

Some of the earliest work on azidothymidine (AZT) monophosphates reported that simple alkyl diesters were inactive in a cellular model of HIV, but that the combination of an alkyl monoester with a phosphoramidate led to surprisingly good cellular potency [117]. In an impressive series of studies, which continue to this day [125], McGuigan and colleagues evaluated lactate-derived systems and diaryl phosphates, before identifying aryloxy phosphoramidates as a highly efficient prodrug approach (reviewed in [118]). Work applying the aryl phosphoramidate technology to nucleoside phosphonates has led to the clinical trials of several such agents.

Most recent work on aryl phosphoramidates has employed protected L-alanine esters for the amine moiety [126]. In several cases where the D-alanine analogues were prepared, they were found to display significantly lower activity [127–129]. One comparison of PMPA prodrugs reported that a phenyl phosphoramidate displayed a tenfold greater activity than the corresponding POC prodrug.

While the amide component of phosphoramidates appears to be centering on L-alanine derivatives, considerable variations in the aryl ester component are still reported. Simple phenyl esters may well be the most extensively employed substructure [130, 131], especially in nucleotide analogues of the general structure **62** [116, 127, 132–136]. However in cases such as the hydroxamic acid **63**, an inhibitor of 6-phosphogluconate dehydrogenase for use in trypanosomiasis, addition of simple substituents such as methyl groups to the phenyl ring has resulted in more than a tenfold increase in potency [137]. In some cases, deployment of a naphthyl ester (e.g., **64**) has accompanied preparation of the phenyl ester [116, 132, 133]. Other studies have focused on the naphthyl esters (Fig. 13) [138–141].

While aryl phosphoramidates based on amino acids are by far the most common examples of this class, further variations in the ester and amine components certainly are known [142]. For example, in place of an aryl ester, a 3-acyloxymethoxypropyl group has been studied (e.g., **65**), and hydrolysis of the ester was found to be ~20 times faster than the corresponding phenyl phosphoramidate [131]. In this work, small differences in the reactivity of the R_P and S_P isomers were noted, but it was not possible to identify which was the more reactive (Fig. 14).

An interesting alternative to amines derived from amino acids is based on the 4-chlorobutyl system as reflected in structure **66**. After reduction and loss of the nitrofuryl group, cyclization of the butyl group delivers the active nucleotide, as shown by the preservation of activity in cells deficient in the kinase that generates FdUMP [143] or AraC monophosphate from the corresponding nucleosides [144].

Finally, a bidentate ligand has also been employed to prepare cyclic phosphoramidate esters such as the cyclic compounds **67** [145]. The (*N*-3-hydroxypropyl)-amino esters could be prepared from commercially available amino acids, ultimately allowing preparation of both the S_P and R_P isomers, and the phosphorus stereochemistry was assigned based on NMR data. Small differences in activity were found between the phosphorus stereoisomers, and the most

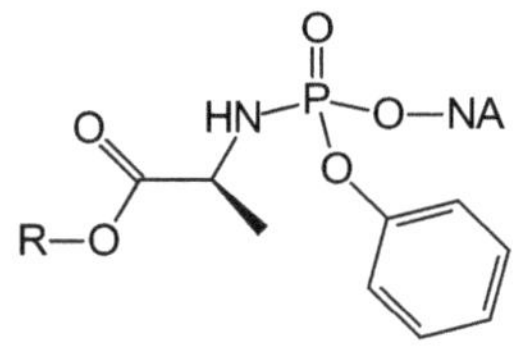

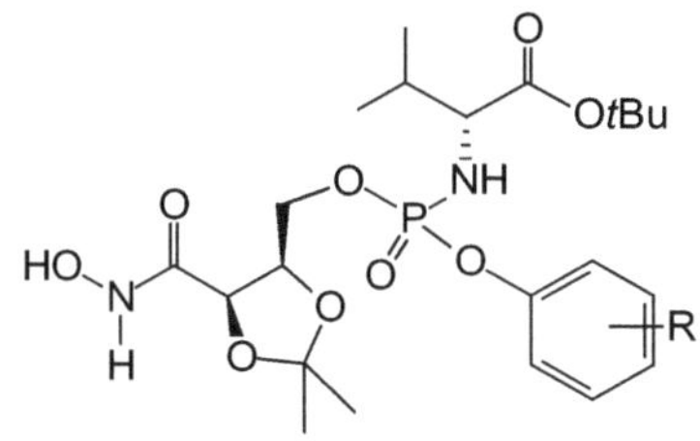

62 R = isopropyl, methyl
NA = AraU, FUDR, PMPA

63 R = H, 4-Me, 2,3-diMe

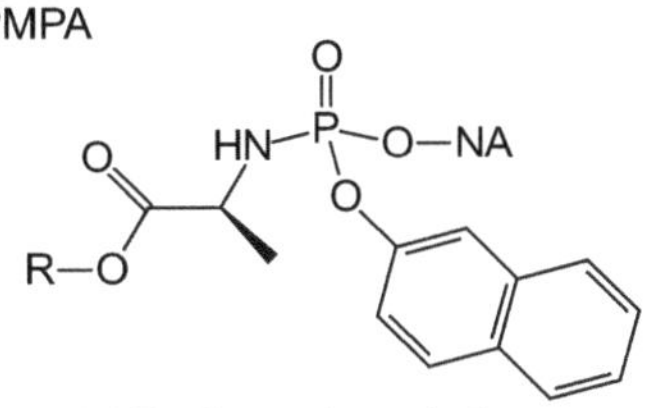

64 R = benzyl, methyl, neopentyl
NA = 2'-MeG

Fig. 13 Aryl phosphoramidate prodrugs

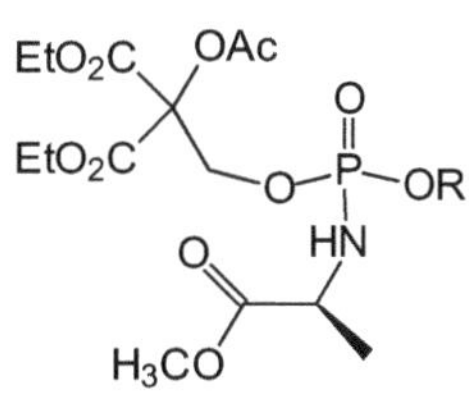

65 R = 2'-*O*-methyl-C

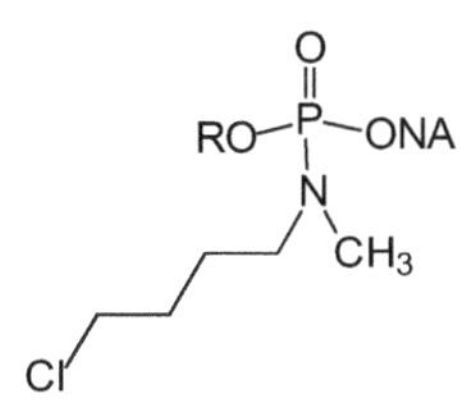

66 R = 5-nitrofuryl
NA = 5FU, araA

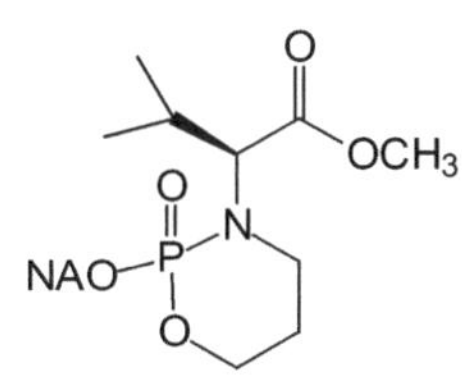

67 NA = PMEA

Fig. 14 Alkyl phosphoramidate prodrugs

active compounds in the series were comparable or slightly greater in potency when compared to the POM analogue.

4.3 *Monoamidates*

Symmetrical diamidates do not introduce a stereocenter at phosphorus, but, in cases such as aryl (or alkyl) phosphoramidates where two different protecting groups are employed, introduction of a stereogenic center at phosphorus can become a concern. Formation of a stereocenter can be avoided if just one amidate is used, as long as the single negative charge in the resulting salt is resonance delocalized; an additional benefit is that drug release does not generate phenol. One recent study of monoamidates prepared compounds such as **68**, where the longer chain alanyl

Fig. 15 Mono phosphoramidate prodrugs

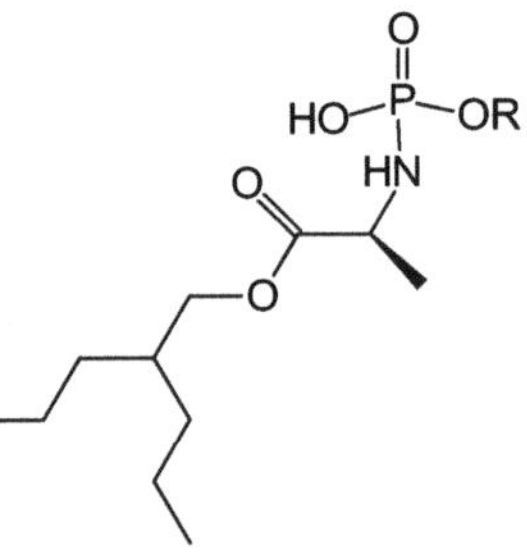

68 R = 2'-*C*-methyl-C

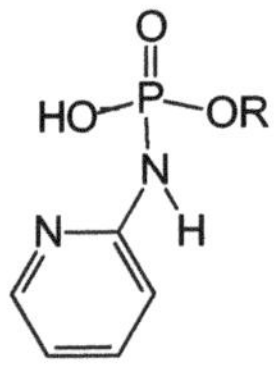

69 R = AZT or ddU

ester was intended to regain some of the lipophilicity sacrificed by removal of the phenyl ester [146]. The resulting compound proved very effective at generating high levels of the nucleoside triphosphate in human hepatocytes. A similar strategy was employed in a series of AZT and ddU analogues (e.g., **69**) [46, 47]. In this work, aryl amines were used to generate the monoamidates, and those based on pyridyl amines displayed notably low cytotoxicity and high antiviral activity. However, because no activity was observed in cell lines deficient in thymidine kinase, further work is necessary to clarify the specific mode of action (Fig. 15).

5 Drug Release

In addition to neutralizing negatively charged phosphorus moieties to allow diffusion across cellular membranes, it has become apparent that the appropriate selection of ester and/or amide prodrugs can be used to impart cell type and tissue type specificity, which can impact the distribution of the free drug. Specifically, by choosing the processes which are entrusted to remove the protecting groups, whether chemical decomposition or enzymatic cleavage, one can control where and when the drug is released. Each drug/prodrug/target must be optimized individually, as some masking groups can be more or less readily removed from different types of drugs [147]. However, we attempt here to describe some general guidelines for planning drug release.

The majority of current prodrugs require enzymatic cleavage prior to biological function. Conceptually speaking, prodrugs designed to undergo chemical decomposition (e.g., simple esters, early cycloSal compounds, or thermolytic prodrugs) offer the least amount of control over tissue distribution. However, by altering the chemical stability to control drug release rates, these compounds may offer advantages in terms of dosing, such as enhanced plasma stability and precisely controlled half-lives of elimination [148]. For example, some thermolytic prodrugs of DNA oligonucleotides have been examined [149]. Based on these studies, it was suggested that an optimal half-life of a thermolytic prodrug for animal work is 100–200 h at 37 °C, and that resistance to enzymatic cleavage allows better distribution of the drug to the cellular targets. In addition, depending on the masking group, thermolytic release has the potential to reduce accumulation of toxic intermediates and to overcome limitations in esterase activity on charged substrates when multiple groups need to be removed [150, 151].

Ester-based compounds such as the POM analogs undergo relatively slow chemical decomposition but are generally susceptible to cleavage by non-specific esterases, which can be found in most cells as well as blood. For example, a representative bis-POM compound was demonstrated to have a 5-minute half-life in plasma [152]. The mechanism of cleavage involves an initial enzymatic step followed by spontaneous release of the linker. Interestingly, while the first POM group is rapidly removed, leading to formation of the monoester, the mono-POM compound has a much longer half-life in plasma, and would likely offer advantages in terms of membrane permeability relative to the free diacid. Ultimately, the most promising use of POM-modified compounds may be to increase oral bioavailability, allowing accumulation of the free drug in the plasma at elevated levels, but not necessarily offering strong advantages in terms of tissue distribution relative to IV dosing.

When selecting which ester groups to use to increase oral absorption of a particular compound, it is important to keep in mind that the majority of neutral protecting groups greatly increase oral absorption. It may therefore be more important to optimize the rate of release (i.e., esterase susceptibility) rather than attempt to maximize the rate of oral absorption. An excellent example of this rational was shown for various ester prodrugs of squalene synthase inhibitors [153]. Here, the authors clearly demonstrate a reduction in oral activity when protecting groups are either removed too quickly or removed too slowly.

In contrast to the POM prodrugs, HDP esters require phospholipase C for removal of the protecting group and tend to have increased stability in the blood. This affords the actual drug with both increased oral availability and increased tissue distribution relative to IV dosing. The phospholipid masking groups tend to increase greatly the duration of exposure to the drugs, which can at times be detected even up to 10 days later [154]. Presumably drug release requires prodrug insertion in the cell membrane, followed by slow cleavage intracellularly via phospholipase C metabolism. This combination of uptake, plasma stability, and long-lasting cellular effects makes the mono phospholipid strategy especially effective for oral delivery of drugs such as brincidofovir, a highly promising antiviral agent.

Cleavage of the prodrugs of cyclic cidofovir and other related nucleotides requires several enzymatic steps to achieve the free drug. For example, the dipeptide prodrugs put forth by McKenna require removal of one of the amino acids by puromycin-sensitive aminopeptidase and spontaneous release of the remaining amino acid [155]. Modulating the enzymatic step which removes the first amino acid might provide an additional means to achieve tissue targeting depending on the expression of that enzyme. This hydrolysis likely must precede ring opening, which is also enzymatically-driven [156]. The specific mechanisms of ring opening are still emerging, with PDE7A recently identified as one enzyme important to this process [157].

The cyclic ester prodrugs have further enhanced tissue specificity of drug release, especially with the HepDirect system. Because of its specificity for metabolism by CYP3A4, accumulation of the free drug in the liver is favored, the predominant site of CYP3A4 expression. Likewise, while early generation cycloSal derivatives lacked selectivity, more recent modifications of these structures have afforded compounds which are stable in the presence of serum but not in cell extracts, thus offering some advantages over POM prodrugs [158].

The phosphoramidates also exhibit increased plasma stability relative to the acyloxyalkyl ester compounds. Whether diamidate (e.g., **70**), or aryl amidate (**71**), these compounds appear to be released through an initial enzymatic hydrolysis of the amino acid carboxyl ester [159]. In the case of diamidates and aryloxy amino acid amidates, this initial step is catalyzed by carboxypeptidase Y (cathepsin A) or carboxylesterase I [160], to afford the corresponding carboxylate (**72** or **73**). By increasing specificity for cathepsin A, one can target the free drug to tissues such as peripheral blood mononuclear cells, which have been found to express this enzyme strongly [161, 162]. The initial hydrolysis leads to spontaneous elimination of one of the protecting groups, which may involve the cyclic intermediate **74**, leaving a monoamide intermediate [163]. The amidate **75** is subsequently cleaved by cellular phosphoramidases such as Hint1 [160, 164] to release the free drug (**76**). Monophosphoramidates have been evaluated for enzymatic cleavage by human Hint1, and a set of guidelines for specific phosphoramidate cleavage by Hint enzymes has been established. For example, purine bases may be preferred but pyrimidine bases can also be used, phosphoramidate oxygen atoms are required, and an electrophilic group is required at the ribose 2′-position, while sterically-crowded amines reduce cleavage (Fig. 16) [165].

A set of phenol-containing phosphonoamidates with various amino acid esters has been tested against a panel of cellular serine and cysteine proteases. This study found large differences in the metabolism of the prodrugs, such that modifying the

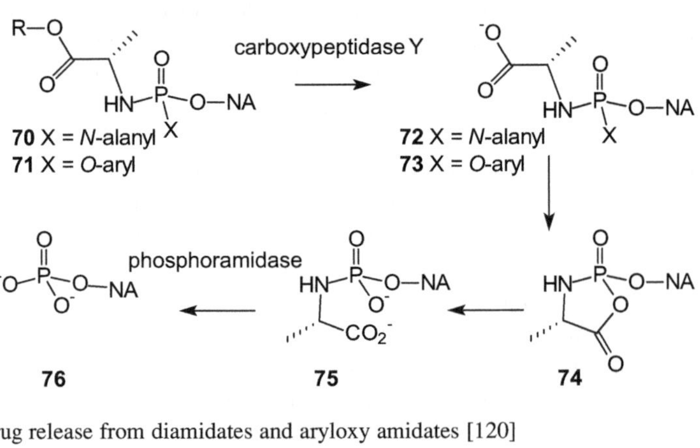

Fig. 16 Drug release from diamidates and aryloxy amidates [120]

amino acid ester group can completely change the enzyme activity. It was noted that the alanine isopropyl amidate bearing a phenyl ester is metabolized by cathepsin A [166]. The authors suggest that, based on differential tissue expression of these proteases, one might be able to design specific phosphonamidate prodrugs that are preferentially metabolized in various tissues.

Some interesting differences in drug release also may occur based on the subcellular localization of the enzymes responsible for removal of the protecting groups. Much like prodrugs which target a specific cell type, prodrugs which target a particular subcellular locale would be best served by undergoing hydrolysis by a singular enzyme with known subcellular distribution. However, it should be noted that this approach is not without risks, as lack of redundant conversion pathways may increase the potential for drug interactions or sensitivity to genetic polymorphisms [167]. Nonetheless, there have been several attempts to develop single enzyme cleavage systems. For example, the phospholipase C-driven release of HDP prodrugs might be expected to occur primarily in the cytoplasm or at the intracellular face of the plasma membrane [168]. Likewise, HepDirect prodrugs (which require CYP3A4) would be similarly released in the cytoplasm near the inner face of the plasma membrane [169]. On the other hand, release of phosphoramidates by Hint1 may occur in multiple compartments, as this enzyme has been reported to exhibit both cytoplasmic and nuclear localization [170]. However, phosphoramidates that are substrates for carboxypeptidase Y may be expected to be released in the lysosome, which would confer an advantage against targets that are lysosomal but would require membrane diffusion or active translocation of the free acid to reach cytoplasmic targets. Indeed, GS-9191, an amidate prodrug of 9-(2-phosphonylmethoxyethyl)-N^6-cyclopropyl-2,6-diaminopurine, was demonstrated to be initially hydrolyzed by cathepsin A in lysosomes followed by pH-dependent translocation to the cytoplasm [171].

Finally, in a few select cases, it is not even necessary to remove the protecting groups in order to observe biological activity. For example, methylated phosphates were able to function as agonists for the adenosine A1 receptor, a GPCR, without removal of the methyl group, resulting in enhanced in vivo bioavailability and pharmacodynamic effect [172]. Likewise, simple esters which block the ATP gated ion channel P2X were able to offer a cardioprotective effect [173]. Findings such as these suggest that, while removal of the masking group is required for activity of antiviral agents and inhibitors of isoprenoid biosynthesis, both of which require coordination of the negatively charged phosphate in an enzyme active site, it is not necessary for certain phosphorus moieties including some receptor agonists. While binding to the receptor may be reduced by the presence of a small masking group, the pharmacokinetic advantages may outweigh a small loss in ligand binding affinity.

6 Current Applications of Prodrug Technology

In recent years, there has been a dramatic increase in the synthesis of phosphate and phosphonate prodrugs. It could be argued that synthesis of prodrug forms is becoming the norm during generation of these drugs. In particular, two classes of prodrugs, the classic acyloxyalkyl phosphonates and the amidates [116, 174], account for the majority of recent examples. Here, we attempt to highlight some important recent developments in three key areas (nucleotides, isoprenoids, and phosphorylated proteins) for which novel applications of prodrugs are revolutionizing their respective fields.

6.1 Nucleosides, Nucleotides, and Nucleic Acid Metabolism

An intense area of research into phosphate and phosphonate prodrugs has centered on the development of effective antiviral agents. The discovery of HIV stimulated tremendous efforts in this regard [59], and the continued importance of the disease encourages continued efforts in this area, including use of prodrugs such as tenofovir disoproxil as preventative agents [175]. Furthermore, increasing concern for viral infections such as HCV [100, 135, 176, 177] and herpes virus [56] also motivates research. Because there are numerous reviews focused on prodrugs of nucleotides and nucleoside phosphonates [178–184], the following paragraphs focus primarily on other emerging applications in the nucleotide area.

Much of the past research into nucleotide and nucleoside phosphonate prodrugs has been designed to afford drugs which include phosphorus and so do not require formation of a monophosphate through a kinase mediated reaction. This strategy, often referred to as a kinase bypass [15–18], can circumvent development of drug-resistant strains through mutations which delete one specific kinase. An intriguing alternative has been suggested through formation of prodrug forms based on nucleoside diphosphates (e.g., **77**) [185, 186]. Compounds with acyl residues larger than C_6H_{13} were found to be highly active in a CEM/TK$^-$ assay, demonstrating that this is an effective strategy for intracellular delivery of nucleoside diphosphates. It remains to be seen whether this or a similar strategy may be extended to other types of diphosphates or even triphosphates (Fig. 17).

At the same time, applications of the more traditional kinase bypass strategy to novel nucleoside analogues continue to appear. In this context, a recent paper described phenyl phosphoroamidate **78**. After efforts to optimize the nucleoside analogue, both phosphorus stereoisomers were isolated and identified by crystallography. The more active S_P isomer, **78**, then became the first C-nucleoside to enter clinical development for treatment of HCV through inhibition of NS5B polymerase [187]. These findings illustrate that one consequence of the aryl phosphoramidates is the introduction of a new stereogenic center at phosphorus [188, 189] which can dramatically impact biological activity. However, the majority of studies make no mention of this stereochemistry, suggesting that assays were conducted on

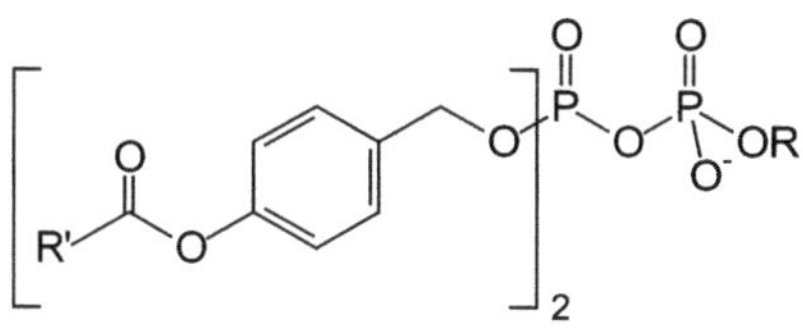

77 R = d4T; R' = methyl, *n*-alkyl, etc.

Fig. 17 A diphosphate prodrug

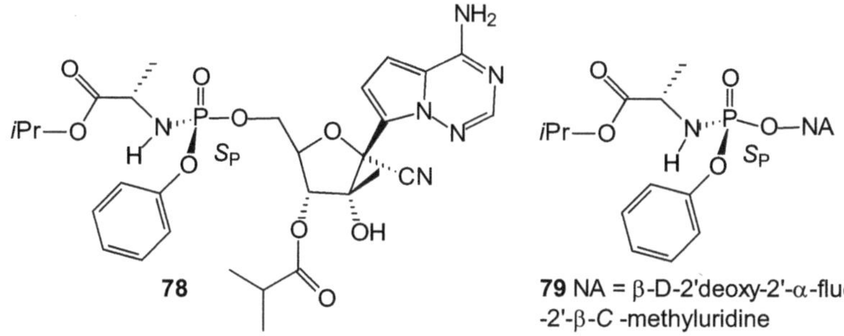

79 NA = β-D-2'deoxy-2'-α-fluoro-
-2'-β-*C* -methyluridine

Fig. 18 Prodrugs assayed as the S_P isomers

diastereomeric mixtures. Consistent with these results, another elegant study separated the phosphorus-centered diastereomers of phosphoramidate **79**, identified them by diffraction analysis, and assayed the individual isomers, reporting that the S_P diastereomer shown displayed significantly greater activity [113, 190]. The prodrug **79** produces high levels of the triphosphate in multiple species following oral dosing. Its toxicity is low with high potency towards HCV, even in resistant cells, and so it was recently approved for treatment of HCV (as Sofosbuvir) [191, 192]. This success suggests that the issue of phosphorus stereochemistry needs to be addressed in other aryl phosphoramidates as well. It is interesting to note that in both cases the S_P isomer was more potent, although neither the underlying cause of this phenomenon nor its prevalence are known at this time (Fig. 18).

In addition, prodrug strategies are continuing to find application in other non-viral applications. For example, several prodrug forms of acyclic immucillin phosphonates have been prepared in an effort to identify antimalarial agents which function through inhibition of purine biosynthesis in plasmodium. While the POM modified phosphonate **80** is able to penetrate erythrocytes, hydrolysis limits drug entry into the parasite. However, lysophospholipid prodrugs (e.g., **81**) are able to deliver the compound to the parasites, where they are cleaved by phospholipase A and C, releasing the free phosphonic acid [193]. This target has also been addressed with bisamidate prodrugs of bisphosphonates such as **82** as potential antimalarial agents [194] and antibiotics [195]. Taken together, these studies reiterate the importance of optimizing each prodrug-drug-target combination (Fig. 19).

Some groups have evaluated prodrugs for their ability to prevent precipitation of phosphorus compounds. It has long been recognized that geminal bisphosphonates

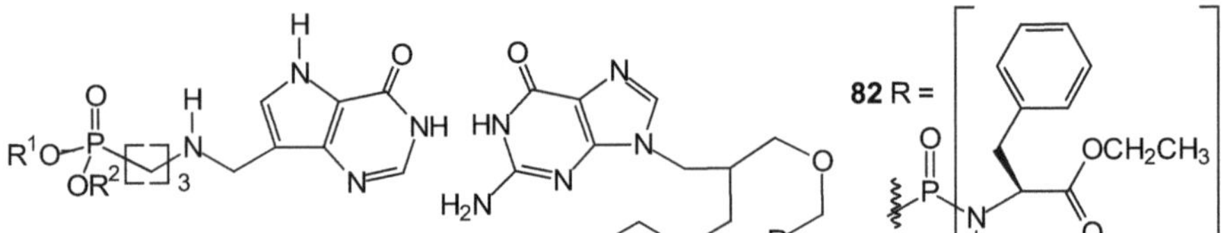

80 $R^1 = R^2 = POM$
81 $R^1 = CH_3CH_2$; $R^2 = CH_3(CH_2)_{14}$

Fig. 19 Prodrugs of some purine biosynthesis inhibitors

can be viewed as analogues of diphosphate which do not readily undergo hydrolysis. Derivatives with small substituents on the methylene carbon, compounds such as etidronate (**83**) and clodronate (**84**) are sometimes referred to as first generation bisphosphonates because of their early clinical use [196]. In vivo, these bisphosphonates are converted to ATP analogues which may directly trigger apoptosis [197, 198], but as small, highly charged molecules they do not readily cross the cell membrane by means other than endocytosis. Various prodrug approaches have been examined to increase their ability to enter cells. For example, several bis-, tri-, and tetra-POM bisphosphonates, including derivatives of clodronate [35] and etidronate [36], have been prepared, while a series of dianhydride derivatives of clodronate has also been made (e.g., **85**) [199]. In the latter study, the acetyl, butyroyl, and benzoyl derivatives display half-lives in human serum of less than 1 min, while the dipivaloyl bisphosphonate has a longer half-life of 3.3 h. The authors demonstrate that, while the solubility of the dianhydrides was less than clodronate, the compounds maintain their solubility in the presence of increasing concentrations of calcium while clodronate does not. Therefore, compounds of this type may increase the oral bioavailability of clodronate and related bisphosphonates. More recently, others have examined a phosphonoamidate prodrug of clodronate [200]. However, while these approaches are likely to increase oral bioavailability of the compounds, their low predicted stability in the plasma may ultimately lead to cleavage of the prodrug and sequestration of the bisphosphonate to the bone rather than allow for distribution to other tissues (Fig. 20).

6.2 *Isoprenoids and Inhibitors of Isoprenoid Metabolism*

A second pathway based upon small molecule diphosphates, isoprenoid biosynthesis, may rival the importance of nucleotide metabolism in the treatment of human disease [201]. Because isoprenoids have important roles in heart disease, bone disease, and cancer, various groups have investigated ways to elevate the concentration and target the delivery of isoprenoid pathway inhibitors using prodrug strategies.

Fig. 20 Examples of first generation bisphosphonates and a prodrug form

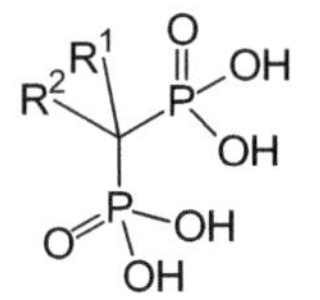

83 R^1 = OH, R^2 = CH_3
84 R^1 = Cl, R^2 = Cl

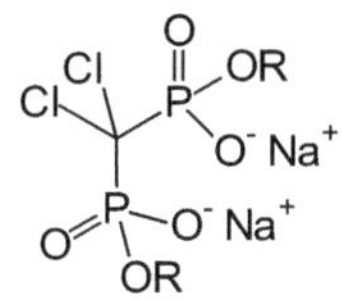

85 R = 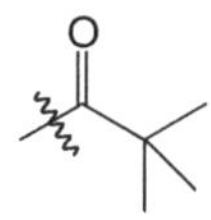

6.2.1 Isoprenoid Diphosphate Synthases

At the heart of cellular isoprenoid biosynthesis pathways lie the isoprenoid diphosphate synthases, including farnesyl diphosphate synthase and geranylgeranyl diphosphate synthase [202]. A number of clinically successful agents used to treat osteoporosis and other bone diseases, collectively known as nitrogenous bisphosphonates and considered to be the second and third generations of methylene bisphosphonate derivatives, are inhibitors of farnesyl diphosphate synthase [196]. The nitrogenous bisphosphonates are highly effective inhibitors in part because their bisphosphonate substructure mimics the functionality of the natural diphosphates. In addition to imparting potent enzyme inhibition, this highly charged bisphosphonate group serves to localize the nitrogenous bisphosphonates to the bone microenvironment, where they achieve long-lasting and local concentrations which are capable of inhibiting osteoclast function. Because bisphosphonate drugs exhibit anticancer effects in the bone, some groups have hypothesized that directing these drugs to other locales through synthesis of bisphosphonate prodrugs would lead to anti-proliferative effects against soft tissue cancers. Simple alkyl and arylbisphosphonates have met with some success [203], but more readily cleaved prodrug forms may be more rewarding.

In 2006, Oldfield and colleagues published the synthesis and evaluation of a tetra-POM bisphosphonate (**86**) which inhibits farnesyl diphosphate synthase. While not evaluated for oral availability, this compound was able to inhibit in vitro growth of cancer cells in a manner that was 20-fold more potent relative to the free acid [204]. This illustrates the anti-cancer potential of the target and also addresses the issue of cell permeability for this class of drugs. In 2008, we reported a series of isoprenoid bisphosphonates which function as geranylgeranyl diphosphate synthase inhibitors. Both the salt (**87**) and tetra-POM derivatives (**88**) were prepared, and growth inhibition and diminished protein geranylgeranylation were greater with the tetra POM compounds by up to 25-fold relative to their respective free acids [205]. The magnitude of the increase varied from compound to compound, with smaller compounds (with higher charge to mass ratios) benefiting more from POM protection. Taken together, these studies suggest that bisphosphonate inhibitors of both farnesyl and geranylgeranyl diphosphate synthases exhibit difficulty entering cells, and neutralization of these compounds with a prodrug approach can increase their in vitro anti-cancer properties (Fig. 21).

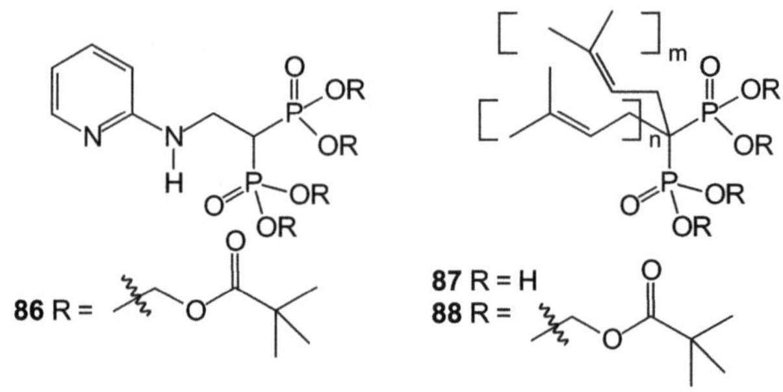

Fig. 21 Prodrugs of farnesyl diphosphate synthase and geranylgeranyl diphosphate synthase inhibitors

6.2.2 Prenyl Transferases

Isoprenoid diphosphates such as farnesyl diphosphate and geranylgeranyl diphosphate are also post-translationally incorporated into proteins, increasing membrane affinity. Because some such prenylated proteins are known oncogenes, (e.g., Ras), there has been a substantial effort to develop inhibitors of the protein prenyl transferases farnesyl transferase and geranylgeranyl transferase I and II. Most efforts have focused on peptidomimetics, but various add back experiments have shown that isoprenoid diphosphates enter cells and several isoprenoid diphosphate analogues also have been reported to show biological activity. While these analogues may suffer from the permeability problems attributed to other phosphates, there have been some reports of prodrug approaches with such compounds [206]. Recently, phosphoramidate prodrugs of the farnesyl [207] and geranylgeranyl [208] monophosphate analogues **89** and **90** have been reported. Because of their similarity to the natural substrates, these compounds are among the most potent isoprenoid analogs reported for inhibition of the prenyltransferases. Their cellular activity was not strong, possibly because of further metabolism of the monophosphate. However, in combination with lovastatin these compounds exhibit some interesting effects on cancer cell viability, which suggests that further examination of the use of masking groups to deliver isoprenoid diphosphate analogues as transferase inhibitors may be warranted (Fig. 22).

6.2.3 Inhibitors of the Non-mevalonate Pathway for Isoprenoid Biosynthesis

In contrast to humans, bacteria and parasites commonly utilize the non-mevalonate pathway of isoprenoid biosynthesis [209]. This provides the opportunity to inhibit their growth selectively by therapeutic targeting of these enzymes. Analogous to statin drugs which inhibit the first enzyme of human isoprenoid synthesis, fosmidomycin (**91**) and its analog FR900098 (**92**) inhibit the first step of the non-mevalonate pathway, 1-deoxy-D-xylulose 5-phosphate (DOXP) reductoisomerase (Dxr). These drugs show some activity against malaria, but also exhibit low oral bioavailability. In 2001, Reichenberg synthesized a series of aryl ester modified FR900098 analogues (**93**) [210]. These compounds exhibited greater

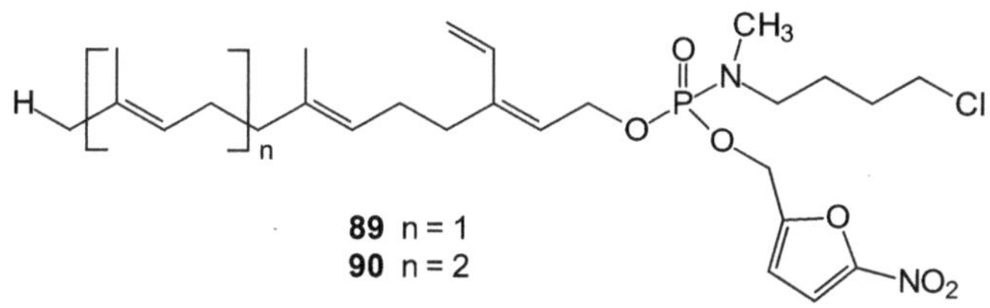

Fig. 22 Monophosphate prodrugs that inhibit prenyl transferases

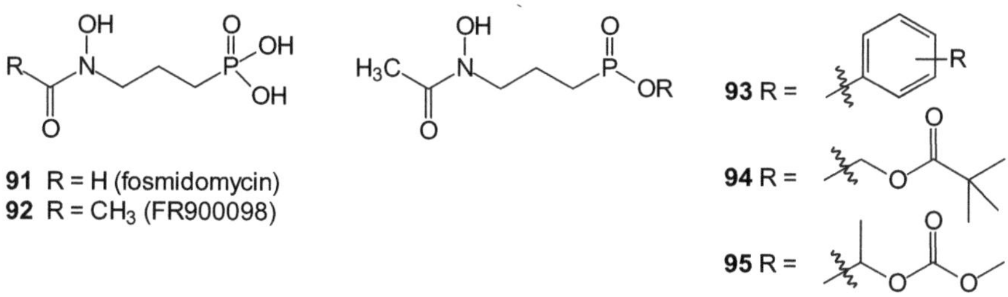

Fig. 23 Fosmidomycin and prodrug analogues

activity in a *Plasmodium* model when dosed orally relative to the free acids. Because of concerns about the potential for toxicity of the phenol metabolite, this group went on to produce a series of acyloxyalkyls (e.g., the POM compound **94**) and then compounds reminiscent of POC derivatives (e.g., **95**) as prodrugs of FR900098 [211, 212]. At least one of these compounds shows improvement in activity over the free acid in the mouse model (Fig. 23).

Later studies on analogues substituted at the C-1 or C-3 positions of fosmidomycin, or constrained by a ring system spanning C-1 and C-2, continued to employ bis-POM prodrugs to explore the SAR of FR900098 analogs against *Plasmodium* Dxr [213–216]. These compounds yielded several more potent inhibitors of *Plasmodium* growth in vitro, but use of the prodrug strategy in this system does not confer the magnitude of increase seen in other cellular systems. It is difficult to say whether the limited efficacy is compound-based or parasite-specific. These results were echoed through studies of a series of fosmidomycin analogs [217], where again the bis-POM protecting strategy was only marginally effective. It is possible that fosmidomycin penetrates cells adequately of itself. The most recent studies reported a series of fosmidomycin analogs with enhanced potency against the purified enzyme and growth of the *Plasmodium* parasite in vitro [218]. While the activity was somewhat enhanced by bis-POM protection, the results parallel findings of limited activity of the tri-POM prodrug squalene synthase inhibitor ER27856 in *Leishmania*. This again suggests either a decreased ability to penetrate the membrane of this organism or lack of ability to release the free drug [219].

Because fosmidomycin and FR900098 are also potent antibacterial agents, a prodrug approach has been examined to determine whether it would increase efficacy in Gram-negative bacteria [220]. Several lipophilic esters were tested on a panel of Gram-positive and Gram-negative bacteria. The largest positive effects were seen in an *M. tuberculosis* strain (H37Rv), where several prodrugs were active at concentrations of 100 μg/mL vs >500 μg/mL for the free acid. Generally, the prodrugs did not enhance efficacy of FR900098 in Gram-negative pathogens, with only one out of six showing a modest increase in potency against *E. coli* k12. It is possible that these bacteria were unable to metabolize the prodrug or that the prodrug is ineffective at crossing the double membrane, because they are too lipophilic to transverse the hydrophilic transmembrane space. However, another group has recently reported activity of the bis-POM derivative **94** in mycobacterium [221]. All in all, future studies which focus on prodrugs of more metabolically stable compounds such as these would be well worthwhile.

6.3 *Protein Phosphorylation and Interactions*

The phosphorylation and dephosphorylation of amino acid residues in proteins is an essential aspect of many signal transduction pathways. The phosphorylated residues often promote protein–protein interactions via Src Homology 2 (SH2) domains and other protein domains that interact specifically with phosphorylated proteins. Therefore, compounds which interfere with protein to phosphorylated protein interactions are potential therapeutic agents.

6.3.1 Prodrugs of Peptidomimetics Which Target the SH2 Domain

Tyrosine phosphorylation is a primary component of signal transduction pathways, such as those regulated by growth factor receptors. As the SH2 domain is the most prevalent of the phosphotyrosine recognition domains, it has emerged as a target for phosphotyrosine mimicking phosph(on)ate compounds and their respective prodrugs. One early study reported a series of SH2 domain blockers targeted at Src and Abl kinases [222]. The authors suggest that efficient SH2 blockers must contain a phosphate group, or a phosphonate substitute, to allow effective ligand binding. They successfully generated several dipeptide analogs, including the phosphonic acid **96**, the ethyl ester **97**, and the di-POM compound **98** as phenylalanine phosphonates. Both esters exhibited much faster cellular uptake than the acid and the prodrug **98** was readily metabolized in cells to release the free acid (Fig. 24).

A similar approach has been taken to prepare aryl phosphoramidates such as **99** [223], as prodrugs of SH2 domain analogs for Src/Lck inhibitors. The compounds exhibit low micromolar growth inhibition in Jurkat T cells, and undergo spontaneous hydrolysis with half-lives of approximately 30 min. The same masking groups have been applied in phosphonate **100** as a suspected SH2 domain blocker through inhibition of mitotic centromere-associated kinesin protein function in a panel of

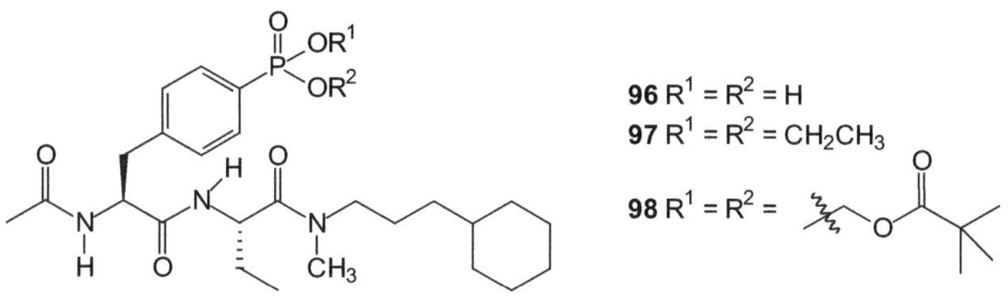

Fig. 24 Bis-POM prodrug of an SH2 targeted peptidomimetic

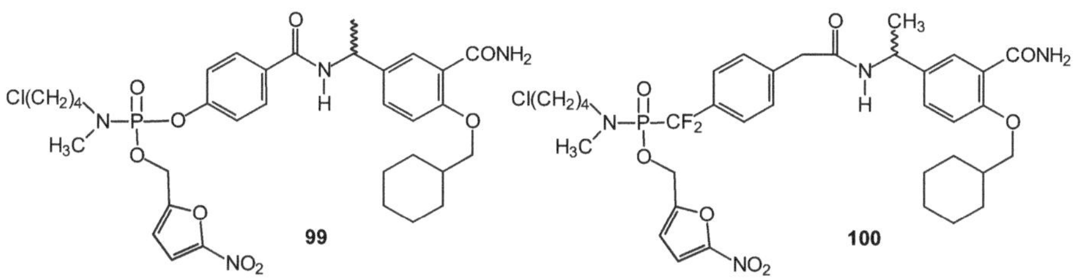

Fig. 25 Aryl phosphoramidate peptidomimetics

cell lines [224]. The compound inhibited cell growth but surprisingly the growth inhibition was not restricted to Src-dependent cells. The same prodrug format was used to investigate targeting protein tyrosine phosphatase 1B (PTP1B) as a means of sensitizing cells to insulin signaling for treatment of diabetes and obesity [225]. In this study, cellular activity of the prodrug was seen at low nanomolar concentrations (Fig. 25).

The goal of a similar study was to develop a phosphopeptide mimic which binds to the SH2 domain of Stat3, preventing its association with growth factor receptors and subsequent phosphorylation and translocation to the nucleus. In this case, a xenograft experiment was performed to assess activity in an MDA-MB-468 tumor model, and a ~30% decrease in tumor volume was observed after 4 weeks of treatment [226]. In contrast to the earlier findings [222], in this study the bis-POM compound **101**, prepared as a phosphotyrosine surrogate, is significantly more effective than the mono-POM analog at producing cellular activity [227]. Subsequent studies therefore relied on the bis-POM approach to identify inhibitors of Stat3 [228, 229] and Stat6 [230] binding interactions.

Finally, while most SH2 analogs have utilized POM or amidate masking groups, one lab has recently generated a cycloSal analog, **102** [231]. In this study, compounds were active in the micromolar range in a cell-free assay system. It is not yet clear how rapidly such compounds are hydrolyzed, and it may be unlikely that they would offer significant enhancement in cellular delivery of the phosphorylated peptide because of the chemical nature of the cycloSal hydrolysis. However, this report clearly demonstrates that many different prodrug strategies can be applied to obtain analogues of peptide phosphates (Fig. 26).

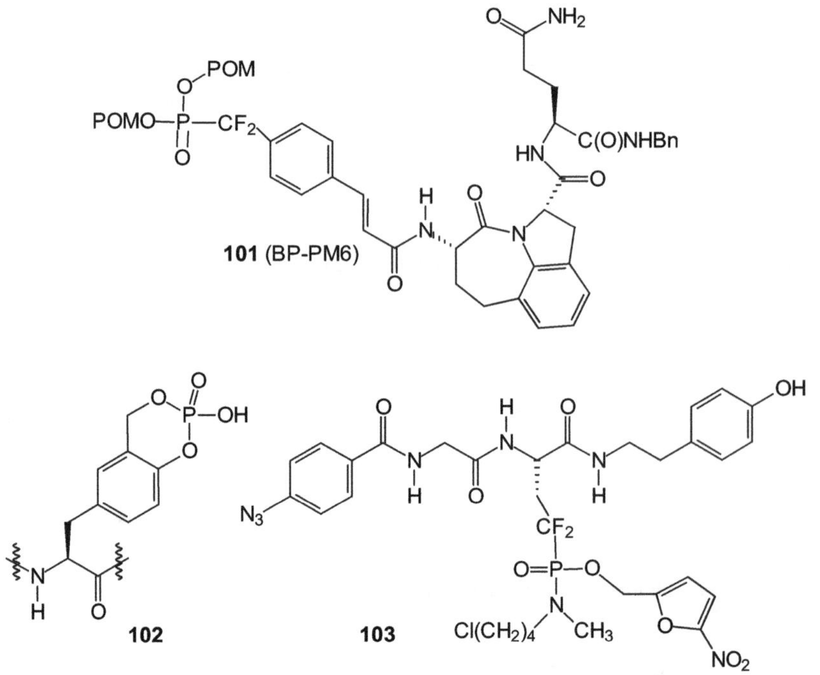

Fig. 26 Some prodrug forms used in peptidomimetics

6.3.2 Serine/Threonine Analogs

Only recently has the prodrug approach enabled cellular studies of phosphoserine and phosphothreonine analogs. After a phosphoserine analog with the ability to block interactions of 14-3-3 proteins was identified, a mixed phosphonoamidate ester prodrug of the corresponding difluoromethylene phosphonate was prepared (**103**). This prodrug exhibited an approximately 20-fold increase in activity over the free acid [232]. A recent report on phosphothreonine analogs targeted towards Plk1 has also relied on a bis-POM strategy [233].

Conclusions

Incorporation of phosphates and phosphonates in potential drugs may once have been viewed as unrewarding, because of the inherent tension between a requirement for high negative charge density for bioactivity and the limited ability of such compounds to traverse the cell membrane. With the development of various prodrug forms, it has now become routine. The understanding of both the chemistry and biological activity of different prodrugs has advanced tremendously since the earliest studies, and the pace of those

(continued)

advancements is increasing. From diesters of simple alcohols and symmetrical phosphorus species, the art has advanced to necessarily asymmetric amidate esters. The phosphorus stereochemistry which once was avoided is now recognized as a potential design element, and strategies for asymmetric synthesis have begun to appear. The increasing complexity of the prodrug forms has allowed evolution of drug release strategies and fosters drug targeting. These advances, together with the growing number of phosphorus-containing drugs in clinical use and clinical trials, make it clear that studies of prodrug forms continue to be a vibrant research area.

Acknowledgements Financial support from the University of Connecticut, Department of Pharmaceutical Sciences (AJW) and from the Roy J. Carver Charitable Trust as a Research Program of Excellence (DFW) and the NIH (R01CA-172070, DFW) is gratefully acknowledged.

References

1. Westheimer FH (1987) Why nature chose phosphates. Science 235:1173–1178
2. Engel R (1977) Phosphonates as analogs of natural phosphates. Chem Rev 77:349–367
3. Metcalf WW, van der Donk WA (2009) Biosynthesis of phosphonic and phosphinic acid natural products. Annu Rev Biochem 78:65–94
4. Williams R, Jencks WP, Westheimer FH (2001) Table of pK_a values. http://research.chem.psu.edu/brpgroup/pKa_compilation.pdf. Accessed 14 Oct 2014
5. Wiemer DF (1997) Synthesis of nonracemic phosphonates. Tetrahedron 53:16609–16644
6. Kornberg RD, McNamee MG, McConnell HM (1972) Measurement of transmembrane potentials in phospholipid vesicles. Proc Natl Acad Sci USA 69:1508–1513
7. Ghosh S, Chan JMW, Lea CR, Meints GA, Lewis JC, Tovian ZS, Flessner RM, Loftus TC, Bruchhaus I, Kendrick H et al (2004) Effects of bisphosphonates on the growth of *Entamoeba histolytica* and *Plasmodium* species in vitro and in vivo. J Med Chem 47:175–187
8. Huttunen KM, Rautio J (2011) Prodrugs – an efficient way to breach delivery and targeting barriers. Curr Top Med Chem 11:2265–2287
9. Plunkett W, Huang P, Xu YZ, Heinemann V, Grunewald R, Gandhi V (1995) Gemcitabine: metabolism, mechanisms of action, and self-potentiation. Semin Oncol 22:3–10
10. McGuigan C, Jones BCNM, Riley PA (1991) Trans-esterification reactions yield novel masked phosphate derivatives of the anti-cancer agent araC. Bioorg Med Chem Lett 1:607–610
11. Maiti M, Persoons L, Andrei G, Snoeck R, Balzarini J, Herdewijn P (2013) Synthesis and anti-herpetic activity of phosphoramidate ProTides. ChemMedChem 8:985–993
12. Krise JP, Stella VJ (1996) Prodrugs of phosphates, phosphonates, and phosphinates. Adv Drug Deliv Rev 19:287–310
13. Cho A (2006) Recent advances in oral prodrug discovery. Ann Rep Med Chem 41:395–407
14. He GX, Krise JP, Oliyai R (2007) Prodrugs of phosphonates, phosphinates, and phosphates. In: Stella VJ, Borchardt RT, Hageman MJ, Oliyai R, Maag H, Tilley JW (eds) Prodrugs: Challenges and Rewards. Part 2, Springer, New York, pp 923–964
15. Meier C (1998) Pro-nucleotides - recent advances in the design of efficient tools for the delivery of biologically active nucleoside monophosphates. Synlett 233–242

16. Mackman RL, Cihlar T (2004) Prodrug strategies in the design of nucleoside and nucleotide antiviral therapeutics. Ann Rep Med Chem 39:305–321
17. Ariza ME (2005) Current prodrug strategies for the delivery of nucleotides into cells. Drug Des Rev 2:373–387
18. Ray AS, Hostetler KY (2011) Application of kinase bypass strategies to nucleoside antivirals. Antiviral Res 92:277–291
19. Bobeck DR, Schinazi RF, Coats SJ (2010) Advances in nucleoside monophosphate prodrugs as anti-HCV agents. Antivir Ther 15:935–950
20. Madela K, McGuigan C (2012) Progress in the development of anti-hepatitis C virus nucleoside and nucleotide prodrugs. Fut Med Chem 4:625–650
21. Sofia MJ, Chang W, Furman PA, Mosley RT, Ross BS (2012) Nucleoside, nucleotide, and non-nucleoside inhibitors of hepatitis C virus NS5B RNA-dependent RNA-polymerase. J Med Chem 55:2481–2531
22. Hecker SJ, Erion MD (2008) Prodrugs of phosphates and phosphonates. J Med Chem 51:2328–2345
23. Greene TW, Wuts PGM (1991) Protective groups in organic synthesis, 2nd edn. Wiley, New York, NY
24. Cho MJ, Kurtz RR, Lewis C, Machkovech SM, Houser DJ (1982) Metronidazole phosphate–a water-soluble prodrug for parenteral solutions of metronidazole. J Pharm Sci 71:410–414
25. Chung MC, Bosquesi PL, dos Santos JL (2011) A prodrug approach to improve the physico-chemical properties and decrease the genotoxicity of nitro compounds. Curr Pharm Des 17:3515–3526
26. Palte MJ, Davis AKF, McGrath NA, Spiegel CA, Raines RT (2012) Ribonucleoside 3-′-phosphates as pro-moieties for an orally administered drug. ChemMedChem 7:1361–1364
27. Ellis GA, McGrath NA, Palte MJ, Raines RT (2012) Ribonuclease-activated cancer prodrug. ACS Med Chem Lett 3:268–272
28. O'Boyle NM, Greene LM, Keely NO, Wang S, Cotter TS, Zisterer DM, Meegan MJ (2013) Synthesis and biochemical activities of antiproliferative amino acid and phosphate derivatives of microtubule-disrupting beta-lactam combretastatins. Eur J Med Chem 62:705–721
29. Sweeny DJ, Li W, Clough J, Bhamidipati S, Singh R, Park G, Baluom M, Grossbard E, Lau DTW (2010) Metabolism of fostamatinib, the oral methylene phosphate prodrug of the spleen tyrosine kinase inhibitor R406 in humans: contribution of hepatic and gut bacterial processes to the overall biotransformation. Drug Metab Dispos 38:1166–1176
30. Gu Y, Atwell GJ, Wilson WR (2010) Metabolism and excretion of the novel bioreductive prodrug PR-104 in mice, rats, dogs, and humans. Drug Metab Dispos 38:498–508
31. Chou LC, Chen CT, Lee JC, Way TD, Huang CH, Huang SM, Teng CM, Yamori T, Wu TS, Sun CM et al (2010) Synthesis and preclinical evaluations of 2-(2-fluorophenyl)-6,7-methylenedioxyquinolin-4-one monosodium phosphate (CHM-1-P-na) as a potent antitumor agent. J Med Chem 53:1616–1626
32. Ferriz JM, Vinsova J (2010) Prodrug design of phenolic drugs. Curr Pharm Des 16:2033–2052
33. Dhareshwar SS, Stella VJ (2010) A novel prodrug strategy for beta-dicarbonyl carbon acids: syntheses and evaluation of the physicochemical characteristics of C-phosphoryloxymethyl (POM) and phosphoryloxymethyloxymethyl (POMOM) prodrug derivatives. J Pharm Sci 99:2711–2723
34. Gotthard G, Hiblot J, Gonzalez D, Elias M, Chabriere E (2013) Structural and enzymatic characterization of the phosphotriesterase OPHC2 from pseudomonas pseudoalcaligenes. PLoS One 8:e77995
35. Niemi R, Vepsalainen J, Taipale H, Jarvinen T (1999) Bisphosphonate prodrugs: synthesis and in vitro evaluation of novel acyloxyalkyl esters of clodronic acid. J Med Chem 42:5053–5058

36. Niemi R, Turhanen P, Vepsalainen J, Taipale H, Jarvinen T (2000) Bisphosphonate prodrugs: synthesis and in vitro evaluation of alkyl and acyloxymethyl esters of etidronic acid as bioreversible prodrugs of etidronate. Eur J Pharm Sci 11:173–180
37. Slatter JG, Feenstra KL, Hauer MJ, Kloosterman DA, Parton AH, Sanders PE, Scott G, Speed W (1996) Metabolism of the bisphosphonate ester U-91502 in rats. Drug Metab Dispos 24:65–73
38. Serafinowska HT, Ashton RJ, Bailey S, Harnden MR, Jackson SM, Sutton D (1995) Synthesis and in vivo evaluation of prodrugs of 9-[2-(phosphonomethoxy)ethoxy]adenine. J Med Chem 38:1372–1379
39. McGuigan C, Tollerfield SM, Riley PA (1989) Synthesis and biological evaluation of some phosphate triester derivatives of the anti-viral drug AraA. Nucleic Acids Res 17:6065–6075
40. Jones BCNM, McGuigan C, Riley PA (1989) Synthesis and biological evaluation of some phosphate triester derivatives of the anti-cancer drug AraC. Nucleic Acids Res 17:7195–7201
41. Lan S, Hsieh DC, Hillyer JW, Fancher RM, Rinehart KJ, Warrack BM, White RE (1998) Metabolism of α-phosphonosulfonate squalene synthase inhibitors. I. Disposition of a farnesylethyl α-phosphonosulfonate and ester prodrugs in rats. Drug Metab Dispos 26:993–1000
42. Dang Q, Liu Y, Rydzewski RM, Brown BS, Robinson E, van Poelje PD, Colby TJ, Erion MD (2007) Bis(para-methoxy)benzyl phosphonate prodrugs with improved stability and enhanced cell penetration. Bioorg Med Chem Lett 17:3412–3416
43. Dang Q, Brown BS, van Poelje PD, Colby TJ, Erion MD (1999) Synthesis of phosphonate 3-phthalidyl esters as prodrugs for potential intracellular delivery of phosphonates. Bioorg Med Chem Lett 9:1505–1510
44. Topczewski JJ, Neighbors JD, Wiemer DF (2009) Total synthesis of (+)-schweinfurthins B and E. J Org Chem 74:6965–6972
45. De Lombaert S, Erion MD, Tan J, Blanchard L, el-Chehabi L, Ghai RD, Sakane Y, Berry C, Trapani AJ (1994) N-Phosphonomethyl dipeptides and their phosphonate prodrugs, a new generation of neutral endopeptidase (NEP, EC 3.4.24.11) inhibitors. J Med Chem 37:498–511
46. Romanowska J, Sobkowski M, Szymanska-Michalak A, Kolodziej K, Dabrowska A, Lipniacki A, Piasek A, Pietrusiewicz ZM, Figlerowicz M, Guranowski A et al (2011) Aryl H-phosphonates 17: (N-aryl)phosphoramidates of pyrimidine nucleoside analogues and their synthesis, selected properties, and anti-HIV activity. J Med Chem 54:6482–6491
47. Romanowska J, Szymanska-Michalak A, Boryski J, Stawinski J, Kraszewski A, Loddo R, Sanna G, Collu G, Secci B, La Colla P (2009) Aryl nucleoside H-phosphonates. part 16: synthesis and anti-HIV-1 activity of di-aryl nucleoside phosphotriesters. Bioorg Med Chem 17:3489–3498
48. De Clercq E, Holy A, Rosenberg I, Sakuma T, Balzarini J, Maudgal PC (1986) A novel selective broad-spectrum anti-DNA virus agent. Nature 323:464–467
49. Khandazhinskaya A, Yasko M, Shirokova E (2006) The synthesis and antiviral properties of acyclic nucleoside analogues with a phosphonomethoxy fragment in the side chain. Curr Med Chem 13:2953–2980
50. Farquhar D, Srivastva DN, Kattesch NJ, Saunders PP (1983) Biologically reversible phosphate-protective groups. J Pharm Sci 72:324–325
51. Srivastva DN, Farquhar D (1984) Bioreversible phosphate protective groups – synthesis and stability of model acyloxymethyl phosphates. Bioorg Chem 12:118–129
52. Iyer RP, Phillips LR, Biddle JA, Thakker DR, Egan W, Aoki S, Mitsuya H (1989) Synthesis of acyloxyalkyl acylphosphonates as potential prodrugs of the antiviral, trisodium phosphonoformate (foscarnet sodium). Tetrahedron Lett 30:7141–7144
53. Saperstein R, Vicario PP, Strout HV, Brady E, Slater EE, Greenlee WJ, Ondeyka DL, Patchett AA, Hangauer DG (1989) Design of a selective insulin receptor tyrosine kinase inhibitor and its effect on glucose uptake and metabolism in intact cells. Biochemistry 28:5694–5701

54. Lee WA, Martin JC (2006) Perspectives on the development of acyclic nucleotide analogs as antiviral drugs. Antiviral Res 71:254–259
55. Atack JR, Prior AM, Fletcher SR, Quirk K, McKernan R, Ragan CI (1994) Effects of L-690,488, a prodrug of the bisphosphonate inositol monophosphatase inhibitor L-690,330, on phosphatidylinositol cycle markers. J Pharmacol Exp Ther 270:70–76
56. Topalis D, Pradere U, Roy V, Caillat C, Azzouzi A, Broggi J, Snoeck R, Andrei G, Lin J, Eriksson S et al (2011) Novel antiviral C5-substituted pyrimidine acyclic nucleoside phosphonates selected as human thymidylate kinase substrates. J Med Chem 54:222–232
57. Naesens L, Bischofberger N, Augustijns P, Annaert P, Van den Mooter G, Arimilli MN, Kim CU, De Clercq E (1998) Antiretroviral efficacy and pharmacokinetics of oral bis(isopropyloxycarbonyloxymethyl)-9-(2-phosphonylmethoxypropyl)adenine in mice. Antimicrob Agents Chemother 42:1568–1573
58. Arimilli MN, Kim CU, Dougherty J, Mulato A, Oliyai R, Shaw JP, Cundy KC, Bischofberger N (1997) Synthesis, in vitro biological evaluation and oral bioavailability of 9-[2-(phosphonomethoxy)propyl]adenine (PMPA) prodrugs. Antivir Chem Chemother 8:557–564
59. De Clercq E, Holy A (2005) Acyclic nucleoside phosphonates: a key class of antiviral drugs. Nat Rev Drug Discovery 4:928–940
60. Ripin DHB, Teager DS, Fortunak J, Basha SM, Bivins N, Boddy CN, Byrn S, Catlin KK, Houghton SR, Jagadeesh ST et al (2010) Process improvements for the manufacture of tenofovir disoproxil fumarate at commercial scale. Org Process Res Dev 14:1194–1201
61. Montagu A, Pradere U, Roy V, Nolan SP, Agrofoglio LA (2011) Expeditious convergent procedure for the preparation of bis(POC) prodrugs of new (E)-4-phosphono-but-2-en-1-yl nucleosides. Tetrahedron 67:5319–5328
62. Peyrottes S, Egron D, Lefebvre I, Gosselin G, Imbach JL, Perigaud C (2004) SATE pronucleotide approaches: an overview. Mini-Rev Med Chem 4:395–408
63. Oh CH, Liu LJ, Hong JH (2010) Design and synthesis of dually branched 5′-norcarbocyclic adenosine phosphonodiester analogue as a new anti-HIV prodrug. Nucleosides Nucleotides Nucleic Acids 29:721–733
64. Liu LJ, Hong JH (2010) Design and synthesis of novel sate derivatives of acyclic isocytosine and 9-deazaadenine C-nucleosides. Nucleosides Nucleotides Nucleic Acids 29:257–266
65. Fu X, Ou Y, Pei J, Liu Y, Li J, Zhou W, Lan Y, Wang A, Wang Y (2012) Synthesis, anti-HBV activity and renal cell toxicity evaluation of mixed phosphonate prodrugs of adefovir. Eur J Med Chem 49:211–218
66. Meier C, Lorey M, De Clercq E, Balzarini J (1997) Cyclic saligenyl phosphotriesters of 2′,3′-dideoxy-2′,3′-didehydrothymidine (d4T) – a new pro-nucleotide approach. Bioorg Med Chem Lett 7:99–104
67. Meier C, Balzarini J (2006) Application of the cycloSal-prodrug approach for improving the biological potential of phosphorylated biomolecules. Antiviral Res 71:282–292
68. Meier C, Knispel T, De Clercq E, Balzarini J (1999) cycloSal-pronucleotides of 2′,3-′-dideoxyadenosine and 2′,3′-dideoxy-2′,3′-didehydroadenosine: synthesis and antiviral evaluation of a highly efficient nucleotide delivery system. J Med Chem 42:1604–1614
69. Meier C, Gorbig U, Muller C, Balzarini J (2005) cycloSal-PMEA and cycloAmb-PMEA: potentially new phosphonate prodrugs based on the cycloSal-pronucleotide approach. J Med Chem 48:8079–8086
70. Meier C, Ruppel MFH, Vukadinovic D, Balzarini J (2004) "Lock-in"-cycloSal-pronucleotides – a new generation of chemical Trojan horses? Mini-Rev Med Chem 4:383–394
71. Gisch N, Balzarini J, Meier C (2007) 5-Diacetoxymethyl-cycloSal-d4TMP – a prototype of enzymatically activated cycloSal-pronucleotides. Nucleosides Nucleotides Nucleic Acids 26:861–864

72. Meier C, Lorey M, De Clercq E, Balzarini J (1998) cycloSal-2′,3′-dideoxy-2′,3′-didehydrothymidine monophosphate (cycloSal-d4TMP): synthesis and antiviral evaluation of a new d4TMP delivery system. J Med Chem 41:1417–1427
73. Morales EHR, Roman CA, Thomann JO, Meier C (2011) Linear synthesis of chiral cycloSal-pronucleotides. Eur J Med Chem 2011:4397–4408
74. Rios Morales EH, Balzarini J, Meier C (2012) Stereoselective synthesis and antiviral activity of methyl-substituted cycloSal-pronucleotides. J Med Chem 55:7245–7252
75. Wolf S, Warnecke S, Ehrit J, Freiberger F, Gerardy-Schahn R, Meier C (2012) Chemical synthesis and enzymatic testing of CMP-sialic acid derivatives. Chembiochem 13:2605–2615
76. Huchting J, Ruthenbeck A, Meier C (2013) Synthesis of cycloSal-(glycopyranosyl-6)-phosphates as activated sugar phosphates. Eur J Org Chem 2013:6907–6916
77. Erion MD, Bullough DA, Lin C, Hong Z (2006) HepDirect prodrugs for targeting nucleotide-based antiviral drugs to the liver. Curr Opin Invest Drugs 7:109–117
78. Erion MD, Reddy KR, Boyer SH, Matelich MC, Gomez-Galeno J, Lemus RH, Ugarkar BG, Colby TJ, Schanzer J, Van Poelje PD (2004) Design, synthesis, and characterization of a series of cytochrome P(450) 3A-activated prodrugs (HepDirect prodrugs) useful for targeting phosph(on)ate-based drugs to the liver. J Am Chem Soc 126:5154–5163
79. Erion MD, van Poelje PD, Mackenna DA, Colby TJ, Montag AC, Fujitaki JM, Linemeyer DL, Bullough DA (2005) Liver-targeted drug delivery using HepDirect prodrugs. J Pharmacol Exp Ther 312:554–560
80. Boyer SH, Sun Z, Jiang H, Esterbrook J, Gomez-Galeno JE, Craigo W, Reddy KR, Ugarkar BG, MacKenna DA, Erion MD (2006) Synthesis and characterization of a novel liver-targeted prodrug of cytosine-1-beta-D-arabinofuranoside monophosphate for the treatment of hepatocellular carcinoma. J Med Chem 49:7711–7720
81. Hecker SJ, Reddy KR, van Poelje PD, Sun Z, Huang W, Varkhedkar V, Reddy MV, Fujitaki JM, Olsen DB, Koeplinger KA et al (2007) Liver-targeted prodrugs of 2′-C-methyladenosine for therapy of hepatitis C virus infection. J Med Chem 50:3891–3896
82. Bookser BC, Raffaele NB, Reddy KR, Fan K, Huang W, Erion MD (2009) Synthesis of 3′-amino-3′-deoxyguanosine and 3′-amino-3′-deoxyxyloguanosine monophosphate HepDirect prodrugs from guanosine. Nucleosides Nucleotides Nucleic Acids 28:969–986
83. Erion MD, Cable EE, Ito BR, Jiang H, Fujitaki JM, Finn PD, Zhang BH, Hou J, Boyer SH, van Poelje PD et al (2007) Targeting thyroid hormone receptor-beta agonists to the liver reduces cholesterol and triglycerides and improves the therapeutic index. Proc Natl Acad Sci USA 104:15490–15495
84. Boyer SH, Jiang H, Jacintho JD, Reddy MV, Li H, Li W, Godwin JL, Schulz WG, Cable EE, Hou J et al (2008) Synthesis and biological evaluation of a series of liver-selective phosphonic acid thyroid hormone receptor agonists and their prodrugs. J Med Chem 51:7075–7093
85. Tsukada T, Tamaki K, Tanaka J, Takagi T, Yoshida T, Okuno A, Shiiki T, Takahashi M, Nishi T (2010) A prodrug approach towards the development of tricyclic-based FBPase inhibitors. Bioorg Med Chem Lett 20:2938–2941
86. Huttunen KM, Tani N, Juvonen RO, Raunio H, Rautio J (2013) Design, synthesis, and evaluation of novel cyclic phosphates of 5-aminosalicylic acid as cytochrome p450-activated prodrugs. Mol Pharm 10:532–537
87. Sun W, Peng W, Li G, Jiang T (2011) Design, synthesis, and sustained-release property of 1,3-cyclic propanyl phosphate ester of 18 beta-glycyrrhetinic acid. Chem Biol Drug Des 77:206–211
88. Remy DC, Sunthankar AV, Heidelberger C (1962) Studies on fluorinated pyrimidines. XIV. The synthesis of derivatives of 5-fluoro-2′-deoxyuridine 5′-phosphate and related compounds. J Org Chem 27:2491–2500
89. Mukherjee KL, Heidelberger C (1962) Studies of fluorinated pyrimidines. XV. Inhibition of the incorporation of formate-C14 into DNA thymine of ehrlich ascites carcinoma cells by 5-fluoro-2′-deoxyuridine-5′-monophosphate and related compounds. Cancer Res 22:815–822

90. De Napoli L, Di Fabio G, D'Onofrio J, Montesarchio D (2005) An efficient solid phase synthesis of 5′-phosphodiester and phosphoramidate monoester nucleoside analogues. Chem Commun 2586–2588
91. Hostetler KY (2009) Alkoxyalkyl prodrugs of acyclic nucleoside phosphonates enhance oral antiviral activity and reduce toxicity: current state of the art. Antiviral Res 82:A84–A98
92. Hostetler KY (2010) Synthesis and early development of hexadecyloxypropyl-cidofovir: an oral antipoxvirus nucleoside phosphonate. Viruses-Basel 2:2213–2225
93. Hostetler KY, Parker S, Sridhar CN, Martin MJ, Li JL, Stuhmiller LM, van Wijk GMT, van den Bosch H, Gardner MF, Aldern KA et al (1993) Acyclovir diphosphate dimyristoylglycerol – a phospholipid prodrug with activity against acyclovir-resistant herpes-simplex virus. Proc Natl Acad Sci USA 90:11835–11839
94. Hostetler KY, Beadle JR, Kini GD, Gardner MF, Wright KN, Wu TH, Korba BA (1997) Enhanced oral absorption and antiviral activity of 1-O-octadecyl-sn-glycero-3-phospho-acyclovir and related compounds in hepatitis B virus infection, in vitro. Biochem Pharmacol 53:1815–1822
95. Ciesla S, Trahan J, Wan W, Beadle J, Aldern K, Painter G, Hostetler K (2003) Esterification of cidofovir with alkoxyalkanols increases oral bioavailability and diminishes drug accumulation in kidney. Antiviral Res 59:163–171
96. Hostetler KY, Aldern KA, Wan WB, Ciesla SL, Beadle JR (2006) Alkoxyakl esters of (S)-9-[3-hydroxy-2-(phosphonomethoxy)propyl] adenine are potent inhibitors of the replication of wild-type and drug-resistant human immunodeficiency virus type 1 in vitro. Antimicrob Agents Chemother 50:2857–2859
97. Dal Pozzo F, Andrei G, Lebeau I, Beadle JR, Hostetler KY, De Clercq E, Snoeck R (2007) In vitro evaluation of the anti-orf virus activity of alkoxyalkyl esters of CDV, cCDV and (S)-HPMPA. Antiviral Res 75:52–57
98. Williams-Aziz SL, Hartline CB, Harden EA, Daily SL, Prichard MN, Kushner NL, Beadle JR, Wan WB, Hostetler KY, Kern ER (2005) Comparative activities of lipid esters of cidofovir and cyclic cidofovir against replication of herpesviruses in vitro. Antimicrob Agents Chemother 49:3724–3733
99. Marty FM, Winston DJ, Rowley SD, Vance E, Papanicolaou GA, Mullane KM, Brundage TM, Robertson AT, Godkin S, Mommeja-Marin H et al (2013) CMX001 to prevent cytomegalovirus disease in hematopoietic-cell transplantation. N Engl J Med 369:1227–1236
100. Valiaeva N, Wyles DL, Schooley RT, Hwu JB, Beadle JR, Prichard MN, Hostetler KY (2011) Synthesis and antiviral evaluation of 9-(S)-[3-alkoxy-2-(phosphonomethoxy)propyl]nucleoside alkoxyalkyl esters: inhibitors of hepatitis C virus and HIV-1 replication. Bioorg Med Chem 19:4616–4625
101. Dong SD, Lin C, Schroeder M (2013) Synthesis and evaluation of a new phosphorylated ribavirin prodrug. Antiviral Res 99:18–26
102. Krylov IS, Kashemirov BA, Hilfinger JM, McKenna CE (2013) Evolution of an amino acid based prodrug approach: stay tuned. Mol Pharm 10:445–458
103. Oliyai R, Shaw JP, Sueoka-Lennen CM, Cundy KC, Arimilli MN, Jones RJ, Lee WA (1999) Aryl ester prodrugs of cyclic HPMPC. I: Physicochemical characterization and in vitro biological stability. Pharm Res 16:1687–1693
104. Oliyai R, Arimilli MN, Jones RJ, Lee WA (2001) Pharmacokinetics of salicylate ester prodrugs of cyclic HPMPC in dogs. Nucleosides Nucleotides Nucleic Acids 20:1411–1414
105. Ruiz J, Beadle JR, Buller RM, Schreiwer J, Prichard MN, Keith KA, Lewis KC, Hostetler KY (2011) Synthesis, metabolic stability and antiviral evaluation of various alkoxyalkyl esters of cidofovir and 9-(S)-[3-hydroxy-2-(phosphonomethoxy)propyl]adenine. Bioorg Med Chem 19:2950–2958
106. Keith KA, Wan WB, Ciesla SL, Beadle JR, Hostetler KY, Kern ER (2004) Inhibitory activity of alkoxyalkyl and alkyl esters of cidofovir and cyclic cidofovir against orthopoxvirus replication in vitro. Antimicrob Agents Chemother 48:1869–1871

107. Lebeau I, Andrei G, Dal Pozzo F, Beadle JR, Hostetler KY, De Clercq E, van den Oord J, Snoeck R (2006) Activities of alkoxvalkyl esters of cidofovir (CDV), cyclic CDV, and (S)-9-(3-hydroxy-2-phosphonylmethoxypropyl)adenine against orthopoxviruses in cell monolayers and in organotypic cultures. Antimicrob Agents Chemother 50:2525–2529
108. McKenna CE, Kashemirov BA, Eriksson U, Amidon GL, Kish PE, Mitchell S, Kim JS, Hilfinger JM (2005) Cidofovir peptide conjugates as prodrugs. J Organomet Chem 690:2673–2678
109. Eriksson U, Peterson LW, Kashemirov BA, Hilfinger JM, Drach JC, Borysko KZ, Breitenbach JM, Kim JS, Mitchell S, Kijek P et al (2008) Serine peptide phosphoester prodrugs of cyclic cidofovir: synthesis, transport, and antiviral activity. Mol Pharm 5:598–609
110. Peterson LW, Sala-Rabanal M, Krylov IS, Serpi M, Kashemirov BA, McKenna CE (2010) Serine side chain-linked peptidomimetic conjugates of cyclic HPMPC and HPMPA: synthesis and interaction with hPEPT1. Mol Pharm 7:2349–2361
111. Zakharova VM, Serpi M, Krylov IS, Peterson LW, Breitenbach JM, Borysko KZ, Drach JC, Collins M, Hilfinger JM, Kashemirov BA et al (2011) Tyrosine-based 1-(S)-[3-hydroxy-2-(phosphonomethoxy)propyl]cytosine and -adenine ((S)-HPMPC and (S)-HPMPA) prodrugs: synthesis, stability, antiviral activity, and in vivo transport studies. J Med Chem 54:5680–5693
112. Krecmerova M, Holy A, Andrei G, Pomeisl K, Tichy T, Brehova P, Masojidkova M, Dracinsky M, Pohl R, Laflamme G et al (2010) Synthesis of ester prodrugs of 9-(S)-[3-hydroxy-2-(phosphonomethoxy)propyl]-2,6-diaminopurine (HPMPDAP) as anti-poxvirus agents. J Med Chem 53:6825–6837
113. Reddy PG, Bao D, Chang W, Chun B, Du J, Nagarathnam D, Rachakonda S, Ross BS, Zhang H, Bansal S et al (2010) 2′-Deoxy-2′-alpha-fluoro-2′-beta-C-methyl 3′,5′-cyclic phosphate nucleotide prodrug analogs as inhibitors of HCV NS5B polymerase: discovery of PSI-352938. Bioorg Med Chem Lett 20:7376–7380
114. Du J, Bao D, Chun B, Jiang Y, Reddy PG, Zhang H, Ross BS, Bansal S, Bao H, Espiritu C et al (2012) Beta-D-2′-alpha-F-2′-beta-C-methyl-6-O-substituted 3′,5′-cyclic phosphate nucleotide prodrugs as inhibitors of hepatitis C virus replication: a structure-activity relationship study. Bioorg Med Chem Lett 22:5924–5929
115. Reddy PG, Chun B, Zhang H, Rachakonda S, Ross BS, Sofia MJ (2011) Stereoselective synthesis of PSI-352938: a beta-D-2′-deoxy-2′-alpha-fluoro-2′-beta-C-methyl-3′,5′-cyclic phosphate nucleotide prodrug for the treatment of HCV. J Org Chem 76:3782–3790
116. Mehellou Y, Balzarini J, McGuigan C (2009) Aryloxy phosphoramidate triesters: a technology for delivering monophosphorylated nucleosides and sugars into cells. Chemmedchem 4:1779–1791
117. Devine KG, McGuigan C, O'Connor TJ, Nicholls SR, Kinchington D (1990) Novel phosphate derivatives of zidovudine as anti-HIV compounds. AIDS 4:371–373
118. Pertusati F, Serpi M, McGuigan C (2012) Medicinal chemistry of nucleoside phosphonate prodrugs for antiviral therapy. Antivir Chem Chemother 22:181–203
119. McGuigan C, Bourdin C, Derudas M, Hamon N, Hinsinger K, Kandil S, Madela K, Meneghesso S, Pertusati F, Serpi M et al (2013) Design, synthesis and biological evaluation of phosphorodiamidate prodrugs of antiviral and anticancer nucleosides. Eur J Med Chem 70:326–340
120. McGuigan C, Madela K, Aljarah M, Bourdin C, Arrica M, Barrett E, Jones S, Kolykhalov A, Bleiman B, Bryant KD et al (2011) Phosphorodiamidates as a promising new phosphate prodrug motif for antiviral drug discovery: application to anti-HCV agents. J Med Chem 54:8632–8645
121. Jansa P, Baszczynski O, Dracinsky M, Votruba I, Zidek Z, Bahador G, Stepan G, Cihlar T, Mackman R, Holy A et al (2011) A novel and efficient one-pot synthesis of symmetrical diamide (bis-amidate) prodrugs of acyclic nucleoside phosphonates and evaluation of their biological activities. Eur J Med Chem 46:3748–3754

122. Erion MD, van Poelje PD, Dang Q, Kasibhatla SR, Potter SC, Reddy MR, Reddy KR, Jiang T, Lipscomb WN (2005) MB06322 (CS-917): a potent and selective inhibitor of fructose 1,6-bisphosphatase for controlling gluconeogenesis in type 2 diabetes. Proc Natl Acad Sci USA 102:7970–7975
123. Dang Q, Liu Y, Cashion DK, Kasibhatla SR, Jiang T, Taplin F, Jacintho JD, Li H, Sun Z, Fan Y et al (2011) Discovery of a series of phosphonic acid-containing thiazoles and orally bioavailable diamide prodrugs that lower glucose in diabetic animals through inhibition of fructose-1,6-bisphosphatase. J Med Chem 54:153–165
124. van Poelje PD, Potter SC, Erion MD (2011) Fructose-1,6-bisphosphatase inhibitors for reducing excessive endogenous glucose production in type 2 diabetes. Handb Exp Pharmacol 203:279–301
125. Pertusati F, Hinsinger K, Flynn ÁS, Powell N, Tristram A, Balzarini J, McGuigan C (2014) PMPA and PMEA prodrugs for the treatment of HIV infections and human papillomavirus (HPV) associated neoplasia and cancer. Eur J Med Chem 78:259–268
126. Saboulard D, Naesens L, Cahard D, Salgado A, Pathirana R, Velazquez S, McGuigan C, De Clercq E, Balzarini J (1999) Characterization of the activation pathway of phosphoramidate triester prodrugs of stavudine and zidovudine. Mol Pharmacol 56:693–704
127. Lee WA, He GX, Eisenberg E, Cihlar T, Swaminathan S, Mulato A, Cundy KC (2005) Selective intracellular activation of a novel prodrug of the human immunodeficiency virus reverse transcriptase inhibitor tenofovir leads to preferential distribution and accumulation in lymphatic tissue. Antimicrob Agents Chemother 49:1898–1906
128. Eisenberg EJ, He GX, Lee WA (2001) Metabolism of GS-7340, a novel phenyl monophosphoramidate intracellular prodrug of PMPA, in blood. Nucleosides Nucleotides Nucleic Acids 20:1091–1098
129. Ballatore C, McGuigan C, De Clercq E, Balzarini J (2001) Synthesis and evaluation of novel amidate prodrugs of PMEA and PMPA. Bioorg Med Chem Lett 11:1053–1056
130. McGuigan C, Hassan-Abdallah A, Srinivasan S, Wang Y, Siddiqui A, Daluge SM, Gudmundsson KS, Zhou H, McLean EW, Peckham JP et al (2006) Application of phosphoramidate ProTide technology significantly improves antiviral potency of carbocyclic adenosine derivatives. J Med Chem 49:7215–7226
131. Leisvuori A, Aiba Y, Lonnberg T, Poijarvi-Virta P, Blatt L, Beigelman L, Lonnberg H (2010) Chemical and enzymatic stability of amino acid derived phosphoramidates of antiviral nucleoside 5′-monophosphates bearing a biodegradable protecting group. Org Biomol Chem 8:2131–2141
132. McGuigan C, Murziani P, Slusarczyk M, Gonczy B, Vande Voorde J, Liekens S, Balzarini J (2011) Phosphoramidate ProTides of the anticancer agent FUDR successfully deliver the preformed bioactive monophosphate in cells and confer advantage over the parent nucleoside. J Med Chem 54:7247–7258
133. Mehellou Y, Valente R, Mottram H, Walsby E, Mills KI, Balzarini J, McGuigan C (2010) Phosphoramidates of 2′-beta-*D*-arabinouridine (AraU) as phosphate prodrugs; design, synthesis, in vitro activity and metabolism. Bioorg Med Chem 18:2439–2446
134. McGuigan C, Derudas M, Gonczy B, Hinsinger K, Kandil S, Pertusati F, Serpi M, Snoeck R, Andrei G, Balzarini J et al (2014) ProTides of *N*-(3-(5-(2′-deoxyuridine))prop-2-ynyl) octanamide as potential anti-tubercular and anti-viral agents. Bioorg Med Chem 22:2816–2824
135. Perrone P, Luoni GM, Kelleher MR, Daverio F, Angell A, Mulready S, Congiatu C, Rajyaguru S, Martin JA, Leveque V et al (2007) Application of the phosphoramidate ProTide approach to 4′-azidouridine confers sub-micromolar potency versus hepatitis C virus on an inactive nucleoside. J Med Chem 50:1840–1849
136. Vanpouille C, Lisco A, Derudas M, Saba E, Grivel JC, Brichacek B, Scrimieri F, Schinazi R, Schols D, McGuigan C et al (2010) A new class of dual-targeted antivirals: monophosphorylated acyclovir prodrug derivatives suppress both human immunodeficiency virus type 1 and herpes simplex virus type 2. J Infect Dis 201:635–643

137. Ruda GF, Wong PE, Alibu VP, Norval S, Read KD, Barrett MP, Gilbert IH (2010) Aryl phosphoramidates of 5-phospho erythronohydroxamic acid, a new class of potent trypanocidal compounds. J Med Chem 53:6071–6078
138. Vernachio JH, Bleiman B, Bryant KD, Chamberlain S, Hunley D, Hutchins J, Ames B, Gorovits E, Ganguly B, Hall A et al (2011) INX-08189, a phosphoramidate prodrug of 6-*O*-methyl-2′-*C*-methyl guanosine, is a potent inhibitor of hepatitis C virus replication with excellent pharmacokinetic and pharmacodynamic properties. Antimicrob Agents Chemother 55:1843–1851
139. McGuigan C, Perrone P, Madela K, Neyts J (2009) The phosphoramidate ProTide approach greatly enhances the activity of beta-2′-C-methylguanosine against hepatitis C virus. Bioorg Med Chem Lett 19:4316–4320
140. McGuigan C, Madela K, Aljarah M, Gilles A, Battina SK, Ramamurty CVS, Srinivas Rao C, Vernachio J, Hutchins J, Hall A et al (2011) Dual pro-drugs of 2′-C-methyl guanosine monophosphate as potent and selective inhibitors of hepatitis C virus. Bioorg Med Chem Lett 21:6007–6012
141. Liederer BM, Borchardt RT (2006) Enzymes involved in the bioconversion of ester-based prodrugs. J Pharm Sci 95:1177–1195
142. Lim SM, Westover KD, Ficarro SB, Harrison RA, Choi HG, Pacold ME, Carrasco M, Hunter J, Kim ND, Xie T et al (2014) Therapeutic targeting of oncogenic K-ras by a covalent catalytic site inhibitor. Angew Chem Int Ed 53:199–204
143. Tobias SC, Borch RF (2001) Synthesis and biological studies of novel nucleoside phosphoramidate prodrugs. J Med Chem 44:4475–4480
144. Tobias SC, Borch RF (2004) Synthesis and biological evaluation of a cytarabine phosphoramidate prodrug. Mol Pharm 1:112–116
145. Lu P, Liu J, Wang Y, Chen X, Yang Y, Ji R (2009) Design, synthesis and evaluation of novel oxazaphosphorine prodrugs of 9-(2-phosphonomethoxyethyl)adenine (PMEA, adefovir) as potent HBV inhibitors. Bioorg Med Chem Lett 19:6918–6921
146. Gardelli C, Attenni B, Donghi M, Meppen M, Pacini B, Harper S, Di Marco A, Fiore F, Giuliano C, Pucci V et al (2009) Phosphoramidate prodrugs of 2 ’-C-methylcytidine for therapy of hepatitis C virus infection. J Med Chem 52:5394–5407
147. Valette G, Pompon A, Girardet JL, Cappellacci L, Franchetti P, Grifantini M, La Colla P, Loi AG, Perigaud C, Gosselin G et al (1996) Decomposition pathways and in vitro HIV inhibitory effects of isoddA pronucleotides: toward a rational approach for intracellular delivery of nucleoside 5′-monophosphates. J Med Chem 39:1981–1990
148. Grajkowski A, Ausin C, Kauffman JS, Snyder J, Hess S, Lloyd JR, Beaucage SL (2007) Solid-phase synthesis of thermolytic DNA oligonucleotides functionalized with a single 4-hydroxy-1-butyl or 4-phosphato-/thiophosphato-1-butyl thiophosphate protecting group. J Org Chem 72:805–815
149. Ausin C, Grajkowski A, Cieslak J, Gapeev A, Beaucage SL (2010) Time-dependent thermocontrol of the hydrophilic and lipophilic properties of DNA oligonucleotide prodrugs. Curr Protoc Nucleic Acid Chem 4:42
150. Kiuru E, Ahmed Z, Lonnberg H, Beigelman L, Ora M (2013) 2,2-Disubstituted 4-acylthio-3-oxobutyl groups as esterase- and thermolabile protecting groups of phosphodiesters. J Org Chem 78:950–959
151. Ora M, Mantyvaara A, Lonnberg H (2011) 3-Acetyloxy-2-cyano-2-(alkylaminocarbamoyl) propyl groups as biodegradable protecting groups of nucleoside 5′-mono-phosphates. Molecules 16:552–566
152. Farquhar D, Khan S, Srivastva DN, Saunders PP (1994) Synthesis and antitumor evaluation of bis[(pivaloyloxy)methyl] 2′-deoxy-5-fluorouridine 5′-monophosphate (FdUMP): a strategy to introduce nucleotides into cells. J Med Chem 37:3902–3909
153. Dickson JK, Biller SA, Magnin DR, Petrillo EW, Hillyer JW, Hsieh DC, Lan SJ, Rinehart JK, Gregg RE, Harrity TW et al (1996) Orally active squalene synthase inhibitors: bis((acyloxy) alkyl) prodrugs of the alpha-phosphonosulfonic acid moiety. J Med Chem 39:661–664

154. Aldern KA, Ciesla SL, Winegarden KL, Hostetler KY (2003) Increased antiviral activity of 1-O-hexadecyloxypropyl-[2-(14)C] cidofovir in MRC-5 human lung fibroblasts is explained by unique cellular uptake and metabolism. Mol Pharmacol 63:678–681
155. Tehler U, Nelson CH, Peterson LW, Provoda CJ, Hilfinger JM, Lee KD, McKenna CE, Amidona GL (2010) Puromycin-sensitive aminopeptidase: an antiviral prodrug activating enzyme. Antiviral Res 85:482–489
156. Mendel DB, Cihlar T, Moon K, Chen MS (1997) Conversion of 1-[((S)-2-hydroxy-2-oxo-1,4,2-dioxaphosphorinan-5-yl)methyl]cytosine to cidofovir by an intracellular cyclic CMP phosphodiesterase. Antimicrob Agents Chemother 41:641–646
157. Schneider E, Kuhn M, Reinecke D, Wolter S, Burhenne H, Kaever V, Seifert R (2013) Fishing for elusive cCMP-degrading phosphodiesterases. BMC Pharmacol Toxicol 14:1–2
158. Gisch N, Balzarini J, Meier C (2007) Enzymatically activated cycloSal-d4T-monophosphates: the third generation of cycloSal-pronucleotides. J Med Chem 50:1658–1667
159. Birkus G, Kutty N, He GX, Mulato A, Lee W, McDermott M, Cihlar T (2008) Activation of 9-[(R)-2-[[(S)-[[(S)-1-(isopropoxycarbonyl)ethyl]amino] phenoxyphosphinyl]-methoxy]propyl]adenine (GS-7340) and other tenofovir phosphonoamidate prodrugs by human proteases. Mol Pharmacol 74:92–100
160. Furman PA, Murakami E, Niu C, Lam AM, Espiritu C, Bansal S, Bao H, Tolstykh T, Steuer HM, Keilman M et al (2011) Activity and the metabolic activation pathway of the potent and selective hepatitis C virus pronucleotide inhibitor PSI-353661. Antiviral Res 91:120–132
161. Ray AS, Vela JE, Boojamra CG, Zhang L, Hui H, Callebaut C, Stray K, Lin KY, Gao Y, Mackman RL et al (2008) Intracellular metabolism of the nucleotide prodrug GS-9131, a potent anti-human immunodeficiency virus agent. Antimicrob Agents Chemother 52:648–654
162. Mackman RL, Ray AS, Hui HC, Zhang L, Birkus G, Boojamra CG, Desai MC, Douglas JL, Gao Y, Grant D et al (2010) Discovery of GS-9131: design, synthesis and optimization of amidate prodrugs of the novel nucleoside phosphonate HIV reverse transcriptase (RT) inhibitor GS-9148. Bioorg Med Chem 18:3606–3617
163. Lonnberg T, Ora M, Lonnberg H (2010) Hydrolytic reactions of nucleoside phosphoramidates: kinetics and mechanisms. Mini-Rev Org Chem 7:33–43
164. Murakami E, Tolstykh T, Bao H, Niu C, Micolochick Steuer HM, Bao D, Chang W, Espiritu C, Bansal S, Lam AM et al (2010) Mechanism of activation of PSI-7851 and its diastereoisomer PSI-7977. J Biol Chem 285:34337–34347
165. Chou TF, Baraniak J, Kaczmarek R, Zhou X, Cheng J, Ghosh B, Wagner CR (2007) Phosphoramidate pronucleotides: a comparison of the phosphoramidase substrate specificity of human and escherichia coli histidine triad nucleotide binding proteins. Mol Pharm 4:208–217
166. Birkus G, Wang R, Liu X, Kutty N, MacArthur H, Cihlar T, Gibbs C, Swaminathan S, Lee W, McDermott M (2007) Cathepsin A is the major hydrolase catalyzing the intracellular hydrolysis of the antiretroviral nucleotide phosphonoamidate prodrugs GS-7340 and GS-9131. Antimicrob Agents Chemother 51:543–550
167. Eichenbaum G, Skibbe J, Parkinson A, Johnson MD, Baumgardner D, Ogilvie B, Usuki E, Tonelli F, Holsapple J, Schmitt-Hoffmann A (2012) Use of enzyme inhibitors to evaluate the conversion pathways of ester and amide prodrugs: a case study example with the prodrug ceftobiprole medocaril. J Pharm Sci 101:1242–1252
168. Rhee SG (2013) Reflections on the days of phospholipase C. Adv Biol Regul 53:223–231
169. Baylon JL, Lenov IL, Sligar SG, Tajkhorshid E (2013) Characterizing the membrane-bound state of cytochrome P450 3A4: structure, depth of insertion, and orientation. J Am Chem Soc 135:8542–8551
170. Weiske J, Huber O (2005) The histidine triad protein Hint1 interacts with pontin and reptin and inhibits TCF-beta-catenin-mediated transcription. J Cell Sci 118:3117–3129

171. Birkus G, Kutty N, Frey CR, Shribata R, Chou T, Wagner C, McDermott M, Cihlar T (2011) Role of cathepsin A and lysosomes in the intracellular activation of novel antipapillomavirus agent GS-9191. Antimicrob Agents Chemother 55:2166–2173
172. Korboukh I, Hull-Ryde EA, Rittiner JE, Randhawa AS, Coleman J, Fitzpatrick BJ, Setola V, Janzen WP, Frye SV, Zylka MJ et al (2012) Orally active adenosine A(1) receptor agonists with antinociceptive effects in mice. J Med Chem 55:6467–6477
173. Kumar TS, Yang T, Mishra S, Cronin C, Chakraborty S, Shen JB, Liang BT, Jacobson KA (2013) 5′-Phosphate and 5′-phosphonate ester derivatives of (N)-methanocarba adenosine with in vivo cardioprotective activity. J Med Chem 56:902–914
174. Mehellou Y (2010) Phosphoramidate prodrugs deliver with potency against hepatitis C virus. Chemmedchem 5:1841–1842
175. Donnell D, Baeten JM, Bumpus NN, Brantley J, Bangsberg DR, Haberer JE, Mujugira A, Mugo N, Ndase P, Hendrix C et al. (2014) HIV protective efficacy and correlates of tenofovir blood concentrations in a clinical trial of PrEP for HIV prevention. J Acquir Immune Defic Syndr 66:340–348
176. Chang W, Bao D, Chun BK, Naduthambi D, Nagarathnam D, Rachakonda S, Reddy PG, Ross BS, Zhang H, Bansal S et al (2011) Discovery of PSI-353661, a novel purine nucleotide prodrug for the treatment of HCV infection. ACS Med Chem Lett 2:130–135
177. Du J, Chun B, Mosley RT, Bansal S, Bao H, Espiritu C, Lam AM, Murakami E, Niu C, Micolochick Steuer HM et al (2014) Use of 2′-spirocyclic ethers in HCV nucleoside design. J Med Chem 57:1826–1835
178. Jordheim LP, Durantel D, Zoulim F, Dumontet C (2013) Advances in the development of nucleoside and nucleotide analogues for cancer and viral diseases. Nat Rev Drug Discovery 12:447–464
179. De Clercq E (2011) The clinical potential of the acyclic (and cyclic) nucleoside phosphonates: the magic of the phosphonate bond. Biochem Pharmacol 82:99–109
180. De Clercq E (2013) Highlights in antiviral drug research: antivirals at the horizon. Med Res Rev 33:1215–1248
181. Poijarvi-Virta P, Lonnberg H (2006) Prodrug approaches of nucleotides and oligonucleotides. Curr Med Chem 13:3441–3465
182. Hurwitz SJ, Schinazi RF (2013) Prodrug strategies for improved efficacy of nucleoside antiviral inhibitors. Curr Opin HIV AIDS 8:556–564
183. Miralles-Lluma R, Figueras A, Busque F, Alvarez-Larena A, Balzarini J, Figueredo M, Font J, Alibes R, Marechal J (2013) Synthesis, antiviral evaluation, and computational studies of cyclobutane and cyclobutene L-nucleoside analogues. Eur J Med Chem 2013:7761–7775
184. Oliveira FM, Barbosa LCA, Ismail FMD (2014) The diverse pharmacology and medicinal chemistry of phosphoramidates - a review. RSC Adv 4:18998–19012
185. Schulz T, Balzarini J, Meier C (2014) The DiPPro approach: synthesis, hydrolysis, and antiviral activity of lipophilic d4T diphosphate prodrugs. ChemMedChem 9:762–775
186. Jessen HJ, Schulz T, Balzarini J, Meier C (2008) Bioreversible protection of nucleoside diphosphates. Angew Chem Int Ed 47:8719–8722
187. Cho A, Zhang L, Xu J, Lee R, Butler T, Metobo S, Aktoudianakis V, Lew W, Ye H, Clarke M et al (2014) Discovery of the first C-nucleoside HCV polymerase inhibitor (GS-6620) with demonstrated antiviral response in HCV infected patients. J Med Chem 57:1812–1825
188. Arbelo Roman C, Wasserthal P, Balzarini J, Meier C (2011) Diastereoselective synthesis of (aryloxy)phosphoramidate prodrugs. Eur J Org Chem 2011:4899–4909
189. Arbelo Roman C, Balzarini J, Meier C (2010) Diastereoselective synthesis of aryloxy phosphoramidate prodrugs of 3'-deoxy-2',3'-didehydrothymidine monophosphate. J Med Chem 53:7675–7681
190. Sofia MJ, Bao D, Chang W, Du J, Nagarathnam D, Rachakonda S, Reddy PG, Ross BS, Wang P, Zhang HR et al (2010) Discovery of a beta-d-2′-deoxy-2′-alpha-fluoro-2′-beta-C-methyluridine nucleotide prodrug (PSI-7977) for the treatment of hepatitis C virus. J Med Chem 53:7202–7218

191. Gane EJ, Stedman CA, Hyland RH, Ding X, Svarovskaia E, Symonds WT, Hindes RG, Berrey MM (2013) Nucleotide polymerase inhibitor sofosbuvir plus ribavirin for hepatitis C. N Engl J Med 368:34–44
192. Sulkowski MS, Gardiner DF, Rodriguez-Torres M, Reddy KR, Hassanein T, Jacobson I, Lawitz E, Lok AS, Hinestrosa F, Thuluvath PJ et al (2014) Daclatasvir plus sofosbuvir for previously treated or untreated chronic HCV infection. N Engl J Med 370:211–221
193. Hazleton KZ, Ho M, Cassera MB, Clinch K, Crump DR, Rosario I, Merino EF, Almo SC, Tyler PC, Schramm VL (2012) Acyclic immucillin phosphonates: second-generation inhibitors of plasmodium falciparum hypoxanthine-guanine-xanthine phosphoribosyltransferase. Chem Biol 19:721–730
194. Keough DT, Spacek P, Hockova D, Tichy T, Vrbkova S, Slavetinska L, Janeba Z, Naesens L, Edstein MD, Chavchich M et al (2013) Acyclic nucleoside phosphonates containing a second phosphonate group are potent inhibitors of 6-oxopurine phosphoribosyltransferases and have antimalarial activity. J Med Chem 56:2513–2526
195. Keough DT, Hockova D, Rejman D, Spacek P, Vrbkova S, Krecmerova M, Eng WS, Jans H, West NP, Naesens LMJ et al (2013) Inhibition of the escherichia coli 6-oxopurine phosphoribosyltransferases by nucleoside phosphonates: potential for new antibacterial agents. J Med Chem 56:6967–6984
196. Ebetino FH, Hogan AL, Sun S, Tsoumpra MK, Duan X, Triffitt JT, Kwaasi AA, Dunford JE, Barnett BL, Oppermann U et al (2011) The relationship between the chemistry and biological activity of the bisphosphonates. Bone 49:20–33
197. Rogers MJ, Ji XH, Russell RGG, Blackburn GM, Williamson MP, Bayless AV, Ebetino FH, Watts DJ (1994) Incorporation of bisphosphonates into adenine-nucleotides by amebas of the cellular slime-mold dictyostelium-discoideum. Biochem J 303:303–311
198. Frith JC, Monkkonen J, Blackburn GM, Russell RGG, Rogers MJ (1997) Clodronate and liposome-encapsulated clodronate are metabolized to a toxic ATP analog, adenosine 5′-(beta, gamma-dichloromethylene) triphosphate, by mammalian cells in vitro. J Bone Miner Res 12:1358–1367
199. Ahlmark M, Vepsalainen J, Taipale H, Niemi R, Jarvinen T (1999) Bisphosphonate prodrugs: synthesis and in vitro evaluation of novel clodronic acid dianhydrides as bioreversible prodrugs of clodronate. J Med Chem 42:1473–1476
200. Webster MR, Zhao M, Rudek MA, Hann CL, Freel Meyers CL (2011) Bisphosphonamidate clodronate prodrug exhibits potent anticancer activity in non-small-cell lung cancer cells. J Med Chem 54:6647–6656
201. Wiemer AJ, Hohl RJ, Wiemer DF (2009) The intermediate enzymes of isoprenoid metabolism as anticancer targets. Anticancer Agents Med Chem 9:526–542
202. Wiemer AJ, Wiemer DF, Hohl RJ (2011) Geranylgeranyl diphosphate synthase: an emerging therapeutic target. Clin Pharmacol Ther 90:804–812
203. Monteil M, Migianu-Griffoni E, Sainte-Catherine O, Di Benedetto M, Lecouvey M (2014) Bisphosphonate prodrugs: synthesis and biological evaluation in HuH7 hepatocarcinoma cells. Eur J Med Chem 77:56–64
204. Zhang Y, Leon A, Song Y, Studer D, Haase C, Koscielski LA, Oldfield E (2006) Activity of nitrogen-containing and non-nitrogen-containing bisphosphonates on tumor cell lines. J Med Chem 49:5804–5814
205. Wiemer AJ, Yu JS, Shull LW, Barney RJ, Wasko BM, Lamb KM, Hohl RJ, Wiemer DF (2008) Pivaloyloxymethyl-modified isoprenoid bisphosphonates display enhanced inhibition of cellular geranylgeranylation. Bioorg Med Chem 16:3652–3660
206. Troutman JM, Chehade KA, Kiegiel K, Andres DA, Spielmann HP (2004) Synthesis of acyloxymethyl ester prodrugs of the transferable protein farnesyl transferase substrate farnesyl methylenediphosphonate. Bioorg Med Chem Lett 14:4979–4982
207. Clark MK, Scott SA, Wojtkowiak J, Chirco R, Mathieu P, Reiners JJ Jr, Mattingly RR, Borch RF, Gibbs RA (2007) Synthesis, biochemical, and cellular evaluation of farnesyl monophosphate prodrugs as farnesyltransferase inhibitors. J Med Chem 50:3274–3282

208. Sane KM, Mynderse M, LaLonde DT, Dean IS, Wojtkowiak JW, Fouad F, Borch RF, Reiners JJ Jr, Gibbs RA, Mattingly RR (2010) A novel geranylgeranyl transferase inhibitor in combination with lovastatin inhibits proliferation and induces autophagy in STS-26 T MPNST cells. J Pharmacol Exp Ther 333:23–33
209. Wiemer AJ, Hsiao CH, Wiemer DF (2010) Isoprenoid metabolism as a therapeutic target in Gram-negative pathogens. Curr Top Med Chem 10:1858–1871
210. Reichenberg A, Wiesner J, Weidemeyer C, Dreiseidler E, Sanderbrand S, Altincicek B, Beck E, Schlitzer M, Jomaa H (2001) Diaryl ester prodrugs of FR900098 with improved in vivo antimalarial activity. Bioorg Med Chem Lett 11:833–835
211. Ortmann R, Wiesner J, Reichenberg A, Henschker D, Beck E, Jomaa H, Schlitzer M (2003) Acyloxyalkyl ester prodrugs of FR900098 with improved in vivo anti-malarial activity. Bioorg Med Chem 13:2163–2166
212. Ortmann R, Wiesner J, Reichenberg A, Henschker D, Beck E, Jomaa H, Schlitzer M (2005) Alkoxycarbonyloxyethyl ester prodrugs of FR900098 with improved in vivo antimalarial activity. Arch Pharm (Weinheim) 338:305–314
213. Kurz T, Schlueter K, Kaula U, Bergmann B, Walter RD, Geffken D (2006) Synthesis and antimalarial activity of chain substituted pivaloyloxymethyl ester analogues of fosmidomycin and FR900098. Bioorg Med Chem 14:5121–5135
214. Kurz T, Behrendt C, Kaula U, Bergmann B, Walter RD (2007) α-Phenylethyl substituted bis (pivaloyloxymethyl) ester analogs of fosmidomycin and FR900098. Aust J Chem 60:154–158
215. Schluter K, Walter RD, Bergmann B, Kurz T (2006) Arylmethyl substituted derivatives of fosmidomycin: synthesis and antimalarial activity. Eur J Med Chem 41:1385–1397
216. Kurz T, Schluter K, Pein M, Behrendt C, Bergmann B, Walter RD (2007) Conformationally restrained aromatic analogues of fosmidomycin and FR900098. Arch Pharm (Weinheim) 340:339–344
217. Haemers T, Wiesner J, Giessmann D, Verbrugghen T, Hillaert U, Ortmann R, Jomaa H, Link A, Schlitzer M, Van Calenbergh S (2008) Synthesis of beta- and gamma-oxa isosteres of fosmidomycin and FR900098 as antimalarial candidates. Bioorg Med Chem 16:3361–3371
218. Brucher K, Illarionov B, Held J, Tschan S, Kunfermann A, Pein MK, Bacher A, Graewert T, Maes L, Mordmueller B et al (2012) α-Substituted β-oxa isosteres of fosmidomycin: synthesis and biological evaluation. J Med Chem 55:6566–6575
219. Granthon AC, Braga MV, Rodrigues JCF, Cammerer S, Lorente SO, Gilbert IH, Urbina JA, de Wanderley S (2007) Alterations on the growth and ultrastructure of leishmania chagasi induced by squalene synthase inhibitors. Vet Parasitol 146:25–34
220. Uh E, Jackson ER, San Jose G, Maddox M, Lee RE, Lee RE, Boshoff HI, Dowd CS (2011) Antibacterial and antitubercular activity of fosmidomycin, FR900098, and their lipophilic analogs. Bioorg Med Chem Lett 21:6973–6976
221. Ponaire S, Zingle C, Tritsch D, Grosdemange-Billiard C, Rohmer M (2012) Growth inhibition of mycobacterium smegmatis by prodrugs of deoxyxylulose phosphate reductoisomerase inhibitors, promising anti-mycobacterial agents. Eur J Med Chem 51:277–285
222. Stankovic CJ, Surendran N, Lunney EA, Plummer MS, Para KS, Shahripour A, Fergus JH, Marks JS, Herrera R, Hubbell SE et al (1997) The role of 4-phosphonodifluoromethyl- and 4-phosphono-phenylalanine in the selectivity and cellular uptake of SH2 domain ligands. Bioorg Med Chem Lett 7:1909–1914
223. Garrido-Hernandez H, Moon KD, Geahlen RL, Borch RF (2006) Design and synthesis of phosphotyrosine peptidomimetic prodrugs. J Med Chem 49:3368–3376
224. Huang R, Oh H, Arrendale A, Martin VA, Galan J, Workman EJ, Stout JR, Walczak CE, Tao WA, Borch RF et al (2013) Intracellular targets for a phosphotyrosine peptidomimetic include the mitotic kinesin, MCAK. Biochem Pharmacol 86:597–611
225. Boutselis IG, Yu X, Zhang ZY, Borch RF (2007) Synthesis and cell-based activity of a potent and selective protein tyrosine phosphatase 1B inhibitor prodrug. J Med Chem 50:856–864

226. Auzenne EJ, Klostergaard J, Mandal PK, Liao WS, Lu Z, Gao F, Bast RC Jr, Robertson FM, McMurray JS (2012) A phosphopeptide mimetic prodrug targeting the SH2 domain of Stat3 inhibits tumor growth and angiogenesis. J Exp Ther Oncol 10:155–162
227. Mandal PK, Liao WSL, McMurray JS (2009) Synthesis of phosphatase-stable, cell-permeable peptidomimetic prodrugs that target the SH2 domain of Stat3. Org Lett 11:3394–3397
228. Mandal PK, Gao F, Lu Z, Ren Z, Ramesh R, Birtwistle JS, Kaluarachchi KK, Chen X, Bast RC, Liao WSL et al (2011) Potent and selective phosphopeptide mimetic prodrugs targeted to the Src homology 2 (SH2) domain of signal transducer and activator of transcription 3. J Med Chem 54:3549–3563
229. Mandal PK, Ren Z, Chen X, Kaluarachchi K, Liao WSL, McMurray JS (2013) Structure-activity studies of phosphopeptidomimetic prodrugs targeting the Src homology 2 (SH2) domain of signal transducer and activator of transcription 3 (Stat3). Int J Pept Res Ther 19:3–12
230. Morlacchi P, Mandal PK, McMurray JS (2014) Synthesis and in vitro evaluation of a peptidomimetic inhibitor targeting the Src homology 2 (SH2) domain of STAT6. ACS Med Chem Lett 5:69–72
231. Chu CY, Chang CP, Chou YT, Handoko HYL, Lo LC, Lin JJ (2013) Development and evaluation of novel phosphotyrosine mimetic inhibitors targeting the Src homology 2 domain of signaling lymphocytic activation molecule (SLAM) associated protein. J Med Chem 56:2841–2849
232. Arrendale A, Kim K, Choi JY, Li W, Geahlen RL, Borch RF (2012) Synthesis of a phosphoserine mimetic prodrug with potent 14-3-3 protein inhibitory activity. Chem Biol 19:764–771
233. Qian WJ, Burke TR Jr, Terrence R (2013) Design and synthesis of a reagent for solid-phase incorporation of the phosphothreonine mimetic (2S,3R)-2-amino-3-methyl-4-phosphonobutyric acid (pmab) into peptides in a bio-reversible phosphonyl-bis-pivaloyloxymethyl (POM) prodrug form. Amino Acids 45:1143–1148

Top Curr Chem (2015) 360: 161–236
DOI: 10.1007/128_2014_564

Published online: 19 October 2014

Recent Advances in Asymmetric Synthesis of *P*-Stereogenic Phosphorus Compounds

Oleg I. Kolodiazhnyi

Abstract This chapter points out significant advances in the asymmetric synthesis of *P*-chiral organophosphorus compounds with many applications in stereoselective synthesis and in asymmetric catalysis, making reference to updated literature findings as well as the author's original research. Asymmetric addition and cycloaddition reactions, oxidation, including metal catalyzed and non-metal biocatalytic methods are described, in addition to synthetic approaches via nucleophilic substitution of appropriately substituted precursors. Use of chiral organophosphorus compounds in some asymmetric transformations such as hydrogenation and alkyl/arylation reactions is also discussed.

Keywords Asymmetric catalysis · Asymmetric synthesis · Biocatalysis · Ephedrine · *P*-Chiral phosphorus compounds · Sparteine

Contents

O.I. Kolodiazhnyi (✉)
Institute of Bioorganic Chemistry and Petrochemistry, National Academy of Sciences of Ukraine, Murmanska Str, 1, Kiev, Ukraine
e-mail: olegkol321@rambler.ru

Abbreviations

Ac	Acetyl
Ad	Adamantyl
All	Allyl
Amano AH	*Pseudomonas cepacia* lipase
Amano AK	*Pseudomonas fluorescens* lipase
Amano PS	*Burkholderia cepacia* lipase
An	Anisyl
Bn	Benzyl
Brn	*endo*-Bornyl
CALB	*Candida antarctica* lipase (Chirazyme®L-2, Novozym 435b)
COD	Cyclooctadiene
Cp	Cyclopentadiene
CRL	*Candida rugosa* lipase
Cy	Cyclohexyl
DABCO	Diazabicyclooctane
DBTA	Dibenzoyltartaric acid
DBU	1,8-Diazabicyclo[5.4.0]undec-7-ene
de	Diastereoisomeric excess
DIAD	Diisopropyl azodicarboxylate
DIPAMP	Ethane-1,2-diylbis[(2-methoxyphenyl)phenylphosphane]
DIPEA	*N*,*N*-Diisopropylethylamine
DMAP	4-Dimethylaminopyridine
DME	Dimethoxyethane
DMF	*N*,*N*-Dimethylformamide
DMSO	Dimethylsulfoxide
DPPE	1,1-Bis(diphenylphosphino)ethane
DPPM	1,1-Bis(diphenylphosphino)methane
dr	Diastereoisomeric ratio
EDC	Ethyl-(*N′*,*N′*-dimethylamino)propylcarbodiimide hydrochloride
ee	Enantiomeric excess
er	Enantiomeric ratio
Fc	Ferrocenyl
Is	2,4,6-Tris(isopropyl)phenyl
L	Ligand
LDBB	Lithium *P*,*P′*-di-*tert*-butylbiphenylide
LPL	*Lipoprotein* lipase
Mes	Mesityl
Mnt	(1*R*,2*S*,5*R*)-Menthyl
Ms	Mesyl (Methanesulfol)
Nphth	Naphthyl
PFL	*Pseudomonas fluorescens* lipase
Piv	Pivaloyl

PLE	Pig (porcine) liver esterase
PPL	Porcine pancreatic lipase
PTE	Phosphotriesterase
PTSA	*p*-Toluenesulfonic acid
Py	Pyridine
RT	Room temperature
TBME	*tert*-Butylmethyl ether
TfO	Triflate
Tol	Tolyl
Ts	Tosyl
WSCI	Water soluble carbodiimide (EDC)
Xyl	Xylyl

1 Introduction

Chiral phosphorus compounds play an important role in many areas of science, including biologically active pharmaceuticals, agrochemicals, and ligands for transition metal complexes [1–3]. Therefore, the availability of chiral organophosphorus compounds has allowed tremendous achievements in chemistry, in particular in the field of asymmetric metal catalysis. However, most of the popular chiral *P*-ligands are based on chiral C-backbones. In comparison, the occurrence of *P*-stereogenic ligands in the literature is less because their synthesis often presents a challenge. Many methods can be used to prepare enantiomerically pure organophosphorus compounds, including classical optical resolution via diastereoisomers, chemical kinetic resolution, enzymatic resolution, chromatographic resolution, and asymmetric catalysis [1–16]. Asymmetric synthesis and asymmetric catalysis have been and remain a primary research field of chemistry [4–10]. Therefore, various methods for the asymmetric synthesis of *P*-chiral organophosphorus compounds have been extensively studied in many academic and industrial research laboratories.

For the last few years, enormous success has been achieved in the asymmetric synthesis of organophosphorus compounds, primarily for phosphine ligands for transition metal-catalyzed asymmetric reactions, and many articles devoted to the synthesis of chiral organophosphorus compounds have been published. It is therefore interesting to analyze and systematize the data dedicated to asymmetric synthesis of *P*-chiral organophosphorus compounds that have been published over the last 5–10 years.

2 Configurational Stability of Chiral Phosphorus Compounds

Chiral compounds bearing a stereogenic phosphorus atom play an important role as reagents, auxiliaries, or catalysts in asymmetric synthesis, and are key components of pharmaceuticals. The configurational stability of chiral phosphorus centers has, therefore, received considerable attention. By analogy with compounds possessing an asymmetric nitrogen center, tetrahedral chiral-at-phosphorus compounds are generally stable to racemization. Chiral phosphines are configurationally stable at room temperature, whereas acyclic amines readily racemize. A trivalent phosphorus atom bonded to three substituents in a pyramidal geometry and possessing one unshared electron pair may spontaneously undergo inversion of configuration. Racemization of trivalent phosphorus compounds depends strongly on their structure, first of all on electron-accepting substituents at the phosphorus atom, which decrease the configurational stability. The barrier of inversion in acyclic phosphines is about 150 kJ/mol, whereas the barrier of inversion for acyclic amines is about 30 kJ/mol and depends on the electronegativities of substituents bonded to the phosphorus atom. However, in many cases, compounds bearing electron-accepting groups at the phosphorus atom are racemized [15]. For instance, chiral chlorophosphines are conformationally labile compounds existing as an equilibrium racemic mixture of (*R*)- and (*S*)-enantiomers. Although calculations indicate substantial pyramidal stability at phosphorus in halophosphines of the type R^1R^2PX, attempts to isolate enantiomerically pure chlorophosphines have been unsuccessful. Jugé et al. [16] carried out experimental and computational studies on the configurational stability and racemization of chlorophosphines. The pure chlorophosphines would be configurationally stable at the phosphorus atom, but traces of acids, e.g., HCl, which are almost unavoidable in the experimental conditions, lead to easy racemization. It was found that HCl acts as a catalyst for inversion at the *P*-center. The mechanism of the racemization is explained by the phosphorus nucleophilic attack on H, with a concerted backside attack by the chlorine on the phosphorus center. The reaction intermediate, as indicated by the gas-phase computation, is an achiral pentacoordinated phosphorus with two chlorine atoms in the axial position (Scheme 1).

The racemization of chiral tertiary phosphines in many cases is an interesting route to unavailable enantiomers [15]. For example, the thermodynamically controlled pyramidal inversion of tertiary phosphines is useful synthetically. The facile acid-catalyzed racemization of secondary phosphines involves formation of an

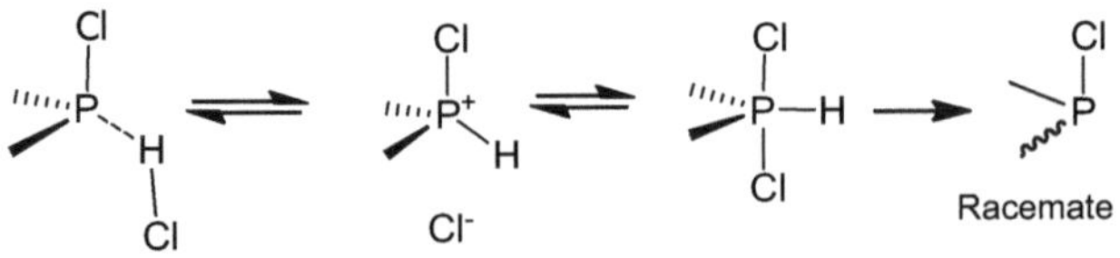

Scheme 1 Thermodynamically controlled racemization of tertiary phosphines

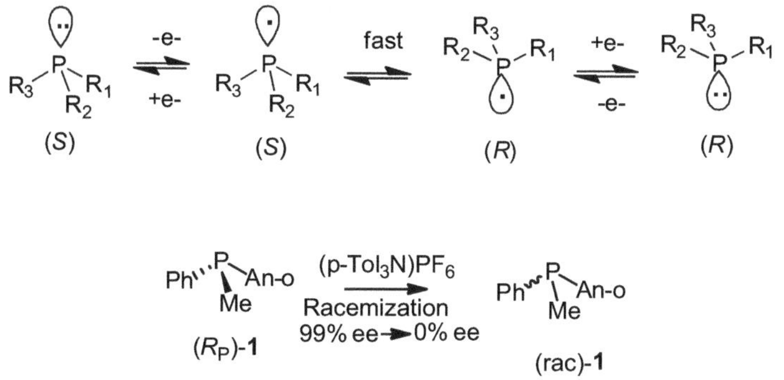

Scheme 2 Examples of catalytic racemization of P(III) compounds

achiral phosphonium ion and often renders isolation of enantiomers difficult [9, 15–17]. Monochlorophosphoramidates and other configurationally unstable phosphorus compounds were employed in DYKAT (Dynamic Kinetic Asymmetric Transformation) and similar stereodynamic synthetic strategies [17]. Radosevich [18] reported that the pyramidal inversion of trivalent phosphines is catalyzed by single-electron oxidation. Specifically, *P*-stereogenic (aryl)methylphenyl phosphines are shown to undergo rapid racemization at ambient temperature when exposed to catalytic quantities of a single-electron oxidant in solution. Under these conditions, transient phosphoniumyl radical cations ($R_3P•^+$) are formed, and computational models indicate that the pyramidal inversion barriers for these open-shell intermediates are on the order of 5 kcal/mol. The pyramidal inversion of trisubstituted (aryl)alkyl phosphines may be catalyzed by single-electron transfer, permitting dynamic stereochemical behavior of *P*-stereogenic phosphines at ambient temperatures. For example, enantioenriched (R_P)-**1** (99% *ee*) is configurationally stable in solution at ambient temperature, with an experimentally determined barrier to thermal inversion of 31.4 kcal/mol. However, according to the proposed outer-sphere ET mechanism, the easy racemization of **1** can be attained by catalytic oxidation with organic aminium oxidant [*P*-An_3N][PF_6], (*P*-Tol_3N)PF_6, (Cp_2Fe) PF_6, Co($OTf)_2$, or copper(II) triflate (Scheme 2)

3 Asymmetric Nucleophilic Substitution at Phosphorus

Asymmetric synthesis utilizing chiral auxiliaries represents the most effective method for the synthesis of chiral organophosphorus compounds. During the last few years this method has seen significant development [17–20]. Nucleophilic displacement at the phosphorus atom is one of the most popular and efficient methods for the preparation of enantiopure *P*-chiral compounds. At present, the most commonly used are chiral secondary alcohols/halophosphines [12, 13, 15, 17], sparteine/dimethylphosphine borane [19, 21], sparteine/secondary phosphine borane [21], and ephedrine/diaminophosphine [1, 5].

3.1 Secondary Alcohols as Chiral Auxiliaries

Optically active secondary alcohols (L-menthol, *endo*-borneol, glucofuranose derivatives, and others) serve as cheap and accessible chiral auxiliaries for the preparation of enantiopure organophosphorus compounds [12, 13, 21–25]. For example, diastereoisomers of menthyl phosphinite boranes [18, 23] are one of the most popular chiral organophosphorus reactants in organic synthesis. The general method to accomplish the resolution of phosphine oxides or quaternary phosphinates is to use chiral secondary alcohols from the chiral pool. As early as 1967, Mislow et al. [22] reported the synthesis of *P*-chiral ligands via resolution of diastereomeric menthylphosphinates and subsequent addition of Grignard reagents, and reduction of the phosphine oxide with trichlorosilane yielded the chiral phosphine (Scheme 3). Representative examples of chiral alkoxyphosphonates were prepared from easily available natural chiral precursors such as L-menthol or glucofuranose, obtained from readily available D-glucose. The unsymmetrically substituted menthyl phosphinates (R_P/S_P)-**2** could be separated into their diastereoisomers by recrystallization or column chromatography. Today, the menthyl phosphinates are useful chiral reactants for organic synthesis. Buono [12, 13] reported that nucleophilic substitution of the alkoxy group of the *H*-phosphinates **2** with organolithiums reagents proceeds stereospecifically with inversion of configurations at phosphorus to give a wide range of *P*-stereogenic tertiary phosphine oxides **3** or secondary phosphine oxides **4** by quenching the reaction mixture with alkyl halides or water, respectively [13, 24]. The enantiopure (S_P)- and (R_P)-

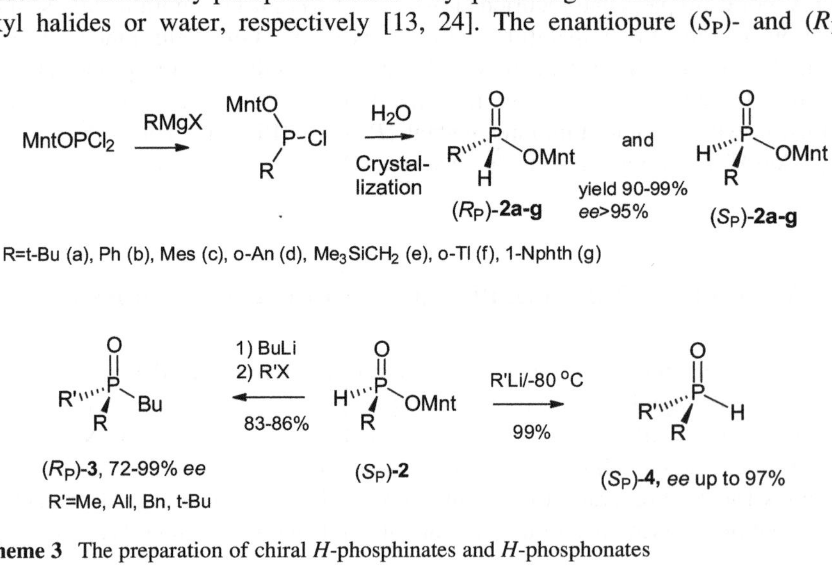

Scheme 3 The preparation of chiral *H*-phosphinates and *H*-phosphonates

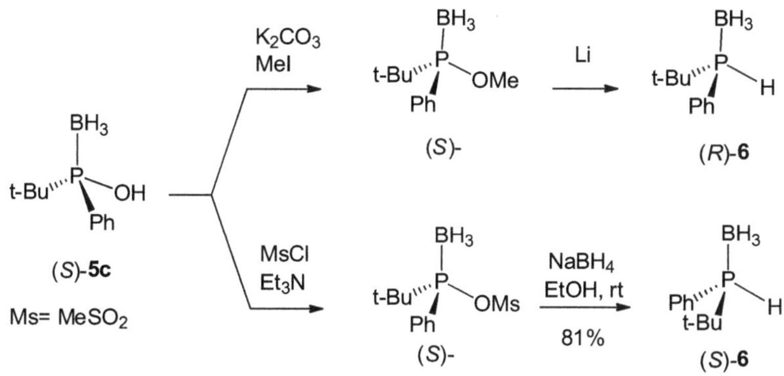

Scheme 4 Enantiopure secondary *H*-phosphines **6**

Table 1 Preparation of **5a–i** from menthyl *H*-phosphinates **2** (Scheme 4)

Products	R	Ar	Config	Yield (%)	*ee* (%)	References
5a	Me	Ph	(*S*)-(−)	78	95	[13]
5b	*n*-Bu	Ph	(−)	70	89	[13]
5c	*t*-Bu	Ph	(*S*)-(−)	70	84	[13]
5d	*t*-Bu	Tol	(*S*)-(−)	70	99	[12]
5e	*t*-Bu	1-Nphth	(*S*)-(−)	62	99	[12]
5f	2-Tol	Ph	(+)	75	97	[13]
5g	2-PhC_6H_4	Ph	(−)	85	95	[13]
5h	1-Nphth	Ph	(*R*)-(−)	75	99	[13]
5i	1-Furyl	Ph	(+)	72	80	[13]

diastereomers of (−)-menthylmesitylphosphine **2c** were isolated by fractional crystallization of an (R_P)/(S_P)-mixture from acetonitrile containing a trace of sodium acetylacetonate as a proton scavenger (yield 66.2%, 97% *de*). The crystal and molecular structure of (S_P)-**2c** (R=Mes) has been defined (Scheme 3).

The transformation of readily available enantiopure *H*-menthylphosphinates **2** into chiral phosphinous acid boranes **5** permits the elaboration of bulky *P*-stereogenic secondary phosphine boranes. Taking advantage of the synthetic potential of these compounds, a broad range of hindered *P*-chiral tertiary phosphine boranes **6** were prepared with excellent enantiomeric excesses [12, 13]. Phosphinous acid **5** can easily be converted into one or the other enantiomer of the secondary phosphines boranes (S_P)- or (R_P)-**6** by stereoselective reduction or substitution of the phosphinite borane derivatives, respectively (Scheme 4 and Table 1).

Buono et al. [11, 13] described the preparation of menthyl phosphinites **2** from $PhPCl_2$ and their separation into diastereoisomers. The subsequent reaction of (S_P)-phosphinite **2** with sodium hydride and with methyl iodide afforded the menthyl methylphenylphosphinite borane (R_P)-**6** with retention of configuration at phosphorus [24]. The palladium-catalyzed ($Pd[PPh_3]_4$) coupling reaction of the (S_P)-**2** with

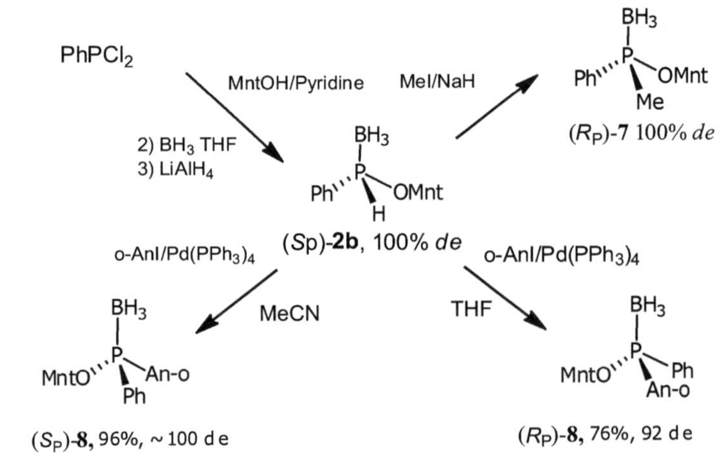

Scheme 5 Preparation of enantiopure (1*R*,2*S*,5*R*)-menthyl phosphinite boranes **6–8**

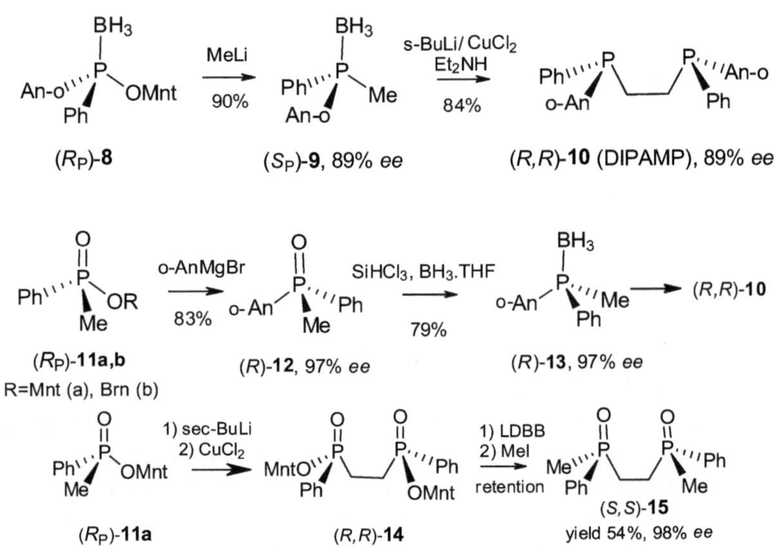

Scheme 6 Synthesis of DIPAMP analogues

o-iodoanisole occurred either with complete retention of the configuration at the phosphorus atom, or with almost complete inversion, depending on the solvent used (MeCN or THF), i.e., both diastereoisomers (R_P)-**8** and (S_P)-**8** can be synthesized from a single starting diastereoisomer (S_P)-**2b** (Scheme 5) [12, 13].

The menthyl phosphonites and menthyl phosphinites were used for the preparation of *P*-chiral monodentate ligands. For example, Imamoto used menthyl phosphinite borane complexes for the synthesis of DIPAMP analogues, as shown in Scheme 6 [23]. The treatment of phosphinite borane (R_P)-**8** with methyllithium,

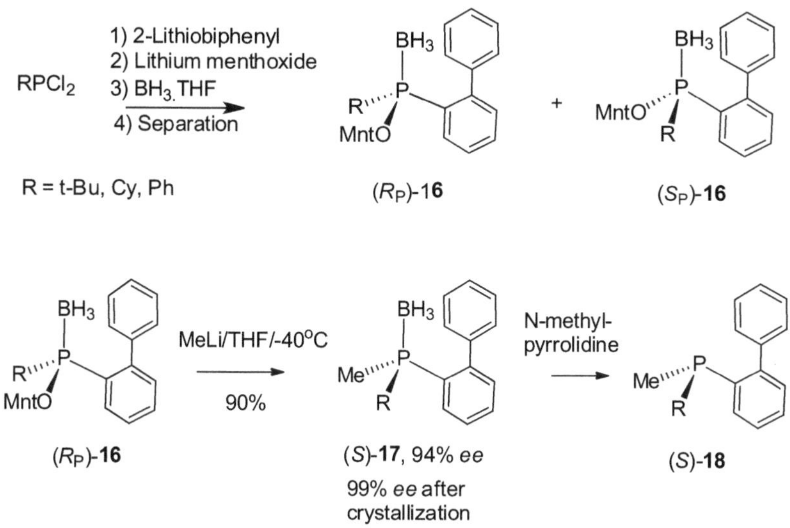

Scheme 7 Synthesis of monodentate phosphine ligands, bearing the 2-diphenyl group

then with *sec*-butyllithium and oxidative coupling with copper(II) chloride, led to the formation of bis-phosphinoethane diborane (*R*,*R*)-**9**. The deboration of **9** by diethylamine afforded the (*R*,*R*)-DIPAMP **10** [22, 23]. Diastereoisomerically pure menthyl- or bornylphosphinates **11a**,**b** react with Grignard reagents to afford tertiary phosphine oxides **12** with inversion of absolute configuration at phosphorus [24, 25]. Phosphine oxides **12** were reduced to tertiary phosphines or their boranes (*R*)-**13** by silanes. Imamoto found that the subsequent treatment of phosphine oxides (*R*,*R*)-**14** with lithium 4,4′-di-*tert*-butyldiphenylide (LDBB) and methyl iodide furnished the diphosphine oxide (*S*,*S*)-**15** with retention of configuration. This method is complementary to the nucleophilic substitution of the menthyl group, in which the substitution at the phosphorus atom occurs with inversion of configuration. Upon examining several reducing agents (alkali metals and Li-NH_3 among others), it was found that LDBB was the reagent of choice because it preserved the stereochemical integrity of the phosphorus atom. Scheme 6 shows the use of this method to prepare diphosphine oxides. The diphosphide obtained upon treatment with LDBB was quenched with methyl iodide to furnish the diphosphine oxide (*S*,*S*)-**15**.

Imamoto et al. improved the menthol methodology by using phosphine boranes instead of phosphine oxides [25]. The use of phosphine boranes avoids the handling of the malodorous, airsensitive, and sometimes corrosive free phosphines until they are needed. The removal of the boranato group is easily achieved by treatment of phosphine borane with excess of diethylamine, DABCO or using certain acids. This step fully retained the configuration at the phosphorus atoms, in contrast to the stereochemical problems associated with reduction of oxides. For example, monodentate phosphines **18**, bearing 2-diphenyl group, were prepared as shown in Scheme 7 and used as effective ligands in complexes of palladium, catalyzing

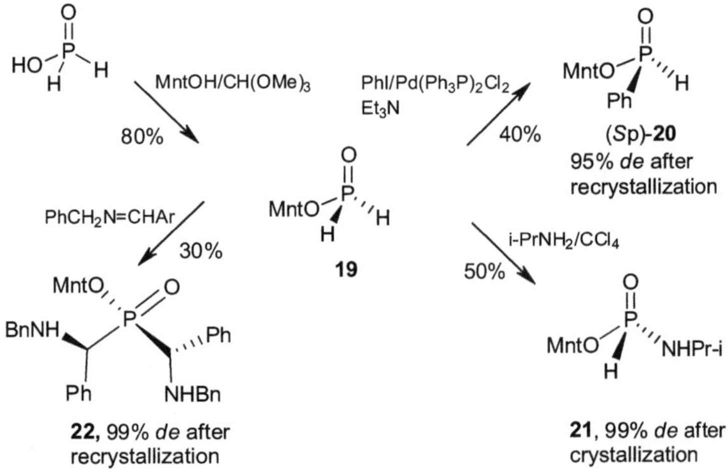

Scheme 8 (*1R,2S,5R*)-Menthyl phosphinate **19** as starting chiral reactant

reactions of carbon-carbon bond formation. Pure stereoisomers **16** and **17** were obtained by HPLC separation or recrystallization, and their structures were proved by X-ray monocrystal analysis.

(1*R*,2*S*,5*R*)-Menthyl-*H*-phosphinate **19** was prepared by reaction of anhydrous hypophosphorous acid with (1*R*,2*S*,5*R*)-menthol in the presence of trimethylorthoformate. It is rather stable, and can be purified by distillation under vacuum. It enters into the Heck reaction with iodobenzene in the presence of a palladium complex to yield the diastereoisomerically enriched (S_P)-(−)-menthyl phenylphosphonite **20**. The Todd–Atherton reaction of compound **19** with CCl_4 and isopropylamine proceeded with the formation of amide **21**. It also reacts with Schiff bases to give bis-amidophosphinates **22** (Scheme 8) [26, 27].

Montchamp has developed methods for the conversion of hypophosphorous acid and alcohols into various enantioenriched *H*-phosphonate diesters: "organophosphorus chemistry without PCl_3" which is interesting from point of view of green chemistry [20, 28, 29]. Compound (R_P)-**24a** was prepared from hypophosphorous acid, paraformaldehyde, and L-menthol in 9% yield, Further cross-coupling of crystalline (R_P)-**24** with bromobenzene gave (R_P)-**25** in 68% yield; on the other hand, the cross-coupling of the mother liquor with bromobenzene and the crystallization of the resulting reaction mixture at room temperature led to (S_P)-**26** in 23% yield and with 97% *de*. The compound (S_P)-**26** was also prepared from phenyl-*H*-phosphinic acid, L-menthol, and paraformaldehyde in 26% yield and with 95% *de*. These phosphorus synthons were functionalized into useful *P*-stereogenic compounds (Scheme 9) [30].

Diastereomer (R_P)-**24** is a versatile *P*-stereogenic building block for the preparation of chiral tertiary phosphines. Cross-coupling of (R_P)-**24** with arylhalogenides and $Pd(OAc)_2$ gives (S_P)-**30** in good yields, and subsequent oxidative cleavage delivers (R_P)-**29** in 81% yield. Compound (S_P)-**30** can be oxidized to form (R_P)-**29**

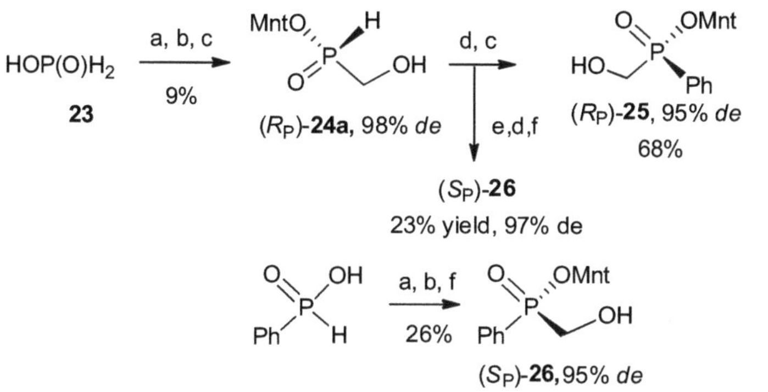

a = $(CH_2O)_n$,75°C; b = L-Menthol, toluene, reflux; c=recrystallization at -18°C
d = $Pd(OAc)_2$, Xantphos, DIPEA, PhBr, DMF/DME, 115°C;
e = mother liquor, f = recrystallization at RT

Scheme 9 Montchamp's preparation of *P*-stereogenic *H*-phosphinates

stereospecifically in 81–95% *de*. Therefore, cross-coupling of **24** followed by oxidation of **30** leads to either P configuration of phosphinates using L-menthol in all cases. The presence of the hydroxymethyl group in compounds **24** and **30** provides opportunities for functionalization since the carbon atom can be preserved if desired. Compound (R_P)-**24** can be viewed as a protected chiral equivalent of alkyl phosphinates $ROP(O)H_2$, since it can be stereospecifically alkylated to form **28**, or cross-coupled to form **30**, and the hydroxymethyl moiety can subsequently be cleaved to form *H*-phosphinates similar to **29**. For example, the Mitsunobu reaction of (S_P)-**30** with phthalimide gives (S_P)-**31** in 70% yield (Scheme 10) [30, 31].

The stereoisomers of ethyl menthyl (methylthio)methylphosphonate **32** were obtained from commercially available diethyl (methylthio)methylphosphonate by subjecting them to a highly diastereoselective hydroperoxide oxidation in the presence of catalytic amounts of a titanium (*R*)- or (*S*)-BINOL complex. The prevailing phosphorus stereoisomer was oxidized with cumene hydroperoxide in the presence of a Sharpless complex between titanium tetra(isopropoxide) and diethyl (*R*,*R*)- or (*S*,*S*)-tartrate, to yield the corresponding (methylsulfinyl) methylphosphonates **33** (76–82% *de*). The sulfoxide **33** was obtained in a diastereomerically pure form (>98% *de*, upon recrystallisation) and was shown to have an (R_P,S_S)-configuration (Scheme 11) [32].

Glucofuranosyl phosphinites are an interesting alternative to menthyl phosphinites. The reaction of chlorophosphines with menthol or borneol proceeds with low diastereoselectivity, and it is therefore necessary to separate the diastereoisomer mixtures by crystallization or chromatography. At the same time, the nucleophilic substitution at trivalent phosphorus of chlorophosphines **34a** or chlorophosphine oxides **34b** with (−)-1,2:5,6-diisopropylidene- or (−)-1,2:5,6-dicyclohexylidene-D-glucofuranose proceeds with good stereoselectivity to afford enantiomerically pure phosphinites **35a** or phosphinates **35b** in good yields

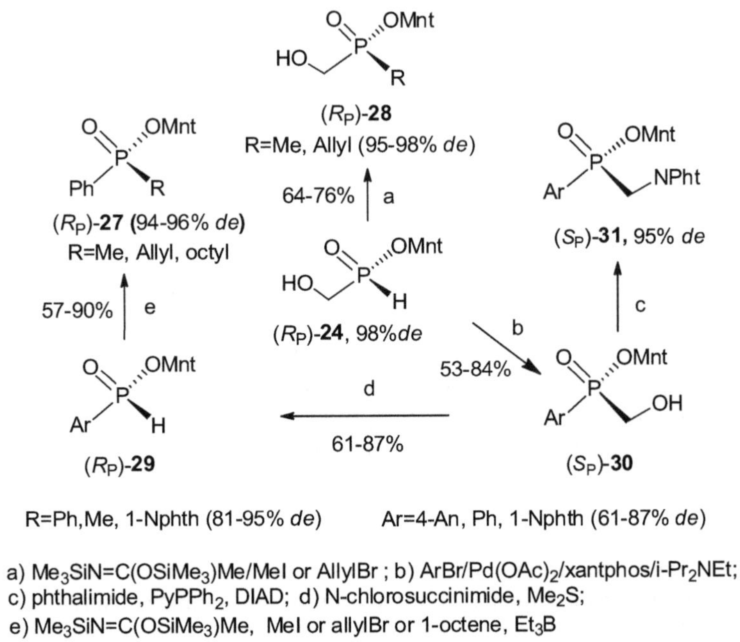

Scheme 10 *P*-Stereogenic *H*-phosphinate **24** as versatile *P*-stereogenic building block

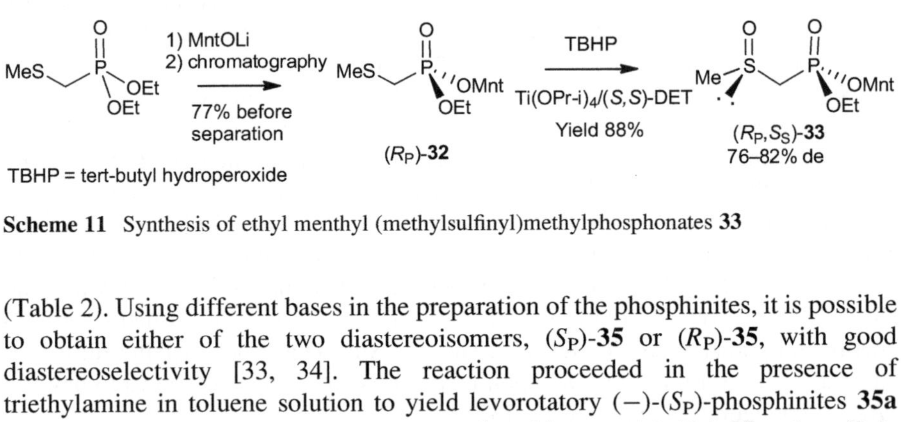

Scheme 11 Synthesis of ethyl menthyl (methylsulfinyl)methylphosphonates **33**

(Table 2). Using different bases in the preparation of the phosphinites, it is possible to obtain either of the two diastereoisomers, (S_P)-**35** or (R_P)-**35**, with good diastereoselectivity [33, 34]. The reaction proceeded in the presence of triethylamine in toluene solution to yield levorotatory (−)-(S_P)-phosphinites **35a** (or (S_P)-phosphinates). Dextrorotatory phosphinates (+)-(R_P)-**35a** (or (R_P)-phosphinates) were obtained by reaction in THF with pyridine as a base. The esters **35a,b** were converted into the corresponding tertiary phosphines (or phosphine oxides) (R_P)- or (S_P)-**36a,b** by reaction with organomagnesium in good yields (Scheme 12). The effect of the achiral base had been identified as the most important factor that determines the stereochemical outcome of the reaction. Indeed, the selectivity was reversed in the presence of pyridine. Hii et al. [34] reported that the diastereoisomeric ratio of **35** remained unchanged when the reactants were used in equimolar amounts, even at lower temperature.

Table 2 Reaction of racemic phenylphosphinic chlorides with glucofuranose (Scheme 12) [6, 33, 34]

Entry	R^1	CR^3_2	Base	Solvent	Yield (%)	(S_P):(R_P) **35**
1	Me	CMe_2	Et_3N	Toluene	70	90:10
2	Et	CMe_2	Et_3N	Toluene	70	96:4
3	*i*-Bu	CMe_2	Et_3N	Toluene	70	95:5
4	Bn	CMe_2	Et_3N	Toluene	75	>99:1
5	Bn	CMe_2	Py	THF	70	25:75
5	Me	c-C_5H_{10}	Et_3N	Toluene	75	95:5
6	Me	c-C_5H_{10}	Et_3N	THF	70	95:5
7	Me	c-C_5H_{10}	Et_3N	CH_2Cl_2	70	87:13
8	Me	c-C_5H_{10}	Py	THF	70	30:70
10	Et	c-C_5H_{10}	Et_3N	Toluene	93	93:7
11	Et	c-C_5H_{10}	Py	THF	94	30:70
12	*i*-Pr	c-C_5H_{10}	Et_3N	Toluene	92	86:14
13	*i*-Pr	c-C_5H_{10}	Py	THF	90	40:60
14	Bn	c-C_5H_{10}	Et_3N	Toluene	95	90:10
15	Bn	c-C_5H_{10}	Py	THF	95	40:60
16	*o*-An	c-C_5H_{10}	Et_3N	Toluene	93	30:70
as17	*o*-An	c-C_5H_{10}	Py	THF	94	55:45
18	1-Nphth	c-C_5H_{10}	Et_3N	Toluene	87	40:60
19	1-Nphth	c-C_5H_{10}	Py	THF	83	55:45

The protected polyhydroxylated 1,2-oxaphosphinane **39** was prepared by a two-step sequence (phenyl-*H*-phosphinate addition on protected mannofuranose followed by intramolecular transesterification) on a gram scale. Deprotection of the di-isopropylidene derivative **39** using acidic cation-exchange resin afforded the free hydroxy organophosphorus heterocycle **40** analogous to C-arylglycosides. The X-ray analysis allowed the absolute configuration of the new created asymmetric centers of the diastereoisomer **39** to be assigned. Recrystallization in ethanol afforded the pure fully deprotected arylphosphinosugar **40**. The phosphinosugar **40** analogous to C-arylated heptopyranose shows a *B* boat structure with P_2(S), C_3(R) absolute configuration (Scheme 13) [35].

An improved strategy for the synthesis of *P*-chiral gluco- and manno-phosphonite boranes **41–43** was developed on the basis of the addition of diethyl phosphonite borane to a glucal-derived aldehyde followed by a cyclization coupled with an ethyl/methyl exchange (Scheme 14) [36].

Montchamp has also described the phosphorylation of the (*R*)-1-(2-naphthyl) ethanol with formation of enantio-enriched *H*-phosphinic acid , precursor to a variety of organophosphorus compounds [31].

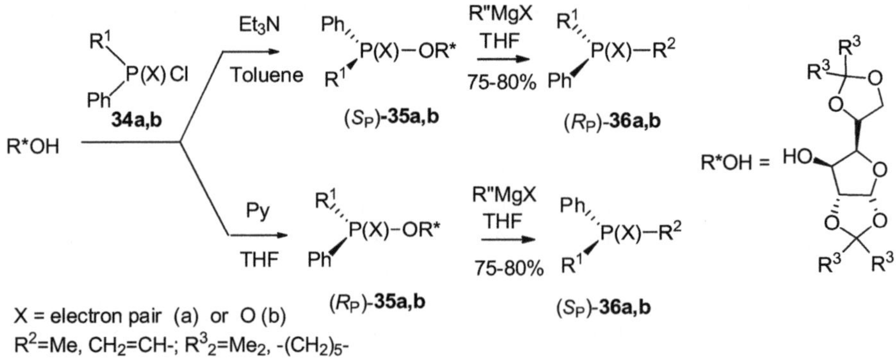

Scheme 12 Glucofuranosyl method for the synthesis of *P*-chiral tertiary phosphine oxides **36**

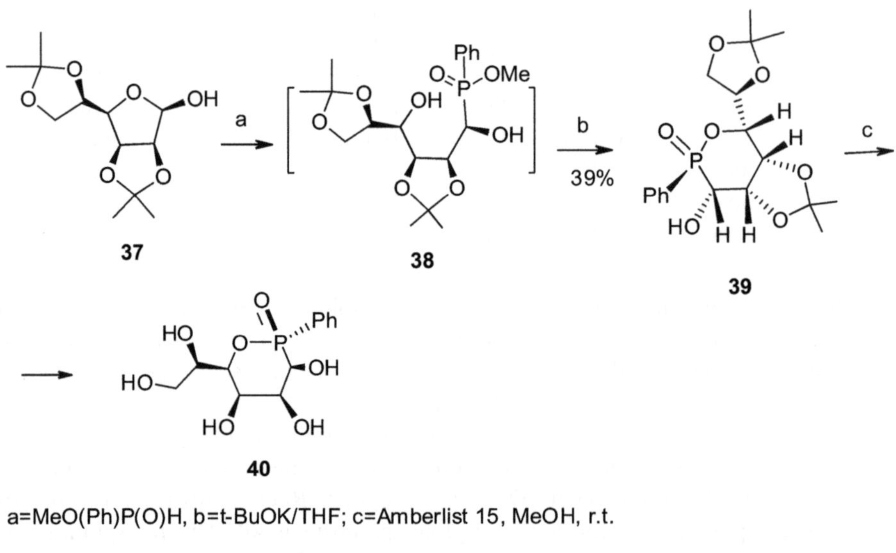

Scheme 13 Preparation of phosphinic analogous to C-arylglycosides

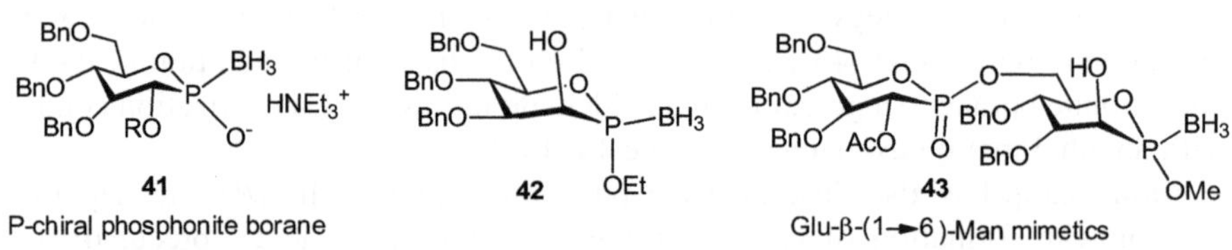

Scheme 14 *P*-Chiral glucophosphonite boranes and phostone–phostone dimer

3.2 *Optically Active Amines as Chiral Auxiliaries*

The unsymmetrically substituted chlorophosphines and chlorophosphine oxides react with chiral primary amines (aminoacid esters or 1-methylbenzylamine) with asymmetric induction at the phosphorus atom (85–90% *de*) to afford enantioenriched aminophosphines **46a–d** and **48a–d** (R=*t*-Bu, R′=Ph), which, after recrystallization, were obtained with 100% diastereoisomeric purity. The absolute configuration of chiral compounds **46** and **48** was established by X-ray analysis and chemical extrapolation. It was found that (*S*)-1-methylbenzylamine generates the (*R*)-configuration at the phosphorus atom, while, in contrast, (*R*)-1-methylbenzylamine gives rise to the (*S*)-configuration. The stereoselectivity of the reaction depends on the reaction conditions, the nature of the base (Et_3N, *i*-Pr_2NEt, DABCO), solvent (toluene, benzene, THF, ether), temperature, and ratio of starting reagents (Scheme 15) [37–40].

The mechanism of asymmetric induction at a trivalent phosphorus atom was explained by formation of pentacoordinated transition state **b**, pseudo-rotation, and exchange of ligands at a pentacoordinated phosphorus atom, resulting in the thermodynamically more stable diastereoisomer. The effect of the reaction conditions on the diastereoisomeric ratio of products **c** shows thermodynamic control. For example, a decrease of the reaction temperature reduces the stereoselectivity, which is impossible with kinetic control, because the temperature lowering leads to a deceleration of the equilibrium establishment of pentacoordinated intermediate complex (Scheme 16) [33].

Aminophosphines **46–49** are useful starting compounds for the preparation of enantiopure compounds **4c** and **50–56** (Scheme 17) [37–40]. The treatment of aminophosphines **48** with borane in THF leads to the formation of the stable

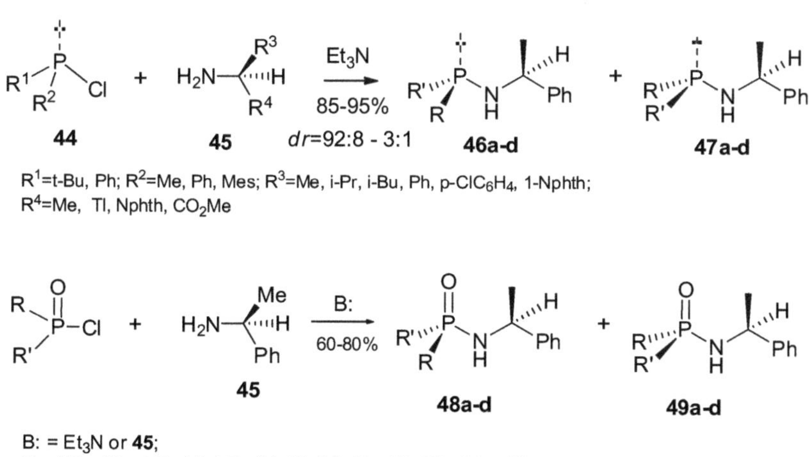

Scheme 15 The reactions of unsymmetrically substituted chlorophosphines and chlorophosphine oxides with chiral primary amines

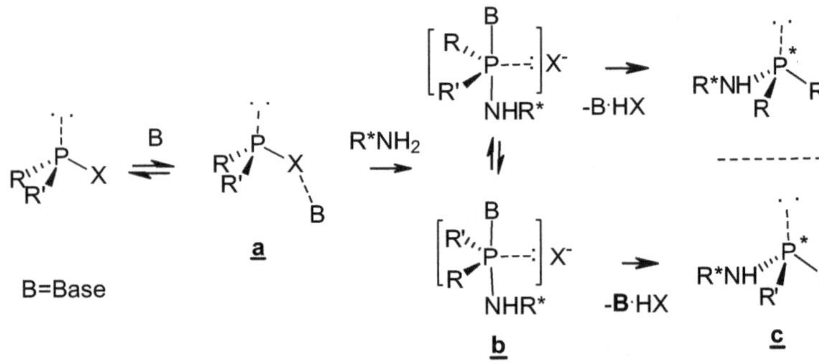

Scheme 16 The mechanism of asymmetric induction at the trivalent phosphorus atom

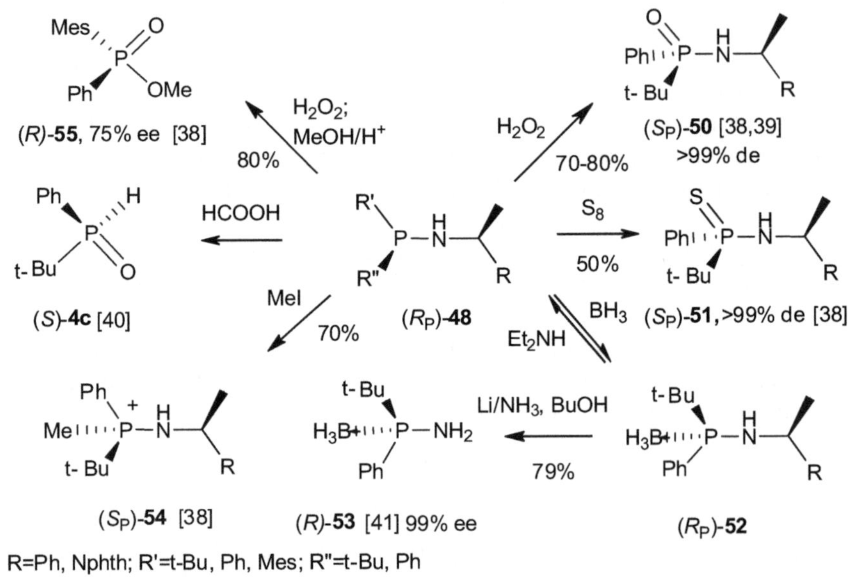

Scheme 17 Aminophosphines **48** as chiral starting reactants

crystalline adduct **52** in quantitative yield. The BH_3 group of the phosphine boranes complex **52** can be easily removed by treatment with diethylamine to furnish the enantiopure aminophosphine (R_P)-**48** with almost quantitative yield. The amino group of compounds was replaced by a methoxy group at the reflux in methanol containing sulfuric acid, with formation of **55**. The acidolysis of **48** afforded enantiomers of *tert*-butylphenylphosphine oxides **4c** [40]. The deprotection of **52** was attained by treatment with lithium amide, leading to the formation of aminophosphine borane **53** (Scheme 17) [41].

The enantiopure aminophosphines **56** were applied as building blocks for the construction of chiral ligands **58** and **59**. The reactivity of the amino group should permit further functionalization which can result in novel structures that preserve

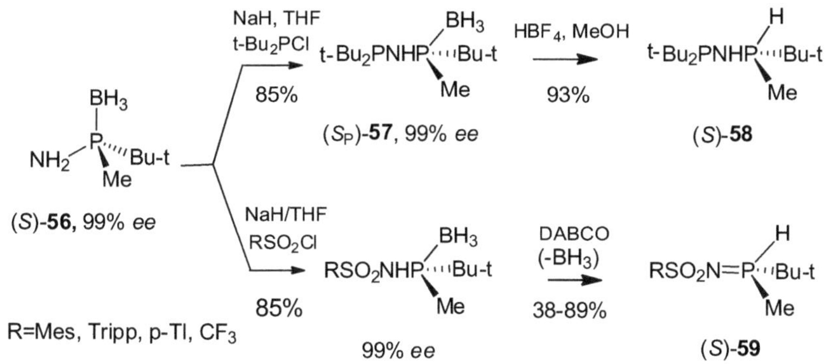

Scheme 18 Synthesis of PNP and PNS ligands **58** and **59**

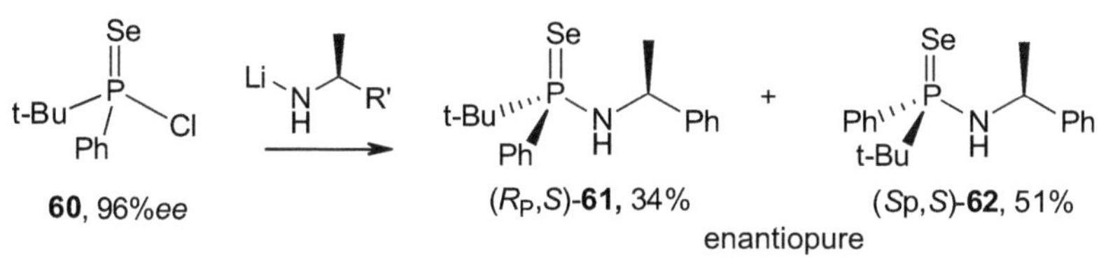

Scheme 19 Preparation of *P*-chiral phosphinoselenoic amides **61** and **62**

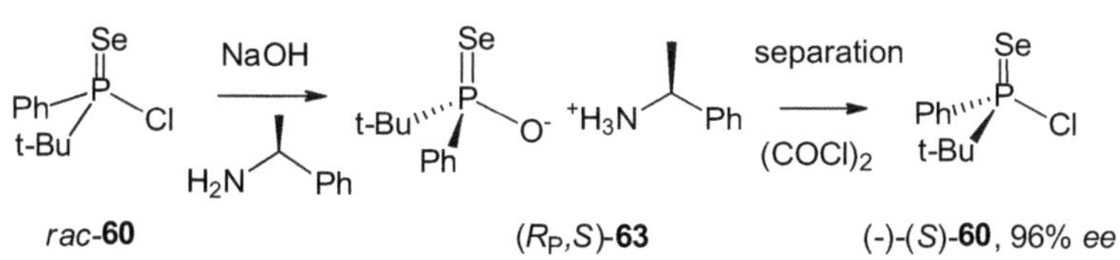

Scheme 20 The preparation of phosphinoselenothioic acid salt (R_P,*S*)-**63** and chloride (*S*)-**60**

the original *P*-chirality. A straightforward application of these *P*-aminophosphines is the preparation of chiral aminodiphosphine (P-N-P) ligands. Thus, Riera and Verdaguer using the aminophosphines **56** have obtained the P-N-P- and P-N-S-ligands **58** and **59**, which were used in asymmetric catalytic hydrogenation (Scheme 18) [41–44].

This method was also used for the preparation of *P*-chiral phosphinoselenoic amides (R_P,*S*)-**61** and (S_P,*S*)-**62** (Scheme 19). Enantiomerically pure amides **61,62** were synthesized by reaction of racemic phosphinoselenoic chlorides *rac*-**60** with optically-active lithium amides. Two diastereomers of (R_P,*S*)-**61** and (S_P,*S*)-**62** were formed in a nearly equal ratio in high yields, and the two diastereomers were successfully separated by column chromatography on silica gel. The absolute configuration of phosphinoselenoic amide (R_P,*S*)-**61** was determined by X-ray analysis. Using this reaction, enantiomerically pure salts of phosphinoselenoic acid **63** and *P*-chiral phosphinoselenoic chlorides (−)-(*S*)-**60** were prepared (Scheme 20) [45, 46].

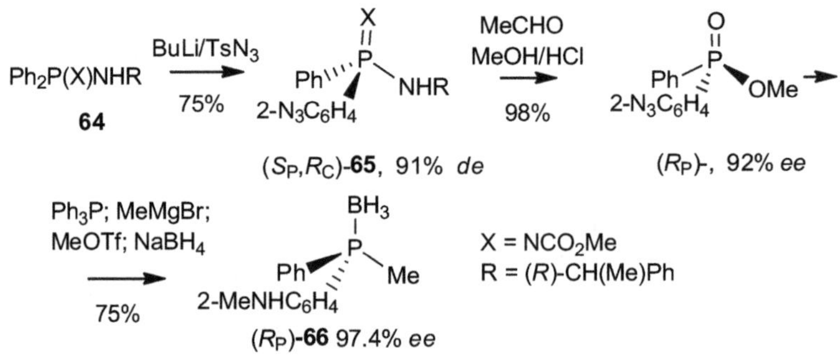

Scheme 21 Synthesis of *P*-chiral compounds **66** via *o*-desymmetrized aminophosphazenes

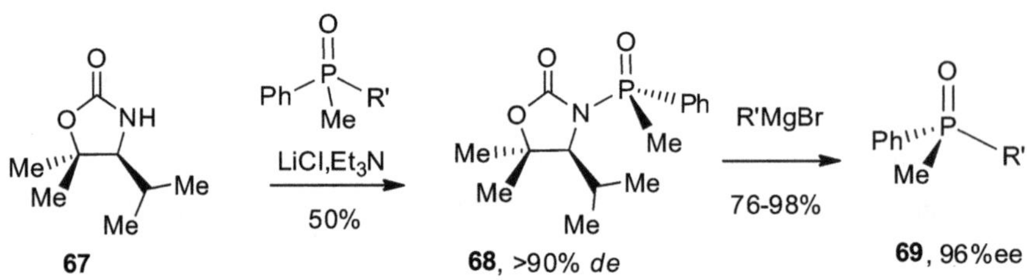

Scheme 22 Synthesis of chiral diaryl-methyl and alkyl-methylphenyl phosphine oxides **69**

Ortiz [47] recently described the *ortho*-directed lithiation of *P*,*P*-diphenylaminophosphazenes **64** followed by electrophilic quenching as an efficient process for the preparation of *P*-chiral *ortho*-functionalized *P*-chirogenic amidophosphinates **65** in good yields and diastereoselectivities. The usefulness of the method was shown with the preparation under mild reaction conditions of a variety of functionalized *P*-chiral compounds **66** in high yield and excellent stereoselectivity, including phosphinic esters, amides, thioamides, phosphine oxides, and (2-aminophenyl)phosphine boranes (Scheme 21).

A range of chiral diaryl-methyl and alkyl-methylphenyl phosphine oxides **69** were synthesized under mild conditions in good yield and excellent enantioselectivity (>98:2 *er*) using the *N*-phosphinoyl oxazolidinone **68** derived from L-valine and methylphenyl phosphinic chloride. The use of lithium chloride and triethylamine allowed the easily phosphorylation of oxazolidinone **67**, leading to the *N*-phosphinoyl oxazolidinone **68** in excellent diastereoselectivity and good yield after isolation of the major diastereoisomer by column chromatography. At the same time, using LiHMDS or MeMgBr for the deprotonation of the oxazolidinone resulted in lower diastereoselectivity of reaction. The methodology involves the highly stereoselective formation of *P*-chiral oxazolidinones **68** which were converted to the interesting phosphine oxides **69** by reaction with Grignard reagents (Scheme 22) [47, 48].

Han [49] reported a diastereoselective method for the synthesis of *P*-chiral phosphine oxides. The sequential nucleophilic substitution of 1,3,2-

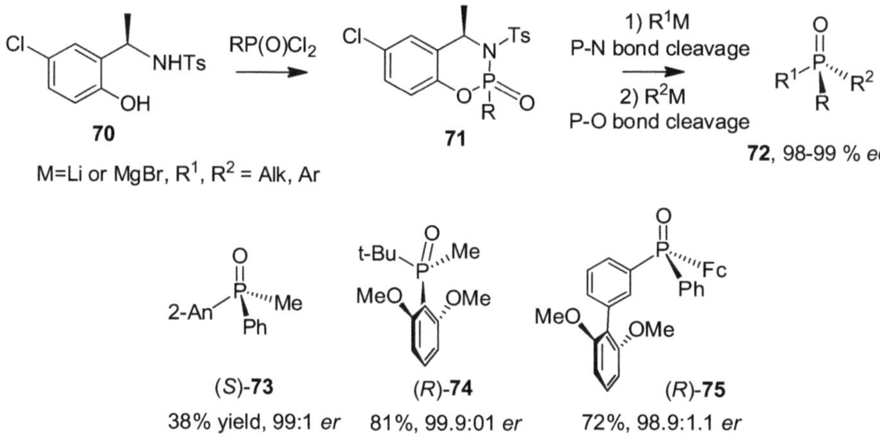

Scheme 23 Synthesis of *P*-chiral phosphine oxides

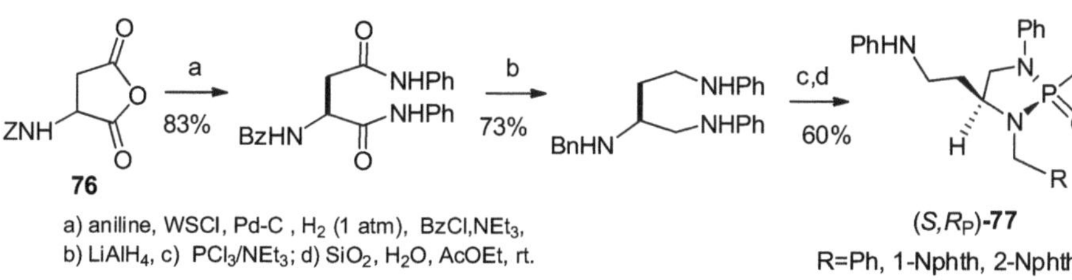

Scheme 24 Synthesis of (S,R_P)-Ph-DIAPHOX **77**

benzoxazaphosphinine-2-oxide **71**, bearing P–N and P–O bonds with metallorganics led to the formation of *P*-chiral phosphine oxides **72**. Cleavage of the P–O bond in **71** by treatment with MeMgCl led to the formation of (*S*)-**73** with inversion of configuration at P with 99:1 *er*. Diastereomerically pure (R_P)-7**4** was obtained by crystallization, and its absolute configuration was confirmed by X-ray crystallographic analysis. The chiral versions of Buchwald-type ligands **74** were prepared effectively using this method. The method was also applied for the asymmetric synthesis of the bulky *P*-chiral phosphine oxides **75**, which are versatile intermediates which can be converted into many important *P*-chiral ligands (Scheme 23).

Nemoto and Hamada [50] has described the development of a new class of chiral phosphorus ligand – aspartic acid-derived *P*-chirogenic diaminophosphine oxides, DIAPHOXs – and their application to several Pd-catalyzed asymmetric allylic substitution reactions. Pd-catalyzed asymmetric allylic alkylation was initially examined in detail using diaminophosphine oxides **77**, resulting in the highly enantioselective construction of quaternary stereocenters. With the use of the Pd-DIAPHOX catalyst system, asymmetric allylic alkylation, asymmetric allylic amination, and enantioselective construction of quaternary carbons were achieved with high *ee* (up to 97–99% in many cases) (Scheme 24).

3.3 Ephedrine as Chiral Auxiliary (Jugé–Stephan Method)

Jugé developed a powerful method (Jugé–Stephan method) [1, 51] for the preparation of *P*-stereogenic phosphines based on the use of ephedrine as a chiral auxiliary. The key reactants in this methodology are 1,3,2-oxazaphospholidine boranes **78**, prepared by a one-pot reaction from bis(diethylamino)phenylphosphine and (−)-ephedrine, followed by protection with BH_3. The cyclization of the (−)-ephedrine takes place stereoselectively, with preferential formation of the (R_P)-diastereoisomer in 90% *de* [52, 53]. The absolute configuration at the phosphorus atom has been determined by chemical correlations and NMR analysis, and proved by X-ray analysis [54]. Oxazaphospholidines react readily with electrophiles or nucleophiles to provide various chiral phosphorus compounds. Enantiomeric antipodes of tertiary phosphines (S_P)-**79** and (R_P)-**81** were obtained from (+)- or (−)-ephedrine, as shown in Scheme 25. The configuration at the *P*-atom is controlled by the configuration at the Ph-substituted C_1 of (+)-pseudoephedrine or (−)-ephedrine, respectively. This was confirmed by X-ray crystal-structure analyses of two intermediate compounds in the synthetic route to the chiral triarylborane-phosphine adducts [54].

Oxazaphospholidine boranes **82** react regio- and stereoselectively with alkyl lithiums or aryl lithiums in THF at −78°C, with formation of acyclic phosphinite boranes **83**. Various substituents R^1=*n*-alkyl, *c*-alkyl, aryl, or ferrocenyl were introduced into aminophosphine boranes **82** in high yield (93–97%) and with high diastereoselectivity (*dr* >98:2). The reaction proceeded with retention of configuration at phosphorus. Recrystallization of aminophosphine boranes **83** in propanol gave the diastereoisomerically pure products [53]. Acid methanolysis of aminophosphine boranes **83** led to the formation of phosphinite boranes **84** with inversion of configuration on the *P*-center to yield the compounds **84** in high

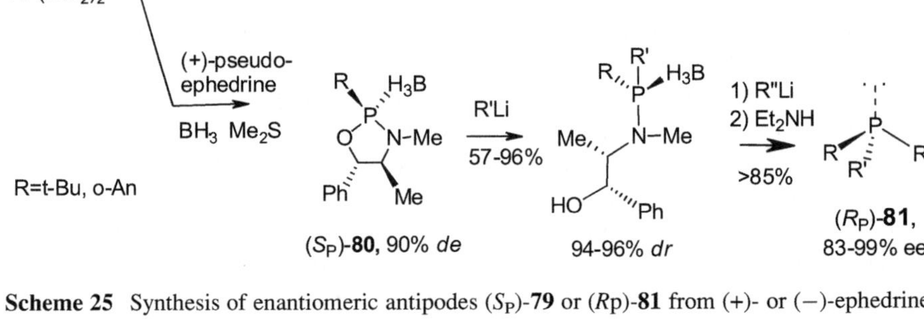

Scheme 25 Synthesis of enantiomeric antipodes (S_P)-**79** or (Rp)-**81** from (+)- or (−)-ephedrine

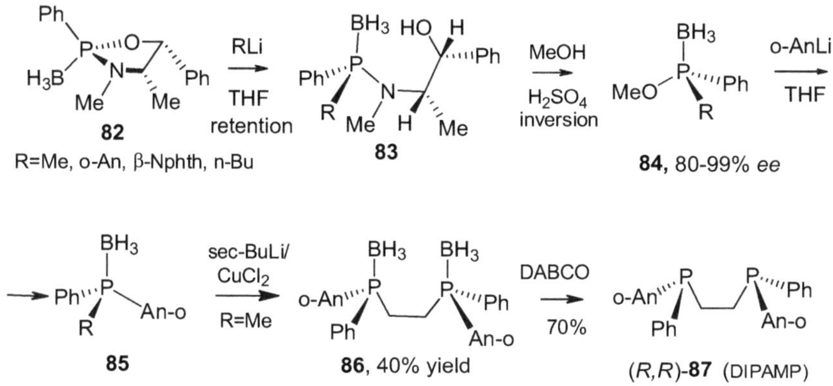

Scheme 26 Syntheses of PAMP·BH_3 **86** and DIPAMP **87** from oxazaphospholidine borane **82**

enantiomeric purity. The phosphinite boranes **84** reacted with organo-lithium compounds at −70°C with inversion of configuration on the *P*-center and with formation of tertiary phosphine borane **85** in good chemical yields and with very good enantiomeric excesses (85–100%). The phosphine borane **85** (R=Me) was coupled to the corresponding diphosphine diborane **86**, which was decomplexed by DABCO to afford the optically pure (*R*,*R*)-DIPAMP **87** in good yield with retention of absolute configuration (Scheme 26) [1, 5, 53, 55].

In addition to the "Jugé−Stephan method" [1, 5],which is based on the nucleophilic ring-opening of ephedrine-derived oxazaphospholidine boranes **82** by reaction with alkyl lithium reagents, Bickelhaupt et al. [56] recently reported the stereodivergent ring-opening of 2-phenyl oxazaphospholidines **88** (R=H or Me) with alkyl lithium reagents. N–H oxazaphospholidines **88** derived from both (+)-*cis*-1-amino-2-indanol and (−)-norephedrine provided inversion products with high stereoselectivity. In contrast, N-Me oxazaphospholidines **88** yield ring-opening products with retention of configuration at the P center. As a result, from a single amino alcohol auxiliary, both enantiomers of key *P*-stereogenic intermediates **89** were synthesized. The acid-catalyzed methanolysis of compounds **89** proceeded with inversion at the P center to give the (*S*)- or (*R*)-methyl phosphinites **90** in very good yields and with high *ee* (Scheme 27).

The *P*-chirogenic aminophosphane-phosphinite ligands (AMP*P) **92**, **93**, supported on the upper rim of a calix[4]arene moiety, were synthesized in two steps using the ephedrine methodology. Ligand **92** was used for the preparation of the corresponding rhodium complex [Rh(COD)-(AMP*P)]BF_4, which was tested for asymmetric catalyzed hydrogenation of various substrates with excellent enantioselectivities up to 98%. For example, the asymmetric hydrogenation of methyl α-acetamidocinnamates catalyzed with these Rh complexes yielded (*S*)-phenylalanine derivatives with 99% *ee*. Investigation of modified *P*-chirogenic aminophosphane-phosphinite ligands **93**, bearing similar substituent on the *P*-chirogenic aminophosphane unit, demonstrates that the calix[4]arene substituent

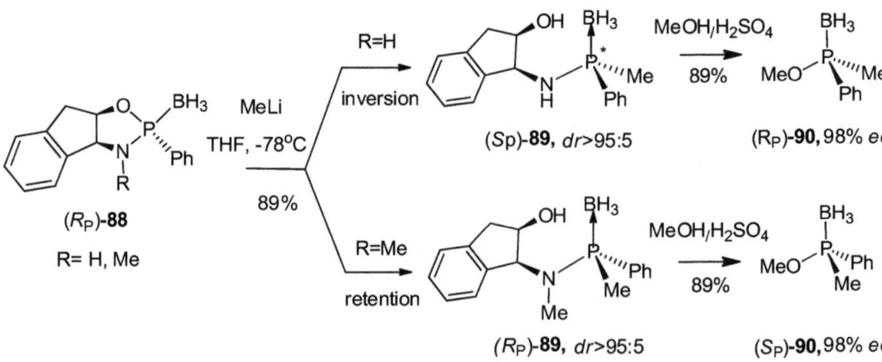

Scheme 27 Stereodivergent ring-opening of 2-phenyl oxazaphospholidines **88**

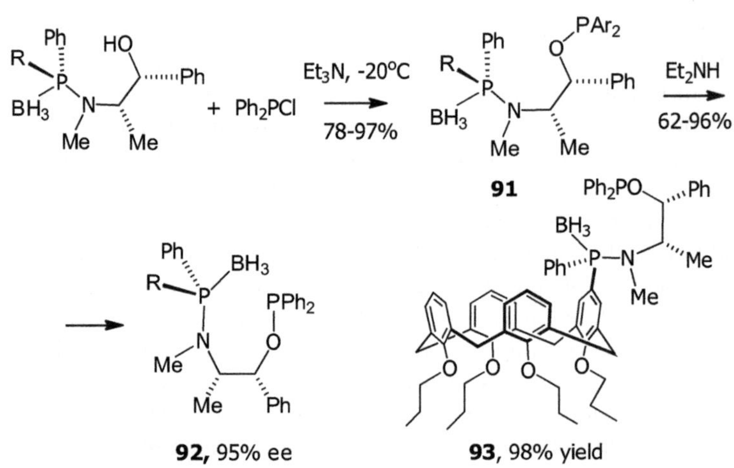

Scheme 28 Preparation of ligands **92** and **93**

of the aminophosphane moiety plays a major role in the better asymmetric induction (Scheme 28) [57].

The stereoselective synthesis of *P*-chirogenic chlorophosphine boranes **95** was achieved by HCl acidolysis of the corresponding aminophosphine boranes **94**. The reaction resulted in P–N bond cleavage with inversion of the configuration at the phosphorus center, leading to the chlorophosphine boranes **95** with high to excellent enantiomeric purities (80–99% *ee*). The enantiomeric purity of the chlorophosphine boranes **95** was determined by reaction with organolithium reagents (Scheme 29) [49, 52, 58].

The stereoselective synthesis of the (*R*,*R*)- (or (*S*,*S*)-ligands) **99** was performed in several steps using the ephedrine methodology with (+)- or (−)-ephedrine, respectively. The key step of the synthesis is the methano bridge formation by reaction of the carbanion derived methyl phosphine borane **98** with the chloro

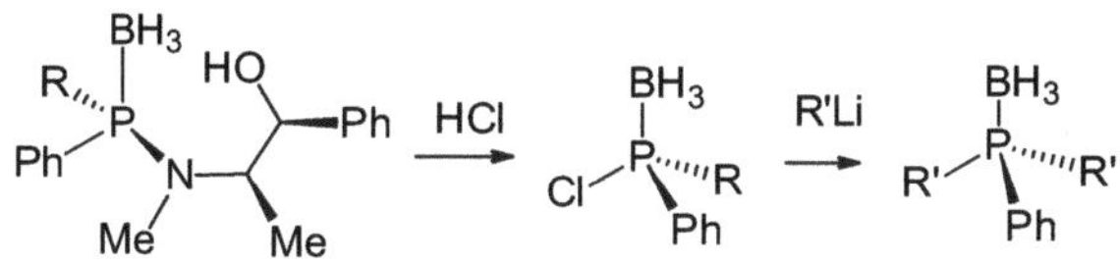

94 from (+)-ephedrine; **95**; **96**

R=Me	R′=o-An	85%	90% ee	(R)
R=o-An	R′=Me	99%	98% ee	(S)
R=o-Tol	R′=Me	87%	98% ee	(S)
R=2-Nphth	R′=Me	61%	85% ee	(S)
R=o-PhPh	R′=Me	45%	99% ee	(S)
R=c-C_6H_{11}	-	74%	80% ee	(R)

Scheme 29 Stereoselective synthesis of *P*-chirogenic chlorophosphine boranes **95**

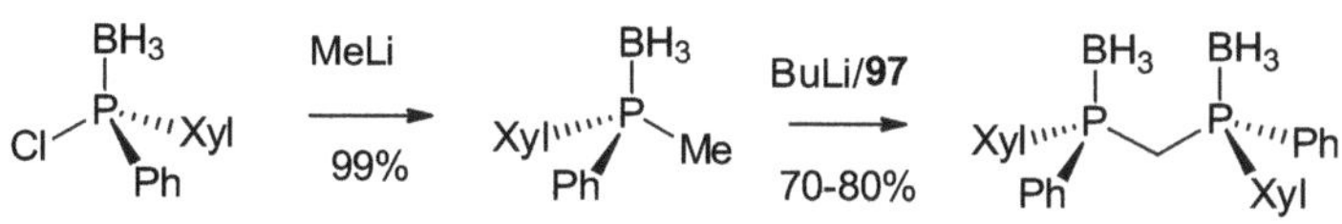

(*S*)-**97**, Xyl=3,5-$Me_2C_6H_3$; (*R,R*)-**98**, 99% *ee*; **99**, (*R,R*)-DPPM 99% *ee*

Scheme 30 Synthesis of the DPPM ligands **99**

phosphine borane **97**. The reaction of the (*S*)-chloro-phenyl-*m*-xylylphosphine borane **97** with MeLi afforded the corresponding (*R*)-methylphosphine borane **95** with inversion of configuration at the *P*-center. After deprotonation of the methylphosphine borane **98** with *n*-BuLi, the reaction with the (*S*)-chloro phosphine borane **97** afforded the protected diphosphine diborane **99** in good yield (Scheme 30) [59].

The chlorophosphine boranes **97** are efficient starting reagents for the synthesis of various classes of *P*-chiral phosphorus compounds. Reactions of chlorophosphine boranes **97** with nucleophiles, such as carbanions, phenoxides, phenylthiolates, or amides, leads to the formation of corresponding organophosphorus compounds **100–103** in yields of 53–99% and with up to 99% *ee*. This method was also used for the preparation of various classes of symmetric and asymmetric *P*-chiral ligands useful for asymmetric reactions, catalyzed by complexes of transition metals (Scheme 31) [52, 60, 61].

The ephedrine/PCl_3 route to *P*-chiral mono- and diphosphines gives easy access to a variety of structures. In this way, the ferrocenyl, 1- and 2-adamantyl *tert*-butylphosphines were formed as borane complexes. For example, in this way Colby and Jamisson [62] synthesized a number of tertiary phosphines **105a–h** in good yields (50–90%) and with enantiomeric purity between 80 and 98% *ee*. The monodentate ferrocenyl phosphines **105a** and **105e** were evaluated as ligands in asymmetric catalytic reductive coupling of alkynes and aldehydes to give chiral

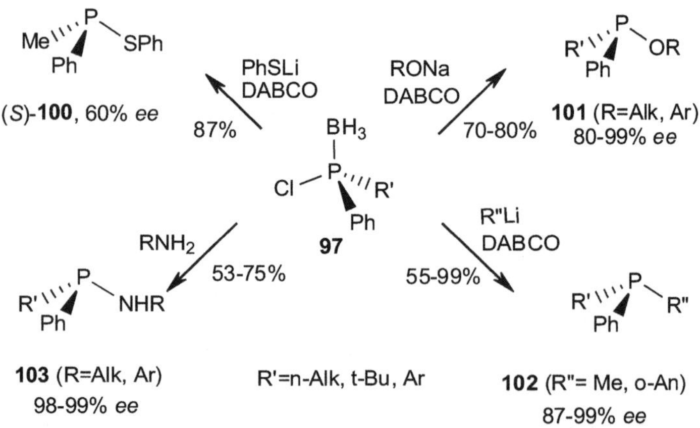

Scheme 31 Chlorophosphine boranes **97** as chiral starting reagents

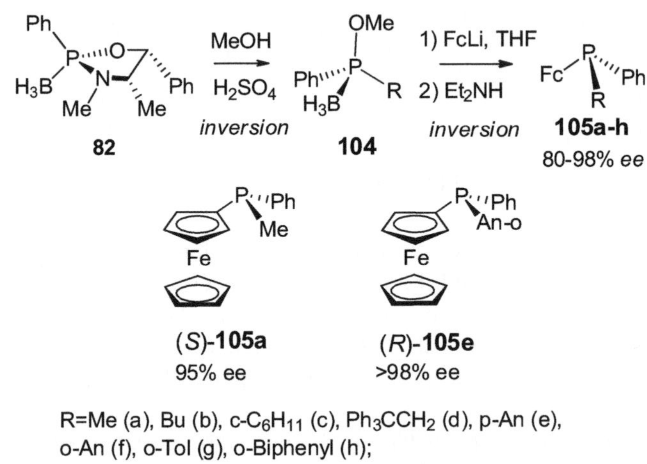

Scheme 32 Monodentate ferrocenyl phosphine ligands **105**

allylic alcohols with good enantioselectivity in many cases and with complete (*E*)-selectivity in all cases (>98:2) (Scheme 32).

Both enantiomers of 2,3-bis(*tert*-butylmethylphosphino)-quinoxaline (QuinoxP*) **109**, 1,2-bis(*tert*-butylmethylphosphino)benzene (BenzP*) **110**, and 1,2-bis(*tert*-butylmethylphosphino)-4,5-(methylenedioxy)-benzene (DioxyBenzP*) **111** were prepared as stereochemically pure compounds in short steps from enantiopure (*S*)- and (*R*)-*tert*-butylmethylphosphine boranes **106** as shown in Scheme 33 [63].

The overall retention of configuration was rationalized by a reaction sequence involving attack of ferrocenyllithium (FcLi) on the electrophilic phosphorus followed by pseudo-rotation and termination by chloride elimination [64]. *Ortho*-lithiation of oxazaphospholidine oxide **113** was carried out with diastereoselectivity

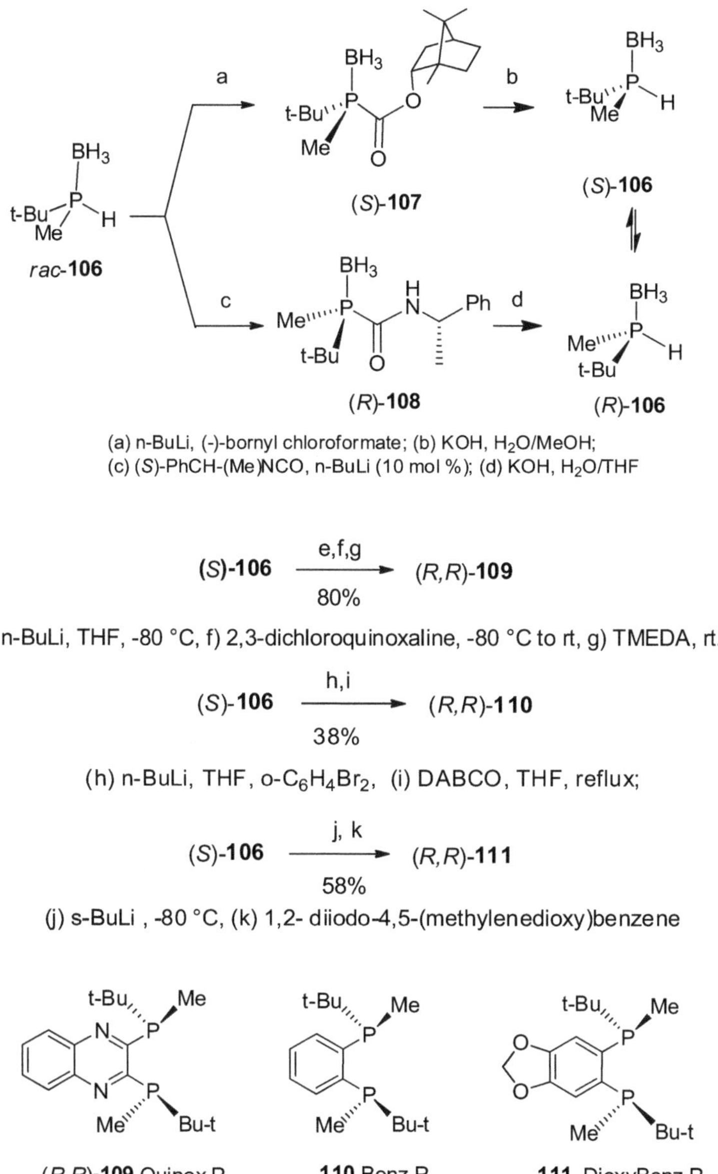

Scheme 33 Synthesis of *P*-chiral ligands **109–111**

of >99%, affording a new and efficient way for introducing planar chirality into the ferrocene backbone. Various electrophiles were used, showing the wide applicability of the new methodology and its potential to generate ligands **114** for use in asymmetric catalysis (Scheme 34).

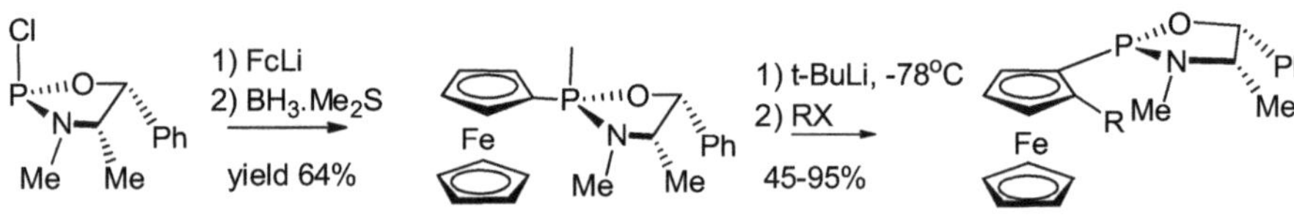

(R_P)-**112** (R_P)-**113**, 99% *de* **114**, 99% *de*

RX= MeI, I_2, Me_3SiCl, Ph_2CO, $B(OMe)_3$, Ph_2PCl, Cy_2PCl, $(p\text{-}CF_3C_6H_4P)_2PCl$, $(p\text{-}An)_2PCl$

Scheme 34 Ferrocene phosphine ligands **114**

The *P*-stereogenic diphosphine ligands **118** were prepared with high diastereo- and enantiomeric purity by a multistep asymmetric synthesis starting from the oxazaphospholidine boranes **82**, via the phosphinite boranes **116**, selective lithiation of the enantiomerically pure phosphine borane (*S*)-**117**, and coupling with Me_2SiCl_2. The Rh, Ru, and Pd complexes bearing diphosphine ligands **118** were prepared and tested as catalyst precursors in hydrogenation reactions with enantioselectivity up to 97.7% *ee*. The Pd complex **119** was tested as catalyst in allylic alkylation reactions (Scheme 35) [21, 63, 65].

The synthesis of *P*-chiral diarylphosphinocarboxylic acids **120** was achieved with excellent enantiopurity starting from the oxazaphospholidine boranes **82**. Amido- and amino-diphosphine ligands **121**, containing an L-proline backbone, were also derived from **82**. The catalytic activities of the ligands **121** were evaluated in the Pd-catalyzed allylic alkylation reaction of 1,3-diphenylpropenyl acetate (Scheme 36) [66].

Buono developed a very effective synthesis of chiral tertiary phosphine oxide starting from oxazaphospholidines. The enantiomerically pure oxazaphospholidine (R_P)-**122** was prepared from $PhP(NMe_2)_2$ and (*S*)-(+)-prolinol. Subsequent treatment of **122** with a variety of acids followed by hydrolysis gave both enantiomers of *tert*-butylphenylphosphine oxide **4c**. It was found that the acid controlled the stereochemistry of the enantiomer obtained. By using acids of high acidity or Amberlyst 15 resin, (+)-(*R*)-**4c** was obtained with good yields and enantioselectivities. When acids of low acidity were used, (−)-(*S*)-**4c** was the preferred enantiomer. For example, *P*-toluenesulfonic acid (PTSA) afforded (*R*)-**4c** in 88% yield and 91% *ee*. After a recrystallization the optically pure compound (*R*)-(+)-**4c** was obtained with >99% *ee* (Scheme 37) [67].

The opening of oxazaphospholidine rings (*R*p)-**122** with *tert*-butyllithium occurred diastereoselectively with retention of absolute configuration on the phosphorus atom, affording the borane complex of aminophosphine (R_P)-**124** [67, 68]. The reaction possibly proceeds via formation of chiral $\sigma^2\lambda^2$-phosphenium cation **126**, which was obtained from (*S*c)-chlorophosphine **125** and then isolated as borane complex **127** (Scheme 38).

Buono et al. [69, 70] have developed the selective syntheses of *P*-chiral diazaphospholidine and triaminophosphine ligands. For example, QUIPHOS-PN_5 **128**, a stable P,N ligand with a stereogenic phosphorus center, was synthesized in

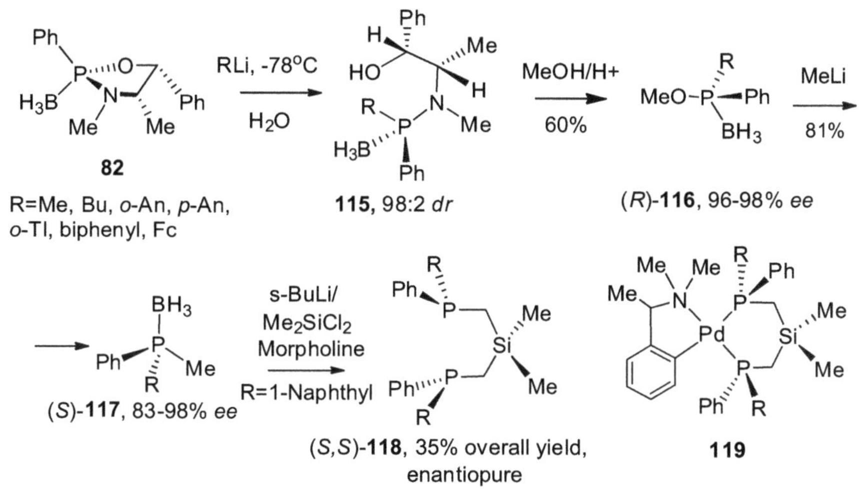

Scheme 35 Preparation of diphosphine ligands **118**

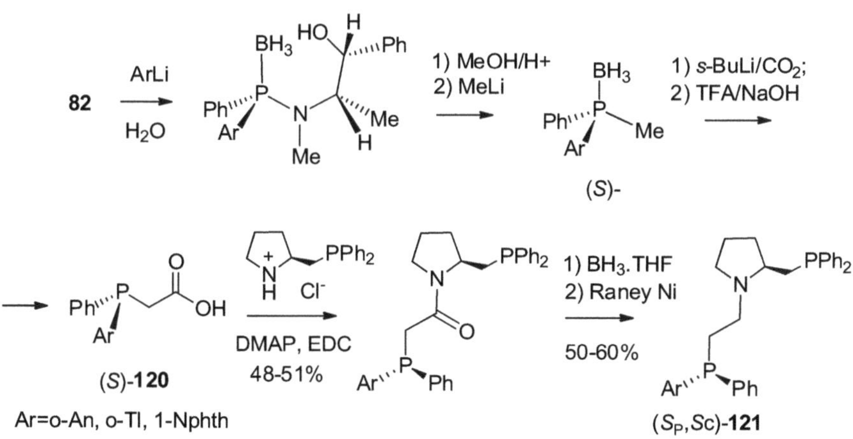

Scheme 36 Synthesis of *P*-chiral amido and amino diphosphines

two steps from 8-bromoquinoline in 61% overall yield. The structure of compound **128** was confirmed by X-ray analysis of a palladium(II) complex **129** bearing this ligand. Pd complex **129** was obtained in quantitative yield by mixing an equimolar amount of $PdCl_2(CH_3CN)_2$ and ligand **128** in methylene chloride. The synthesis, structural, and conformational studies of *P*-chiral triaminophosphines **130**, which feature an indolidine and a 1,2,3,4-tetrahydroquinolidine pattern, was also reported (Scheme 39).

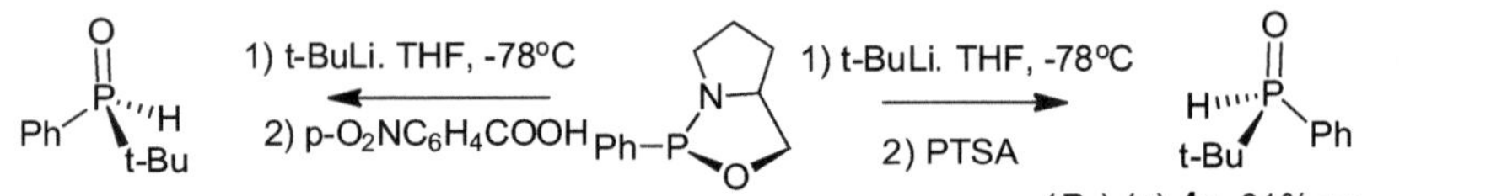

(S_P)-(-)-**4c**, 85% *ee* (R_P)-**122** (R_P)-(+)-**4c**, 91% *ee*, 99% *ee* after recrystallization

Acid	Yield (%)	ee(%)	Config
Amberlyst 15	70	74	(R)-(+)
Triflic acid	20	87	(R)-(+)
PTSA	88	91	(R)-(+)
Formic acid	77	77	(S)-(-)
Acetic acid	84	71	(S)-(-)
P-Nitrobenzoic acid	72	85	(S)-(-)

Scheme 37 Preparation of (*R*)- and (*S*)-*tert*-butylphenylphosphine oxides **4c** from oxazaphospholidine (R_P)-**122**

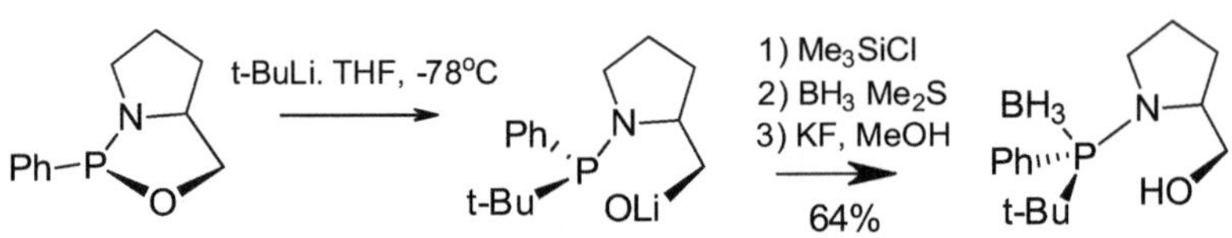

(R_P)-**122** (R_P)-**123** (R_P)-**124**

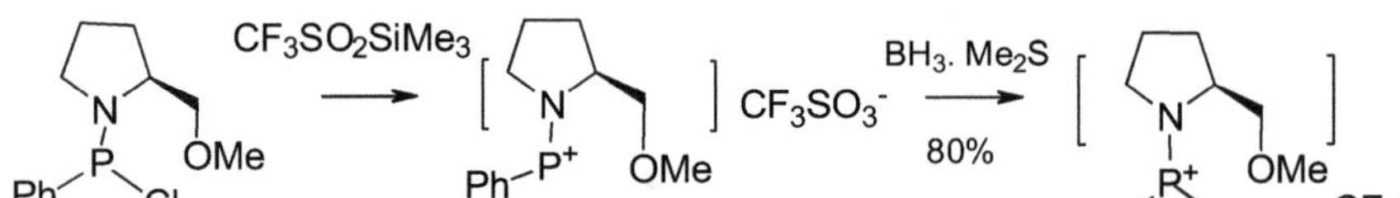

δ_P +166.7 ppm

125 **126** **127**, δ_P +134.2; +138.3ppm (*dr* 9:1)

Scheme 38 Chiral $\sigma^2\lambda^2$-phosphenium cation **127**

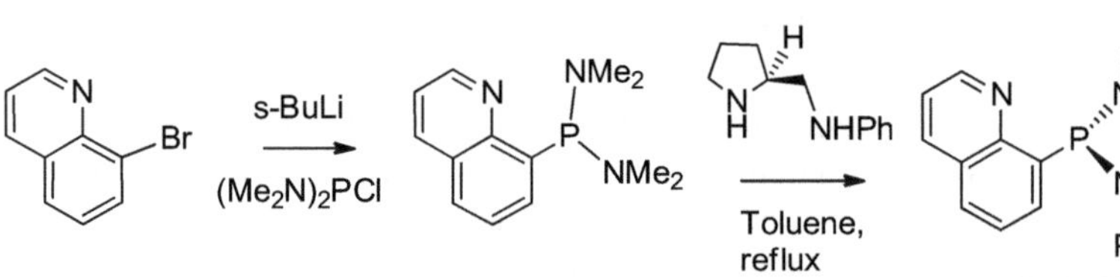

(S_P)-**128**, 61%, enantiopure

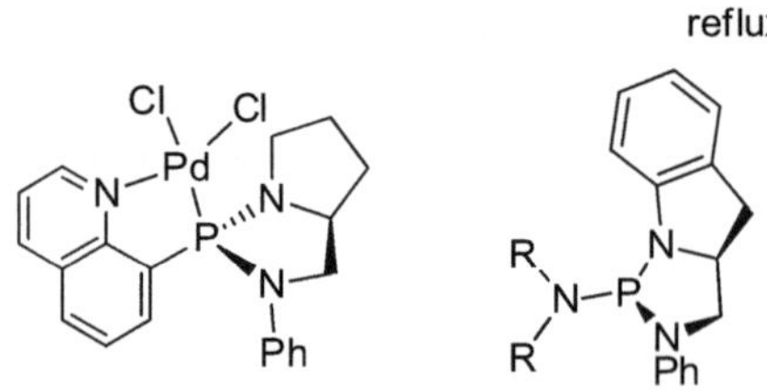

(S_P)-**129**, mp 253° **130**

Scheme 39 *P,N*-Ligand QUIPHOS-PN$_5$ **129**

4 Asymmetric Oxidation of P(III) Compounds

One of attractive methods for preparation of enantiopure phosphine oxides is asymmetric oxidation of phosphines. Developing oxidation methods that involve simply treating a phosphine with an oxidizing reagent/complex to form the enantiopure phosphine oxide has been, and still is, the goal of several research groups.

Thus the enantioselective oxidation of *P*-racemic tertiary phosphines and aminophosphines was successfully attained by treatment with tetrahalomethanes and an alcohol. Oxidation of *N*-phosphorylated amino acids by these reactants proceeded diastereoselectively with formation of aminophosphinates **131** in 80–85% yields and with 50–98% *de*. The compounds were purified by crystallization and obtained in optically pure form (>99% *de*) (Scheme 40) [33, 39].

Gilheany et al. [71–76] reported the oxidation of tertiary phosphine **133** with polyhaloalkanes in the presence of chiral proton-donating auxiliary L-menthol (the asymmetric Appel reaction). As a result, chiral phosphine oxides **134** were prepared with good enantiomeric excess. The latter were then treated with LDA and copper chloride to afford the bis-phosphine oxide which is a useful chiral ligand for asymmetric catalysis. The chiral bis-phosphine oxide (*R*,*R*)-**135** was produced in 98% *ee* and the minor amount of *meso*-isomer formed was easily removed by recrystallization from benzene, which yielded enantiopure (>99.9% *ee*) bis-phosphine oxide **135** in an isolated yield of 73% from the racemic phosphine **133** (Scheme 41).

Although hexachloroacetone gave high selectivity, it has the disadvantage that its by-product can become involved in the process leading to erosion of selectivity. Meantime, the reaction of oxalyl chloride with tertiary phosphine allows the process to run with higher selectivity, without complicating by-products and without the need for extensive purification of the starting material [72].

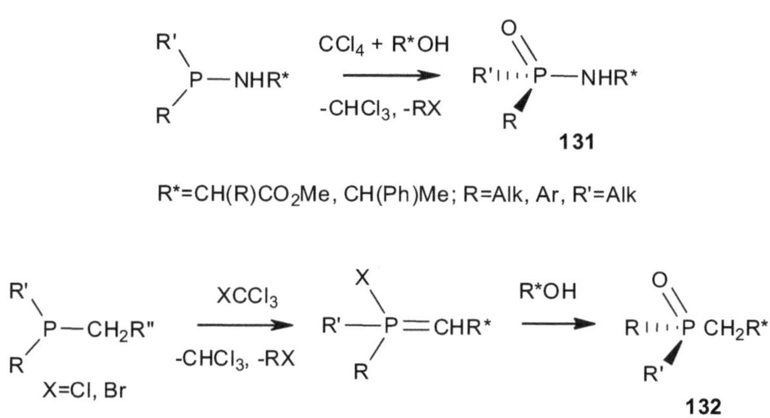

Scheme 40 Asymmetric oxidation of phosphines **131** and **132** with methane tetrahalides

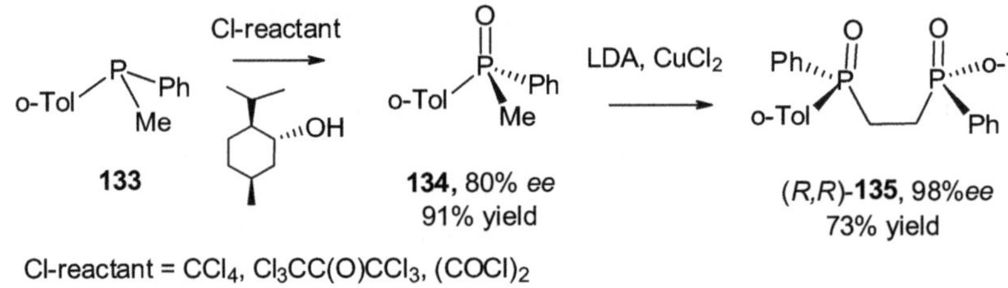

Scheme 41 Asymmetric Appel reaction

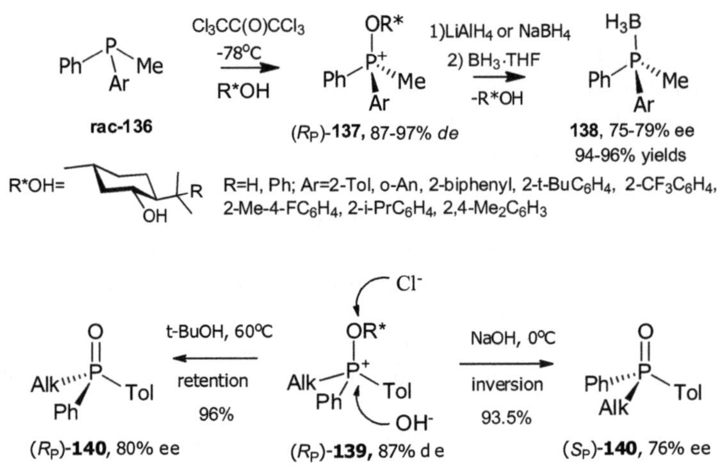

Scheme 42 DKR in oxidation of racemic phosphanes **136**

Gilheany [73, 74] has also developed the preparation of diastereomeric alkoxyphosphonium salts **137**, formed from dynamic resolution of racemic phosphanes **136** by reaction with hexachloroacetone, and menthol. The reaction of **137** with $LiAlH_4$ or $NaBH_4$ gives the corresponding enantioenriched *P*-stereogenic phosphanes **138** in good yields and with moderate enantiomeric excesses. The methodology allows one-pot conversion of racemic phosphanes into enantioenriched phosphanes Both enantiomeric oxides were prepared from a single intermediate, (R_P)-alkoxyphosphonium chloride **139**, which is formed in the course of a selective dynamic kinetic resolution using a single enantiomer of menthol as the chiral auxiliary. The origin of the dual stereoselectivity lies in bifurcation of the reaction pathway of this intermediate, which works as a stereochemical railroad switch. Under controlled conditions, Arbuzov-type collapse of this intermediate proceeds through C–O bond fission with retention of the configuration at the phosphorus center to give tertiary phosphine (R_P)-**140**. Conversely, alkaline hydrolysis of the P–O bond leads to the opposite enantiomer (S_P)-**140** (Scheme 42) [75–77].

5 Asymmetric Catalysis with Chiral Diamines

Since Evans et al. [78] has discovered that prochiral alkyl(dimethyl) phosphine boranes can undergo the enantioselective deprotonation of one methyl group, using butyllithium and (−)-sparteine **141**, these compounds have been widely used for the synthesis of *P*-chirogenic borane phosphines [79–105]. Lithium alkyls form chiral complexes **142** with sparteine **141** and related chiral diamines, which were investigated by single crystal X-ray analysis (Scheme 43) [82–87].

Prochiral dimethylarylphosphine boranes **143** react with a chiral sparteine–alkyllithium complex to undergo enantioselective deprotonation of one of the methyl groups with formation of lithium derivatives **144**, which react with electrophiles to give *P*-chiral compounds **145** (Scheme 44). Sparteine effectively complexes the lithium atom while deprotonation takes place and, in this chiral environment, *sec*-BuLi differentiates between the two enantiotopic methyl groups. *P*-Chirogenic phosphine ligands **146** were prepared via desymmetrization of prochiral phosphine boranes **143** using an *s*-BuLi/(−)-sparteine complex [82–87]. The reaction of lithium derivative **144** with benzophenone led to the formation of alcohol (*S*)-**147** in satisfactory yields and enantioselectivity (Scheme 45) [83].

Enantioselective deprotonation of tertiary dimethylphosphines **143** can be achieved with *s*-BuLi in the presence of accessible and cheap derivatives of alkaloid (−)-cytisine **148**. The derivatives of cytisine **148** are useful sparteine

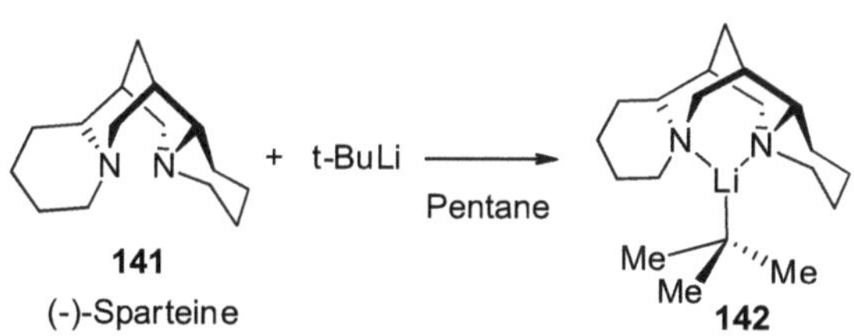

Scheme 43 Complex of (−)-sparteine with *tert*-butyllithium

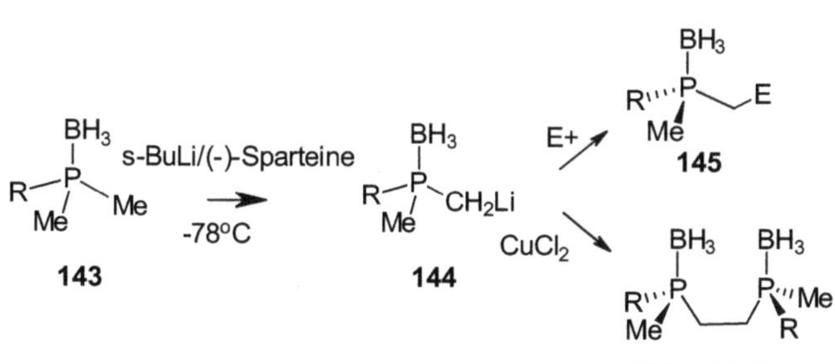

Scheme 44 Enantioselective deprotonation of dimethylarylphosphine boranes **143** with a chiral sparteine-alkyllithium complex

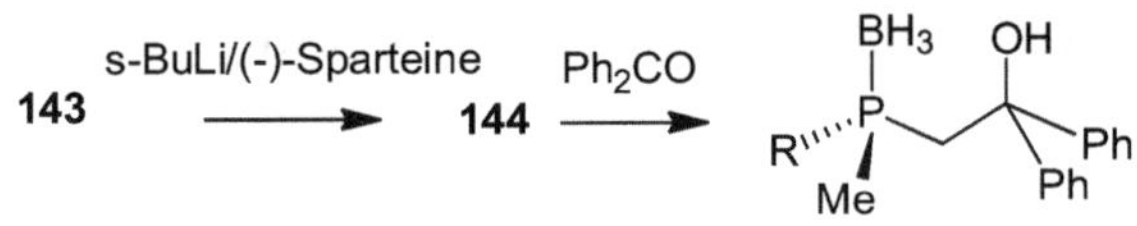

(*S*)-**147**

R	yield (%)	ee (%)
Ph	88	79
o-An	81	83
o-Tol	84	87
1-Nphth	86	82

Scheme 45 The reaction of lithium derivative **144** with benzophenone

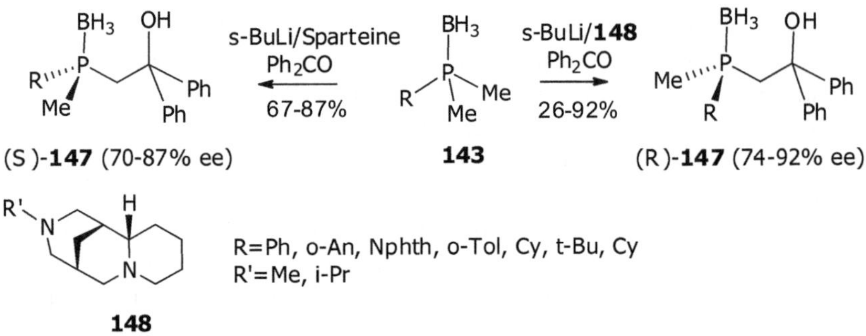

Scheme 46 Enantioselective deprotonation of tertiary dimethylphosphines with (−)-sparteine or cytisine derivatives **148**

surrogates for the desymmetrization of prochiral phenyl-, cyclohexyl-, and *tert*-butyl dimethyl phosphine boranes, yielding chiral phosphine boranes in up to 92% *ee* [88, 89]. Genet reported that increase of stoichiometric amount of chiral diamine augments the enantioselectivity of reaction (Scheme 46) [82].

Enantioselective deprotonation of alkylidenedimethylphosphine boranes **143** by an *s*-BuLi/(−)-sparteine complex, and subsequent oxidation with molecular oxygen in the presence of triethyl phosphite, led to the formation of alkyl(hydroxymethyl) methylphosphine boranes **150** in 91–93% *ee* in case of bulky alkyl groups and 75–81% *ee* in case of cyclohexyl or phenyl groups [87–94]. The treatment of the carbanion with CO_2, led to the formation of phosphorylated carboxylic acids. The reduction of the carboxyl group with borane and reaction with tosyl chloride provided the (*R*)-tosylates **149** in 90% *ee* and good yields. *P*-Chirogenic phosphine-sulfide borane ligands **152** were prepared by reaction of **143** with *s*-BuLi/(−)-sparteine and phenyl disulfide (Scheme 47) [83, 84]. The syntheses of α-alkoxyphosphine boranes as potential ligands for asymmetric organometallic reactions were developed via deprotonation of chiral hydroxymethylphosphine precursors **150**, followed by alkylation with various electrophiles and quenching

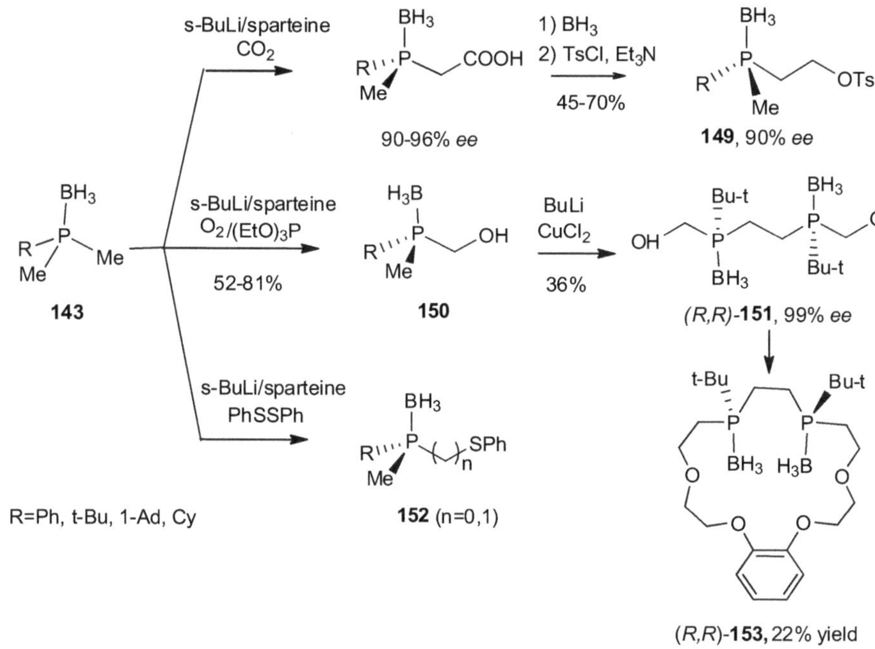

Scheme 47 Asymmetric synthesis of alkylidenedimethylphosphine borane derivatives **149–153**

with a polymer-bound scavenger [87]. Starting from chiral hydroxymethylphosphine **150**, enantiopure *P*-stereogenic secondary bisphosphines **151** were prepared, which were used as key building blocks for the preparation of *P*-stereogenic benzodiphosphacrowns **153** [90, 92].

Johansson et al. [95] reported the preparation of *P*-chirogenic aldehydes **154** via the desymmetrization of prochiral phosphine boranes **143**. The enantio-enriched formyl phosphine boranes were obtained in good yields by reaction of asymmetric lithium derivatives **144** with DMF at −78°C or by deprotonation of phenyl-*m*-anisyl-methylphosphine borane (borane protected PAMP) with *s*-BuLi and subsequent quenching with DMF. The formyl phosphine boranes **154** were transformed to the β-aminophosphine boranes **156** employing reductive amination under microwave irradiation. This methodology gives access to *P*-chirogenic compounds **155–159** which are versatile building blocks for the design and construction of new chiral phosphine ligands. For example, the ligands **155–159** were evaluated in the asymmetric conjugate addition of diethylzinc to *trans*-nitrostyrene (Scheme 48).

Enantiomerically pure bisphosphine (BisP*) **161** [83] and tris-phosphine (MT-Siliphos) **162** [21] ligands were obtained in high yields and used for preparation of various complexes of transition metals (Pd, Pt, Cu, Rh, and Ru) (Scheme 49). Cationic rhodium complexes **162** of bis-phosphines **161** were used as catalysts in asymmetric hydrogenation of (acylamino)acrylates with enantioselectivities up to 99.9% *ee*.

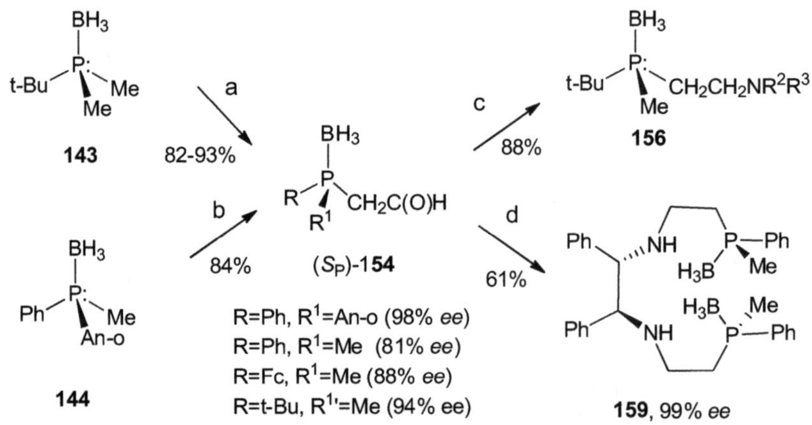

a) s-BuLi/(-)-Sparteine/DMF, -78°C; b) s-BuLi/DMF, -78°C;
c) NR^2R^3=PhCH(Me)NH_2-(*S*); $NaBH(OAc)_3$, SCX-2, DCE, MW (Reductive amination)
d) C_2-symmetric diamine, $NaBH(OAc)_3$, SCX-2, DCE

Scheme 48 Versatile α-formyl phosphine intermediate

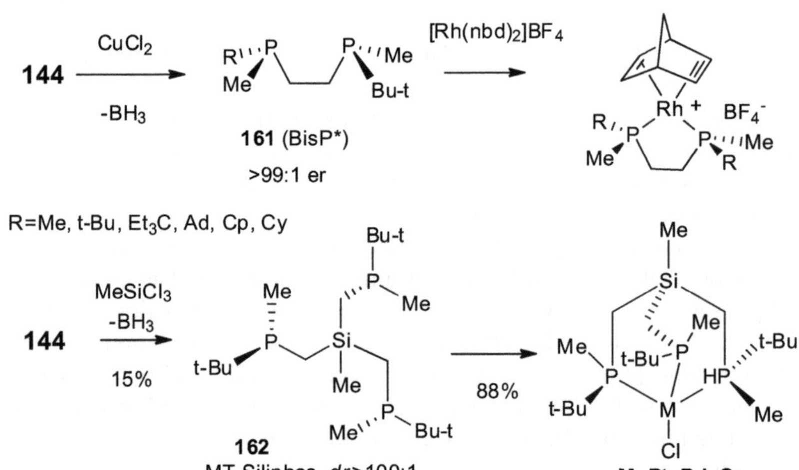

Scheme 49 Enantiomerically pure (BisP*) ligands **161** and MT-Siliphos with bulky substituents

Both *cis*- and *trans*-1,4-diphosphacyclohexanes **163** were synthesized by stereospecific intramolecular coupling reaction of bis-phosphines **146**. The coupling reaction of optically active **146** resulted in the *trans*-isomer **163**; meantime a *cis*-isomer **163** was prepared along with the *trans*-isomer from a mixture of *rac*- and *meso*-bisphosphines **146** (Scheme 50) [96].

P-chirogenic BisP* ligands **165** were prepared from alkyl(hydroxymethyl) methylphosphine boranes **150** in good isolated yield and with high optical purity

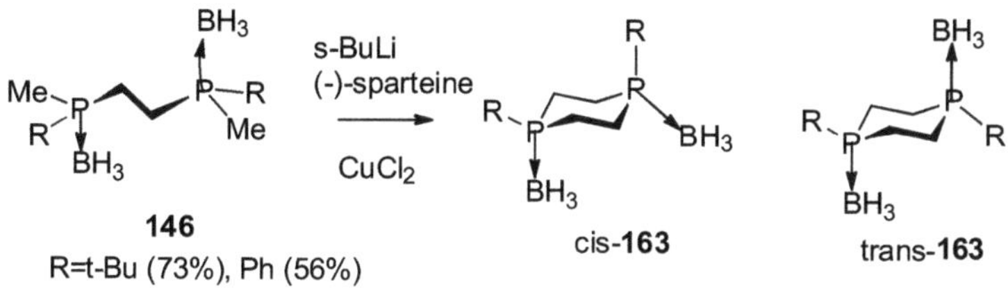

Scheme 50 *cis*- and *trans*-1,4-Diphosphacyclohexanes **163**

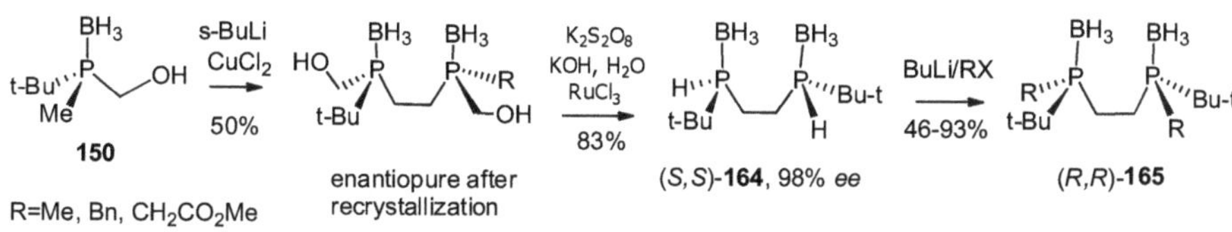

Scheme 51 Synthesis of BisP* borane ligands **165**

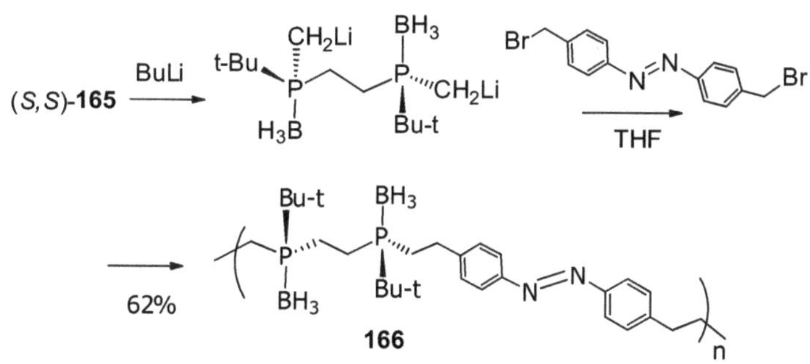

Scheme 52 Synthesis of photoresponsive polymers

as shown in Scheme 51. Rhodium catalysts with BisP* ligands **165** have demonstrated high enantioselectivities (up to 98% *ee*) in the hydrogenation of α-dehydroamino acid derivatives (Scheme 51) [92].

Photoresponsive polymers **166** having chiral phosphine in the main chain were prepared by lithiation of (*S*,*S*)-**165** with BuLi and subsequent reaction of prepared dilithium derivative with azobenzene derivative. According to the GPC analysis, the number-average molecular weight (Mn) and the average molecular weight (Mw) of **166** were found to be 3,000 and 5,000, respectively The polymer isomerized from the *trans*- to the *cis*-form upon UV irradiation and reverted to the *trans*-form reversibly. The polymer was able to coordinate to platinum, and the resulting polymer complex exhibited the Cotton effect owing to the chirality of the phosphorus atoms. The polymer chain was induced to rotate helically when complexed with transition metals through the chiral phosphorus atoms (Scheme 52) [97–99].

Imamoto et al. [100] used diastereoselective deprotonation of dimethylferrocenyl borane for the preparation of ethylene bridged *P*-chirogenic

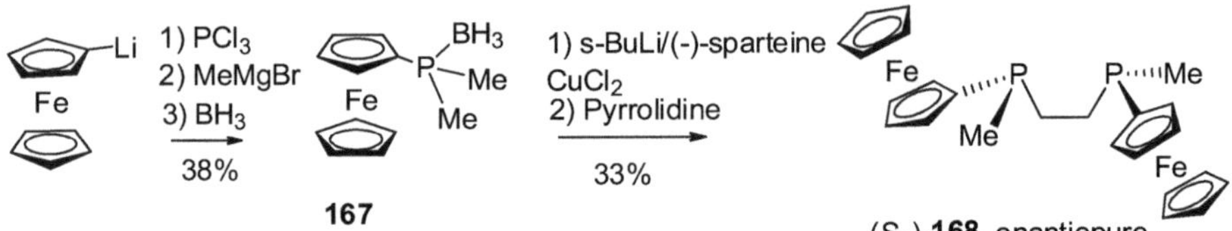

167

(S_P)-**168**, enantiopure

Scheme 53 Preparation of ethylene bridged *P*-chirogenic diferrocene diphosphines (S_P)-**168**

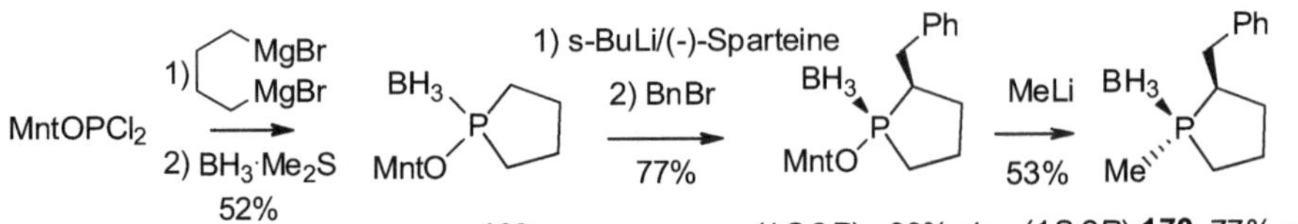

169 (1*S*,2*R*)-, 80% *de* (1*S*,2*R*)-**170**, 77% *ee*

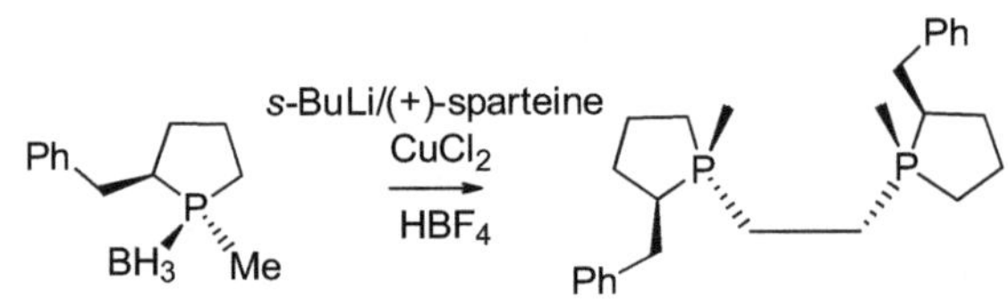

(1*S*,2*R*)-**170**, 77% *ee* (1*R*,2*R*)-**171**, 99% *ee*

Scheme 54 *P*-Chiral ligands **169** bearing phospholane cycles

diphosphines (S_P)-**168** containing a ferrocenyl moiety. One of the enantiotopic methyl groups of **167** was deprotonated with a (−)-sparteine/*sec*-BuLi complex and the resulting carbanion subjected to oxidative dimerization by treatment with copper(II) chloride to give the chiral diphosphine borane with small impurity of a *meso*-product. Recrystallization from toluene allowed the removal of the *meso*-isomer and the production of enantiomerically pure product (S_P)-**168** in 33% yield. The ligand was used in the rhodium-catalyzed asymmetric hydrogenation of dehydroamino acid derivatives (up to 77% *ee*) and in the palladium-catalyzed asymmetric allylic alkylation of 1,3-diphenyl-2-propenyl acetate (up to 95% *ee*) (Scheme 53).

Hoge reported *P*-chiral ligands bearing phospholane rings [101]. The *P*-chiral phospholane ligand **169** was prepared starting from the (1*R*,2*S*,5*R*)-menthyl dichlorophosphite. The treatment of (1*R*,2*S*,5*R*)-menthyl dichlorophosphite with a bis-Grignard reagent generated from 1,4-dibromobutane followed by complexation of the free phosphine with borane-dimethyl sulfide complex led to the formation of phosphinite borane **169**, which was then converted to enantiomerically enriched phospholane borane **170**. The subsequent reaction of methyl phospholane **170** with a *sec*-butyllithium/sparteine complex and copper chloride led to the formation of *P*-chiral diphosphine **171**, which, after recrystallization and deprotection with fluoroboronic acid, was obtained with 99% *ee* (Scheme 54). Asymmetric

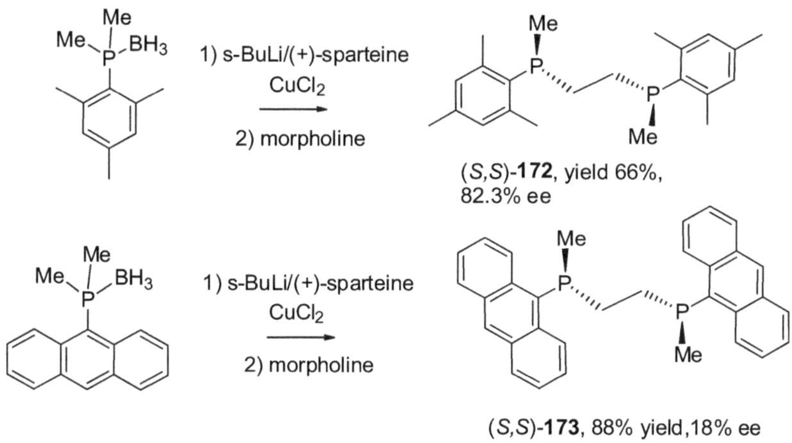

Scheme 55 Sterically hindered diphosphines **171** and **173**

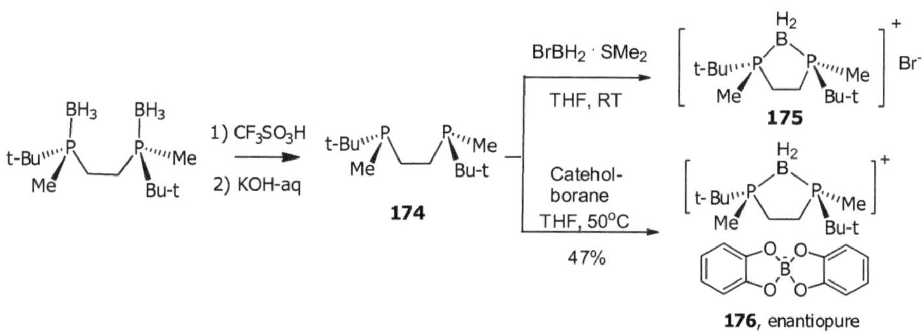

Scheme 56 (*S*,*S*)-1,2-Bis(*tert*-butylmethylphosphino)ethane boranes **175**,**176** (*t*-Bu-BisP*)

hydrogenation of acetamidoacrylic acid derivatives using Rh catalyst with diphosphine ligand **171** provided an enantioselectivity of 77–95% *ee* under low H_2 pressure [101].

Sterically hindered tertiary diphosphines **172** and **173** were synthesized with moderate diastereoselectivity (Scheme 55) [51].

Dihydroboronium derivatives of *t*-Bu-BisP* with different counter anions were prepared, as shown in Scheme 56. The reaction of BisP* with BH_2Br afforded the boronium salt **175** which possessed a bromide ion. The dihydroboronium derivative of (*S*,*S*)-1,2-bis(*tert*-butylmethylphosphino)ethane **175** (*t*-Bu-BisP*) was prepared by the reaction of *t*-Bu-BisP* **174** with catecholborane and used as chiral diphosphine ligand precursor in Rh-catalyzed asymmetric hydrogenated of methyl (*Z*)-acetamidocinnamate to afford the hydrogenation product in up to 94% *ee*. Complexes of iron(III) and *P*-chiral phosphine oxides **176** are catalysts for the asymmetric Diels–Alder reaction of *N*-acrylamide dienophiles [106].

Imamoto has reported the synthesis of *P*-chiral diphosphines with a methylene bridge and bulky alkyl groups on each phosphorus atom. These ligands named MiniPHOS **177** were synthesized using a phosphine borane **151** as an intermediate. The subsequent reaction of **144** with $RPCl_2$, methylmagnesium bromide and borane afforded the diphosphine boranes (*R*,*R*)-**178** and *meso*-**178**. The purification of reaction mixtures by crystallization and deboration resulted in the pure MiniPHOS **181** in yields of 13–28% and with 99% *ee* [19, 105, 106]. Improved synthetic routes to methylene-bridged *P*-chiral diphosphine ligands **181** via tertiary phosphine boranes **143** without the formation of *meso*-isomers was also reported [19]. The use of (−)-sparteine or (+)-sparteine surrogate as chiral catalysts facilitates access to *P*-stereogenic phosphines with opposite configuration. The method was exemplified by the catalytic asymmetric synthesis of each enantiomer of precursors to *t*-Bu-QuinoxP* **179**, Trichickenfootphos **180**, and Mini-PHOS **181** (R=*t*-Bu). The ligand **179** exhibited very good asymmetric induction in Pd-catalyzed asymmetric allylic substitution of 1,3-diphenyl-2-propenyl acetate (up to 98.7% *ee*) and in Ru-catalyzed asymmetric hydrogenation of ketones (up to 99.9% *ee*) (Scheme 57) [83, 84].

Interesting examples of the asymmetric synthesis of *P*-stereogenic vinylic phospholene boranes **183** using sparteine catalysis and Grubbs catalyzed ring-closing metathesis have been described (Scheme 58) [85].

BisP* ligands contain two methylene groups in the backbone and therefore their metal ligand complexes are conformationally flexible. Zhang and Tang prepared TangPhos ligand **186** having a well-defined rigid conformation through two additional five-membered rings on the backbone, which exhibited a more rigid chiral environment than the BisP* ligands. The TangPhos ligand was prepared in three steps using phosphine sulfides **184** and **185** as intermediates (Scheme 59). This ligand can be used for the Rh-catalyzed asymmetric hydrosilylation of α-(acylamino)acrylic acids and α-arylenamides to give optically active amides with enantiomeric excesses 98–99% [102–105].

Zhang et al. [102] have developed a convenient method for the synthesis of *P*-chiral phospholane–oxazolines ligands **188** based on phosphine sulfides as intermediate compounds. Selective deprotonation of **184** by *n*-butyllithium in the presence of (−)-sparteine followed by reaction with CO_2 provided acid **187** with 72% *ee*. Recrystallization of the acid from ethanol yielded the enantiomerically pure (*R*,*R*)-**187** in moderate yield. The condensation of **187** with chiral amino alcohols by using EDC/HOBu-*tert* proceeded smoothly to yield the coupling products, which were subsequently treated with MsCl to form the oxazoline compounds. Desulfurization of oxazoline compounds using Raney Ni provided phospholane–oxazoline ligands **188** in excellent yields. (*S*,*S*)-**188** was used for Ir-complex **189**-catalyzed asymmetric hydrogenation of β-methylcinnamic esters and methylstilbene derivatives. A variety of chiral 3-arylbutyric esters and diaryl (methyl)ethanes were obtained from moderate to very high enantioselectivity (up to 99% *ee*) (Scheme 59).

Imamoto and Crepy [23] obtained access to enantiomerically pure diphosphine dioxides **191** from oxidative dimerization of *rac*-1-*tert*-butylbenzophosphine oxide

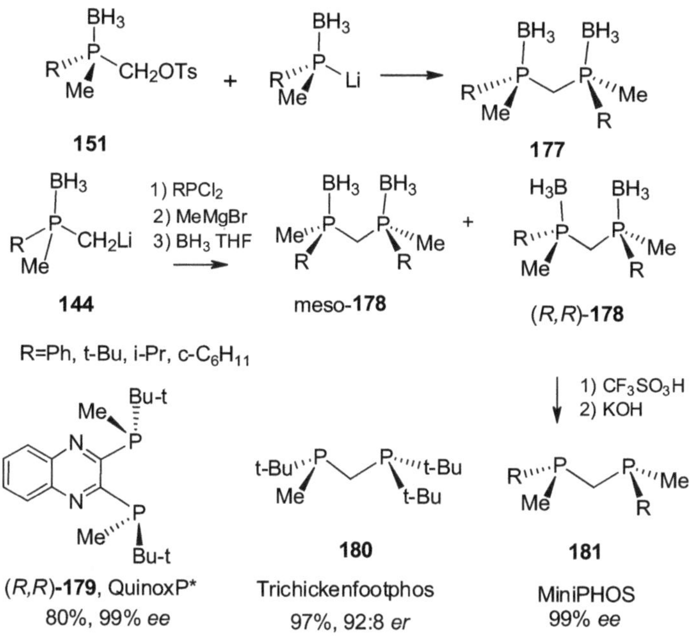

Scheme 57 *P*-Chiral diphosphines with methylene bridge

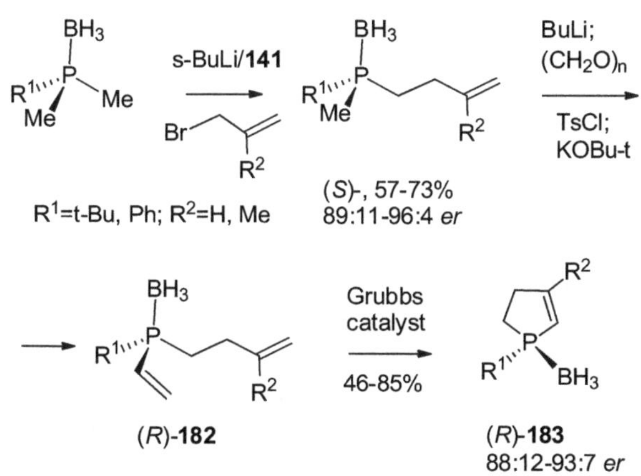

Scheme 58 *P*-Stereogenic vinylic phospholene boranes **183**

190 by treatment with *s*-BuLi/$CuCl_2$, and subsequent resolution with (+)- or (−)-DBTA. The reduction of **191** with hexachlorosilane led to the formation of diphosphine **192** with retention of absolute configuration. The ligand was used as a rhodium complex directly after reduction, for the hydrogenation of α-acetamidocinnamate with 96% *ee* (Scheme 60) [107].

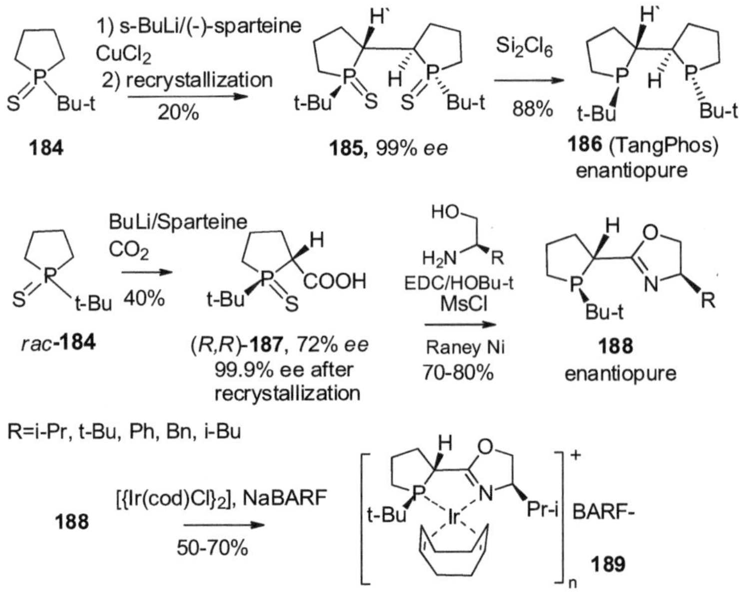

Scheme 59 Preparation of TangPhos ligand **186** and phospholane–oxazoline ligands **188**

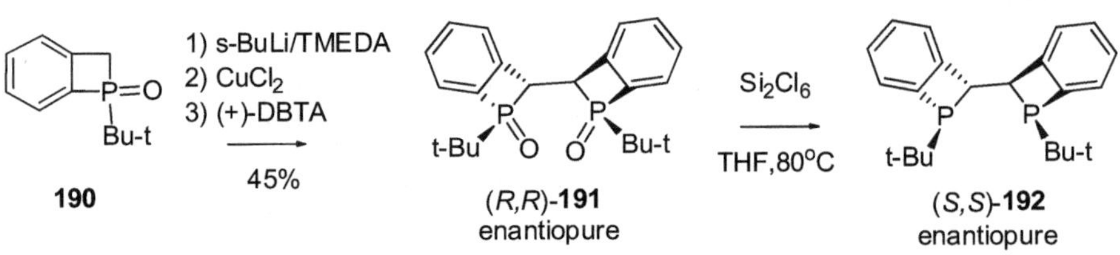

Scheme 60 Synthesis of enantiomerically pure diphosphine dioxides **192**

Starting from a trimethylsilyl-substituted phosphine sulfide **194** (generated by *n*-BuLi/(−)-sparteine-mediated asymmetric lithiation of a dimethylphosphine sulfide **193**), a two-step process of regioselective lithiation-trapping and silyl group removal has been used to prepare a range of *P*-stereogenic compounds, including precursors to diphosphine ligands (e.g., Mini-PHOS). This two-step protocol delivers products **195** with the opposite configuration to that obtained by direct asymmetric lithiation-trapping of a dimethylphosphine sulfide **193** using *n*-BuLi/(−)-sparteine (Scheme 61) [108].

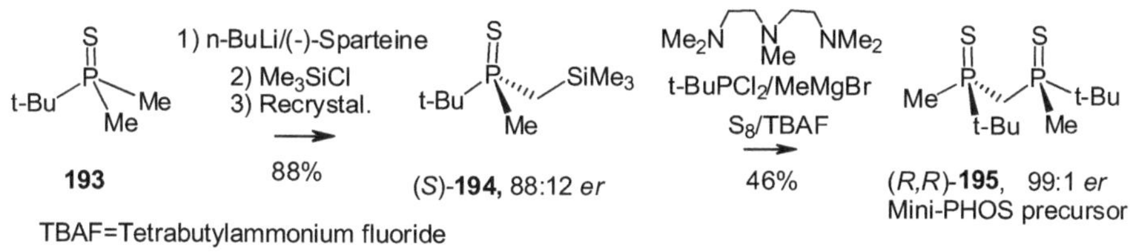

Scheme 61 Preparation of Mini-PHOS

6 Asymmetric Catalysis with Chiral Complexes of Transition Metals

For the last few year the catalytic asymmetric synthesis of tertiary phosphines has attracted the attention of many chemists. Interesting results were published in many articles and reviews [109–115]. Catalyzed by transition metals, asymmetric phosphination of secondary, racemic phosphines with aryl halides or triflates to prepare a tertiary *P*-stereogenic phosphines with control of the stereochemistry at the phosphorus atom is shown in Scheme 62.

One of the routes leading to *P*-stereogenic phosphines is electrophilic substitution at the phosphorus atom of secondary phosphines, as a result of asymmetric catalysis in which a catalyst activates a phosphorus nucleophile or a carbon electrophile, creating an asymmetric environment, i.e., creating preference for one of *Si* or *Re* face sides at the reactive center [103–113]. Upon reaction with chiral metal complexes, racemic secondary phosphines are converted into diastereomeric metal–phosphide complexes **A** or **B**, which interconvert rapidly through the inversion at phosphorus. If the equilibrium **A** ⇆ **B** is faster than the reaction of **A** or **B** with an electrophile **E**, then *P*-stereogenic phosphines **196**, in which pyramidal inversion is slow, can be formed enantioselectively. The product ratio in this dynamic kinetic asymmetric transformation depends both on K_{eq} and on the rate constants k_S and k_R (Scheme 63).

Glueck [116–121] came to the conclusion that the racemic secondary phosphines **197** form, with a platinum complex Pt(Me-Duphos)(Ph)(Br) and $NaOSiMe_3$ in toluene, an adduct **198**, which interconvert rapidly by *P*-inversion (S_P)-**198** ⇆ (R_P)-**198** [118]. Adduct **198** was isolated and studied by low-temperature NMR and X-ray monocrystal analysis. The crystal structure of the adduct showed that the major enantiomer of **198** has an (R_P)-absolute configuration [112].

The treatment of adduct **198** with benzyl bromide led to the formation of tertiary phosphine (R_P)-**199** with 72–78% *ee* and to initial catalyst Pt(Me-Duphos)(Ph)(Br) which confirms the proposed mechanism. The substitution at the tricoordinated phosphorus atom of the secondary phosphine **198** proceeded with retention of absolute configuration at phosphorus, according to classical representations. On the basis of these results, the authors established that the enantioselectivity was determined mainly by the thermodynamic preference for one of the interconverting diastereomers of (S_P)-**198** ⇆ (R_P)-**198**, although their relative rates of alkylation were also important (Curtin–Hammett kinetics) (Scheme 64) [122].

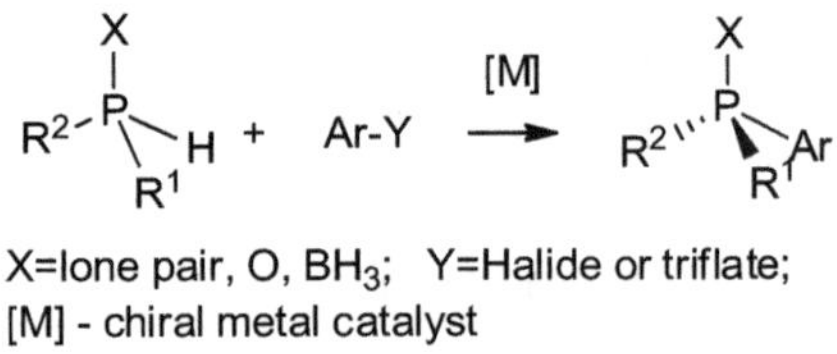

Scheme 62 Catalytic asymmetric synthesis of tertiary phosphines

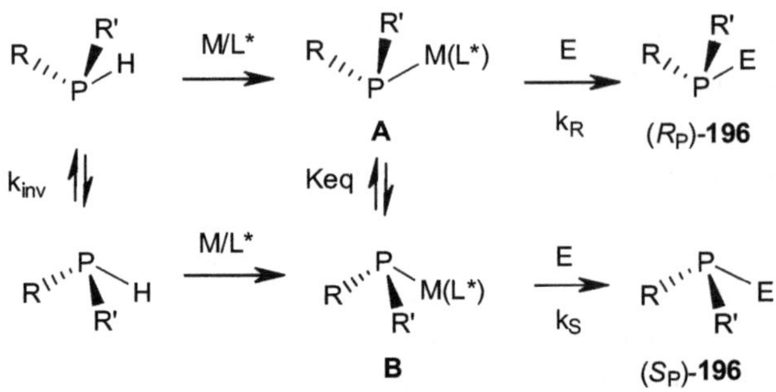

Scheme 63 Mechanism of catalytic electrophilic substitution at trivalent phosphorus

The asymmetric arylation or alkylation of racemic secondary phosphines catalyzed by chiral Lewis acids in many cases led to the formation of enantiomerically enriched tertiary phosphines [120–129]. Chiral complexes of ruthenium, platinum, and palladium were used. For example, chiral complex Pt(Me-Duphos)(Ph)Br catalyzed asymmetric alkylation of secondary phosphines by various RCH_2X (X=Cl, Br, I) compounds with formation of tertiary phosphines (or their boranes) **200** in good yields and with 50–93% *ee* [121]. The enantioselective alkylation of secondary phosphines **201** with benzyl halogenides catalyzed by complexes [RuH (*i*-Pr-PHOX **203**)$_2$]$^+$ led to the formation of tertiary phosphines **202** with 57–95% *ee* [123, 125]. Catalyst [(*R*)-Difluorophos **204**)(dmpe]Ru(H)][BPh_4] was effective at asymmetric alkylation of secondary phosphines with benzyl bromides, whereas (*R*)-MeOBiPHEP **205**/dmpe was more effective in the case of benzyl chlorides (Schemes 65, 66, and 67) [125–127].

The arylation of secondary phosphines with aryl halogenides, catalyzed by chiral complexes of platinum [116–126], ruthenium [127–129], and palladium [130–139], in many cases proceeded with good enantioselectivity and can be considered as one of methods for preparation of enantiomerically enriched tertiary phosphines [109–113]. For example, the reaction of aryl iodides with secondary arylphosphines **201**, catalyzed by chiral complex Pd((*R*,*R*)-Me-Duphos) (*trans*-stilbene), furnished tertiary phosphines **206** with enantioselectivities up to 88% *ee* [112, 115, 118, 130].

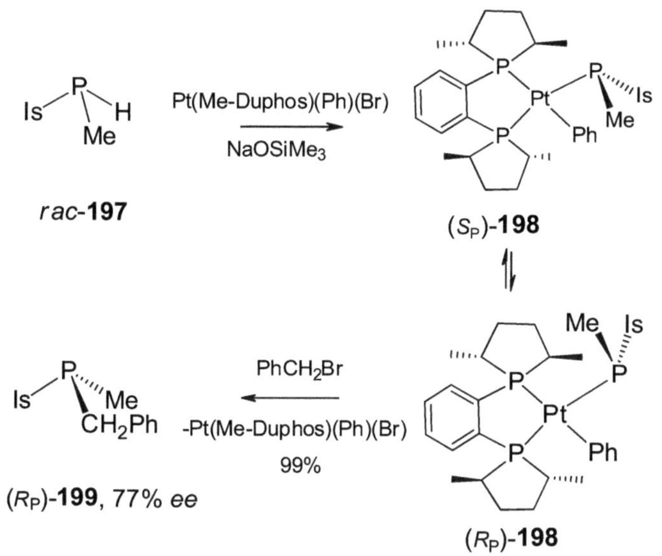

Scheme 64 Asymmetric alkylation of secondary phosphines **196** catalyzed by Pt(II) complexes

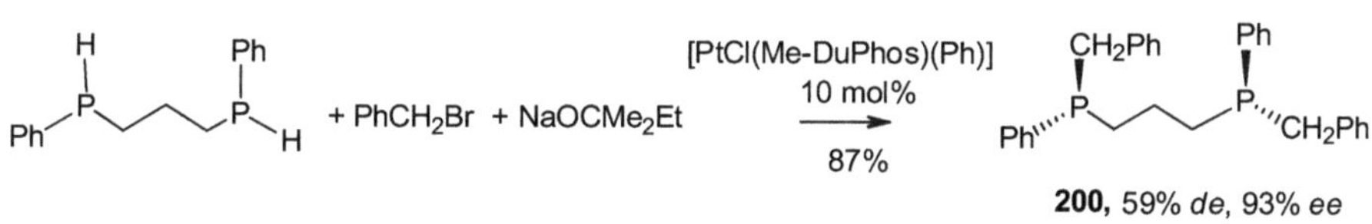

Scheme 65 DuPhos catalyzed asymmetric alkylation of secondary diphosphines [121, 126]

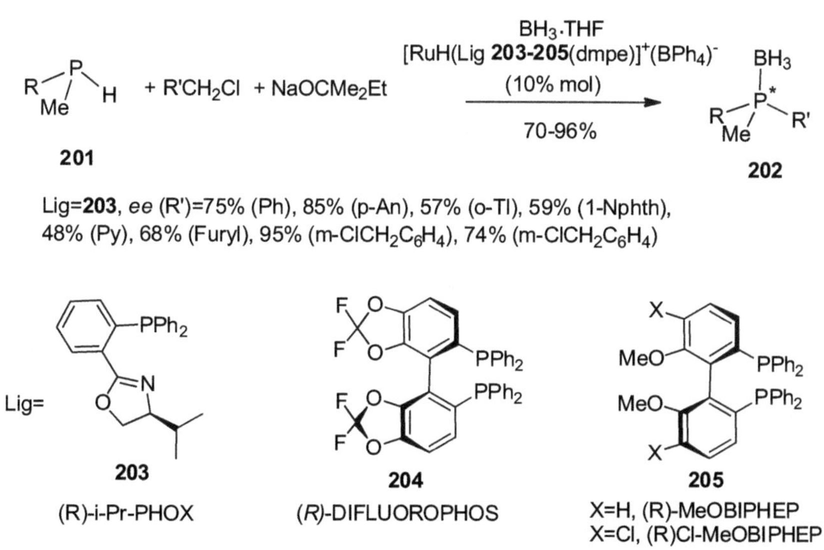

Scheme 66 Asymmetric alkylations of *sec*-phosphines **201**, catalyzed by chiral Ru complexes [127]

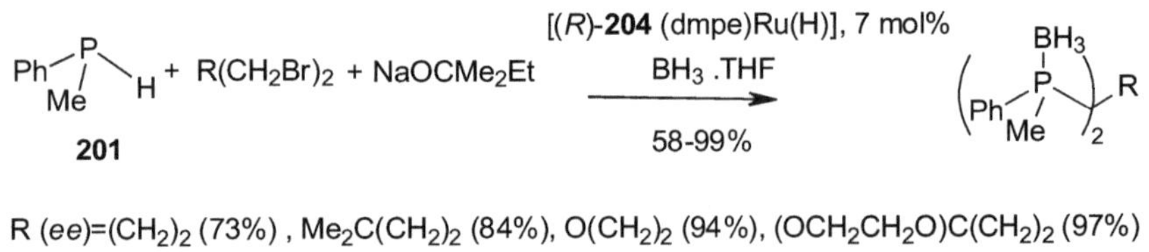

Scheme 67 Difluorophos/dmpe catalyzed asymmetric alkylation of secondary diphosphines **201** [128]

The arylation of secondary phosphines **201** with *ortho*-aryl iodides, catalyzed by generated in situ complex Pd_2 $(dba)_3 \times CHCl_3$, containing chiral ligand Et,Et-FerroTANE **207** and LiBr, led to the formation of corresponding tertiary phosphines with enantioselectivity of 90% *ee* [132, 137]. The palladium complex **209** also showed high enantioselectivity in arylation of secondary phosphines [131, 132]. Some examples of arylation reaction of secondary phosphines with low *ee* were described. The asymmetric arylation of phosphine boranes with anisyl iodide, catalyzed by chiral complex of oxazoline phosphine **208**, led to the formation of enantiomerically enriched tertiary phosphines **206** with 45% *ee* [134]. The Pd complex **210** of (*R*,*S*)-*t*-Bu-JOSIPHOS ligand catalyzed arylation of $PH(Me)(Ph)(BH_3)$ by *o*-anisyl iodide with the formation of PAMP-BH_3 with 10% *ee* (Table 3) [112].

The reaction of secondary phosphine boranes **211** with anisyl iodide, catalyzed by chiral Pd complex with (*S*,*S*)-Chiraphos, proceeded with retention of absolute configuration at phosphorus [135]. Addition of Pd((*S*,*S*)-Chiraphos)(*o*-An) to enantioenriched secondary phosphine **211** in the presence of $NaOSiMe_3$ led to the formation of complex **212**, stable at ambient temperature. This complex at +50°C in excess of diphenylacetylene allowed the formation of (R_P)-**213** in yield of 70% and with enantiomeric purity of 98% *ee* (Scheme 68).

Alkylation or arylation of silylated alkylarylphosphines **214** instead of P–H phosphines for the preparation of chiral tertiary phosphines in some cases led to an appreciable increase of enantioselectivity. For example, as reported by Toste and Bergman [133], the reaction of arylsubstituted iodides with silylphosphines **214** catalyzed by Pd(Et-FerroTANE)Cl_2, in the presence of *N*,*N*'-dimethyl-*N*,*N*,*N*-propylene (DMPU) led to the formation of *P*-chiral tertiary phosphines **215** with 55–98% *ee* (Scheme 69).

Enantioselective intramolecular cyclization of secondary phosphines **216** or their boranes, catalyzed by chiral palladium(diphosphine) complexes, afforded *P*-stereogenic benzophospholanes **217** with moderate stereoselectivity (59–70% *ee*) and yields. However, the absolute configuration of compounds has not been established. This reaction allowed chiral phospholanes to be obtained, which are valuable ligands in asymmetric catalysis (Scheme 70) [115].

Table 3 Arylation of secondary phosphines **201**, catalyzed by chiral palladium complexes (*typical examples*)

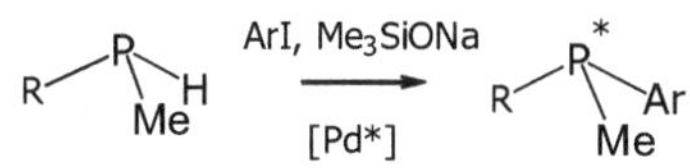

R	ArI	[Pd*]=Catalyst	Yield (%)	*ee* (%)	References
2-PhC_6H_4	2-*t*-BuOCOC_6H_4I	**207**/$CHCl_3$/Pd_2dba_3/LiBr/NEt_3	76	90 (*S*)	[133]
2-An	2-*t*-BuOCOC_6H_4I	**208**/$CHCl_3$/Pd_2dba_3/LiBr/N-Me-piperidine	43	86 (*S*)	[133]
2-$CF_3C_6H_4$	2-*t*-BuOCOC_6H_4I	**207**/$CHCl_3$/Pd_2dba_3/LiBr/NEt_3	39	93 (*R*)	[137]
2-PhC_6H_4	2-MeOCOC_6H_4I	**207**/$CHCl_3$/Pd_2dba_3/LiBr/NEt_3	69	85 (*S*)	[137]
2-PhC_6H_4	2-OCHC_6H_4I	**207**/$CHCl_3$/Pd_2dba_3/LiBr/NEt_3	71	63 (*S*)	[137]
t-Bu	3-AnI	**208**/MeCN/PdL_2/K_2CO_3	–	45	[134]
2-An	3-AnI	**208**/MeCN/PdL_2/K_2CO_3	–	45	[134]
Is	PhI	**209**	84	78 (*S*)	[131]
Is	PhI	**209**	89	75 (*S*)	[131]
Is	4-AnI	**209**	96	82 (*S*)	[132]
Is	*P*-PhOC_6H_4I	**209**	89	88 (*S*)	[132]
Ph	AnI	**210**	76	10	[112]

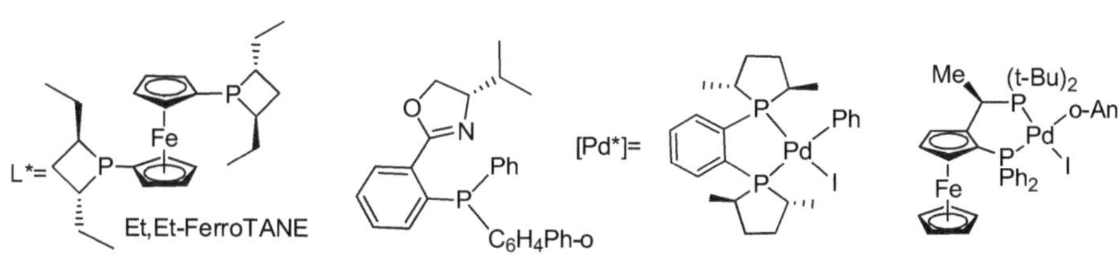

207, Pd_2dba_3 (S_CR_P)-**208**, Pd_2dba_3 **209** **210**

Examples of electrophilic addition of secondary phosphines to alkenes or alkynes were described. [114, 124, 125, 135]. Glueck [124–126] reported enantioselective tandem reaction of alkylated/arylation of primary phosphines catalyzed by platinum complex, proceeding with formation of chiral phosphaacenaphthenes. Palladium-catalyzed hydrophosphination of alkynes **219** under kinetic resolution conditions gave access to 1,1-disubstituted vinylphosphine boranes **220**. However, despite screening several chiral ligands, temperatures, and solvents, the

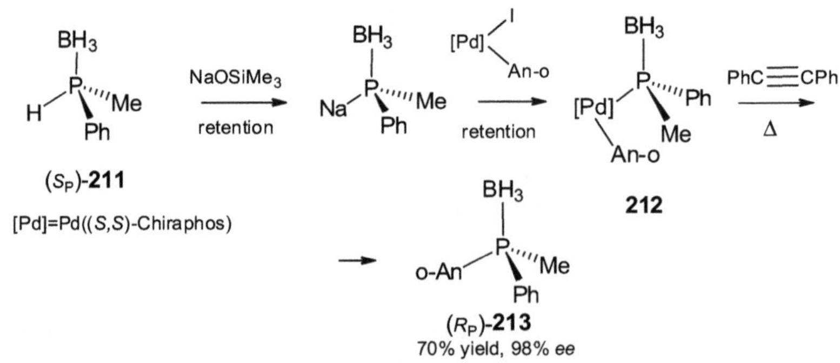

Scheme 68 Reaction of **211** with anisyl iodide catalyzed by chiral Pd (*S*,*S*)-Chiraphos complex

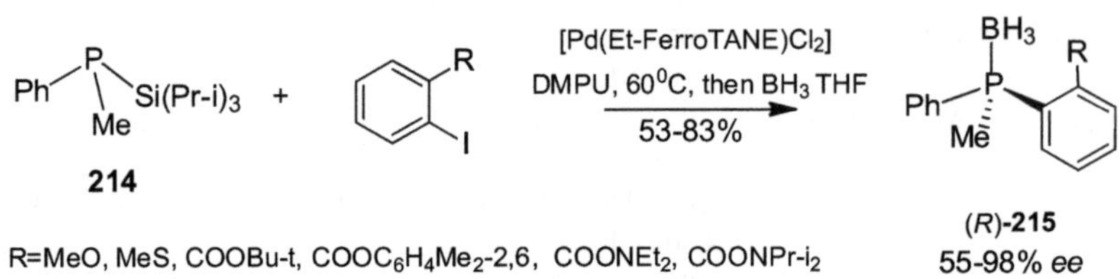

Scheme 69 Reaction of aryl iodides with silylphosphines **214** catalyzed by Pd complex

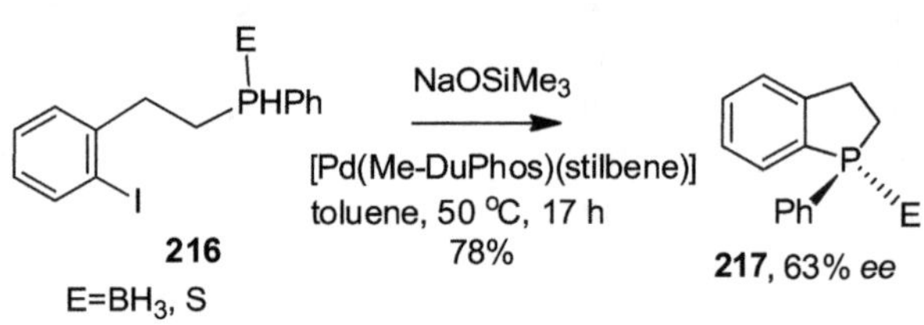

Scheme 70 Enantioselective intramolecular cyclization of secondary phosphines **216**

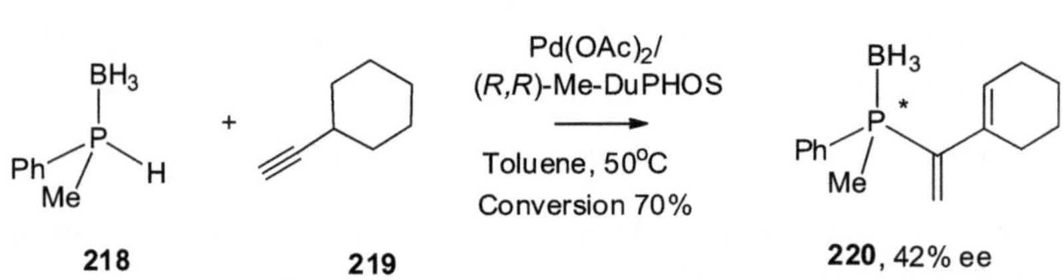

Scheme 71 Electrophilic addition of secondary phosphines to alkynes

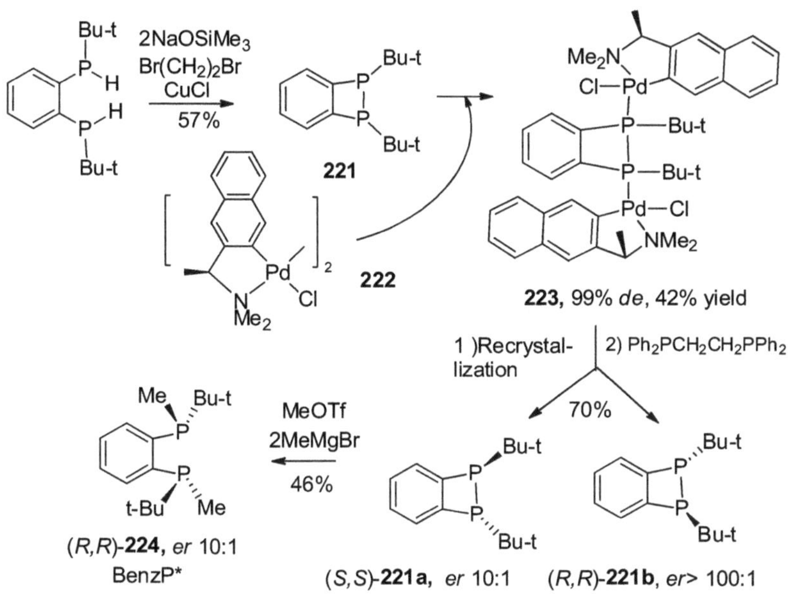

Scheme 72 Synthesis of nonracemic BenzP* **224**

ee of the product was moderate (Scheme 71) [138]. At a conversion of 70%, enantiomeric excess up to 42% was obtained. Mechanisms of metal-catalyzed hydrophosphination of alkenes and alkynes catalyzed by metal complexes with respect to the various potential roles of the metal center was examined by Rosenberg [140]. An attempt to apply chiral ammonium salts as interphase catalyst in the asymmetric alkylation of racemic secondary phosphines was also undertaken, although with low *ee* [141].

Glueck et al. have reported the synthesis, reactivity, and resolution of the benzodiphosphetane *trans*-1,2-(t-Bu_2P)C_6H_4 [142]. Treatment of **221** with 2 equiv. of chiral Pd complex **222** gave dinuclear complex **223**. Recrystallization of **223** from CH_2Cl_2/pentane and separation of diastereomers gave the less soluble (R_P,R_P) isomer **223** in >100:1 *dr*, and (S_P,S_P)-**223** isomer in 14:1 *dr*. Separate treatment of enantioenriched samples of diastereomers **223** with the bidentate bis (phosphine) dppe ($Ph_2PCH_2CH_2PPh_2$) liberated enantiomerically enriched (*S*,*S*)-**221a** and (*R*,*R*)-**221b** which did not epimerize when heated to 105°C in toluene. Sequential electrophilic and nucleophilic alkylation gave C_2-symmetric *P*-stereogenic bis(phosphine) BenzP* (*R*,*R*)-**224** (Scheme 72).

Ferrocene ligands represent an important group of chiral phosphines, among which the most interesting are discussed in Josiphos et al. [143]. Most of these ligands incorporate both carbon-centered chirality and planar chirality, and they have been proven to be effective in numerous asymmetric reactions. Catalysts based upon ferrocene ligands are efficient in various asymmetric reactions, and have attracted considerable interest over the last few years [144–149]. For example, Chen et al. [145] have described a method for the incorporation of *P*-chirogenic

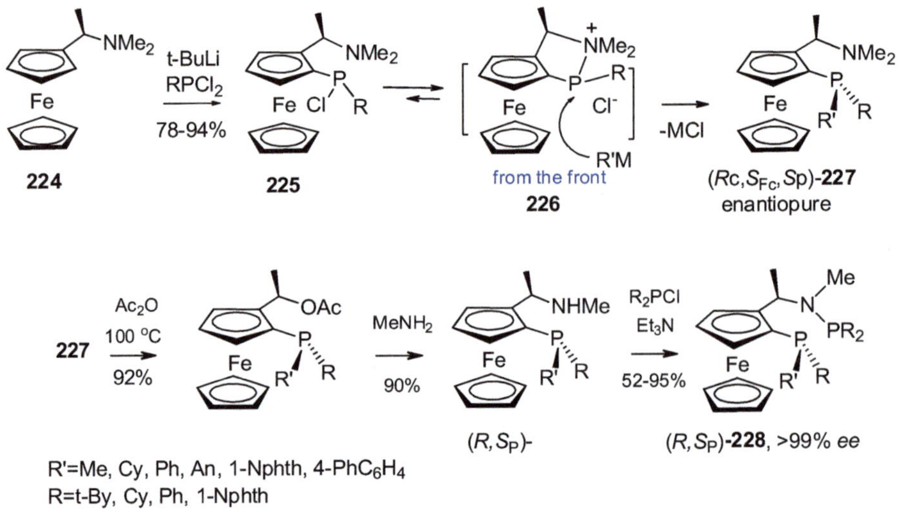

Scheme 73 Chiral ferrocene-phosphine ligands **227** and **228**

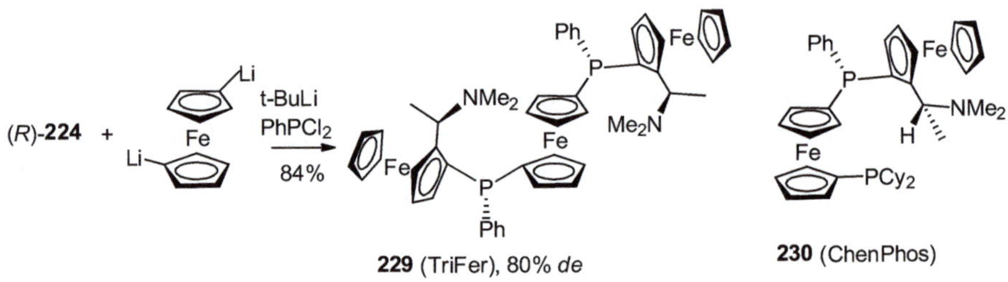

Scheme 74 C_2 Symmetric diphosphine triferrocene ligand **229**

phosphines into ferrocene by means of chiral substituent, as shown in Scheme 73. Slow addition of $RPCl_2$ and Grignard reagent to lithiated Ugi's amine **224** yielded phosphines (R_C,S_{Fc},S_P)-**227** in high yields and with high stereoselectivity. The high stereoselectivity of this reaction was explained by the formation of tertiary ammonium salt **226** as an intermediate. Boaz et al. [146] converted without epimerization the phosphines **227** into **228** ligands of Josiphos or BoPhoz type. These ligands have afforded exceedingly high activity and enantioselectivity in the rhodium-catalyzed asymmetric hydrogenation of dehydro-α-amino acid derivatives (up to 99.5% *ee*), itaconic acids (99% *ee*), and α-ketoesters (97.2% *ee*). Pfaltz [147], using the diastereoselective *ortho*-lithiation of (*R*)-*N*,*N*-dimethyl-1-ferrocenylethylamine **224** and a stereoconvergent intramolecular hydrophosphination, developed access to *P*-chiral ferrocenephospholanes. Starting from Ugi's amine **224**, the ligands TriFer **229** [148] and ChenPhos **230** [150] were prepared in 84% yield and with a diastereoselectivity of 80% (Scheme 74). Ugi's amine (*R*)-**224** was lithiated with BuLi and then treated with $PhPCl_2$, followed by 1,1′-dilithioferrocene generated from 1,1′-dibromoferrocene, by lithiation to afford the TriFer with some formation

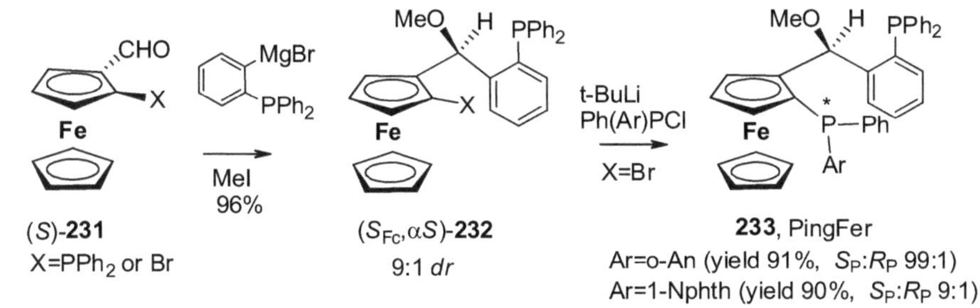

Scheme 75 Synthesis of PingFer ferrocene ligand **233**

of *meso*-isomer, which was removed by crystallization from MeOH. C_2-Symmetric diphosphine ligand **229** contains the central chiral carbon, planar carbon and central chiral phosphine. Use of TriFer as a ligand in rhodium complexes has allowed the hydrogenation of some unsaturated acids with very high enantioselectivity of 98.4–99.6% *ee*. PingFer **233** [149], a *P*-chiral version of the second generation of Taniaphos, was synthesized, proceeding from planar chiral aldehyde **231**. The reaction of ferrocenecarboxaldehyde (*S*)-**231** with Grignard reagent gave the product (S_{Fc},α*S*)-**232** in 96% yield and with a ratio of 9:1 for the αS/αR diastereoisomers. Recrystallization from hexane gave the enantiomerically pure (S_{Fc},α*S*)-PingFer in good yield. These compounds were applied in the highly enantioselective hydrogenation of cinnamic acid derivatives with 99.6% *ee* (Scheme 75).

7 Asymmetric Cycloaddition Reactions

The asymmetric hetero-Diels–Alder reaction is among the most powerful available methodologies for the construction of optically active heterocycles, with extensive synthetic applications in natural or unnatural products synthesis [151, 152]. In the last few years, the scope of this reaction has been extended to organophosphorus compounds. For example, Kumawat and co-workers developed a diastereoselective hetero-Diels–Alder reaction between isoprene and [1,4,2]-diazaphospholo-[4,5-*a*]-pyridines in the presence of sulfur or selenium, leading to the formation of **234** in 50% yield and with 4:1 *dr* (Scheme 76) [151].

Leung et al. [152–167] proposed the Diels–Alder reaction of phospholes with various dienophiles catalyzed by palladium complexes **235** as a method for the synthesis of chiral phosphines. The chiral compounds containing (*S*)-*ortho*-(1-dimethylaminoethyl)-naphthalene palladium **235** were complexed with diene, for example with 3,4-dimethyl-1-phenylphosphole, which then entered into the Diels–Alder reaction with dienophiles (*N*,*N*-dimethylacrylamide, styrene, and others) to result in diastereoisomers of *endo*-amidophosphanorbornenon complexes **236** and **237**. After separation, purification, and decomplexation, the

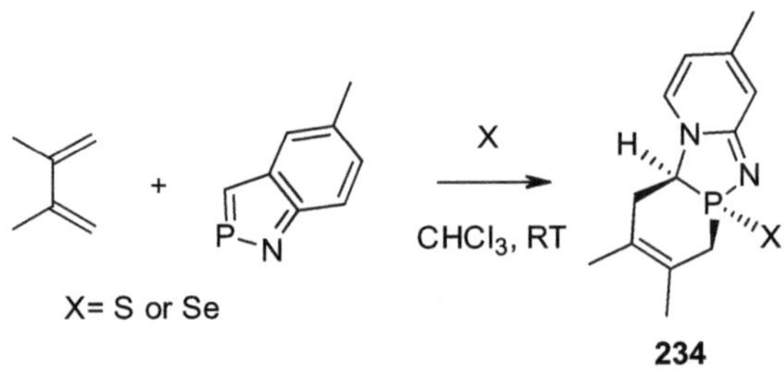

Scheme 76 Hetero-Diels–Alder reaction

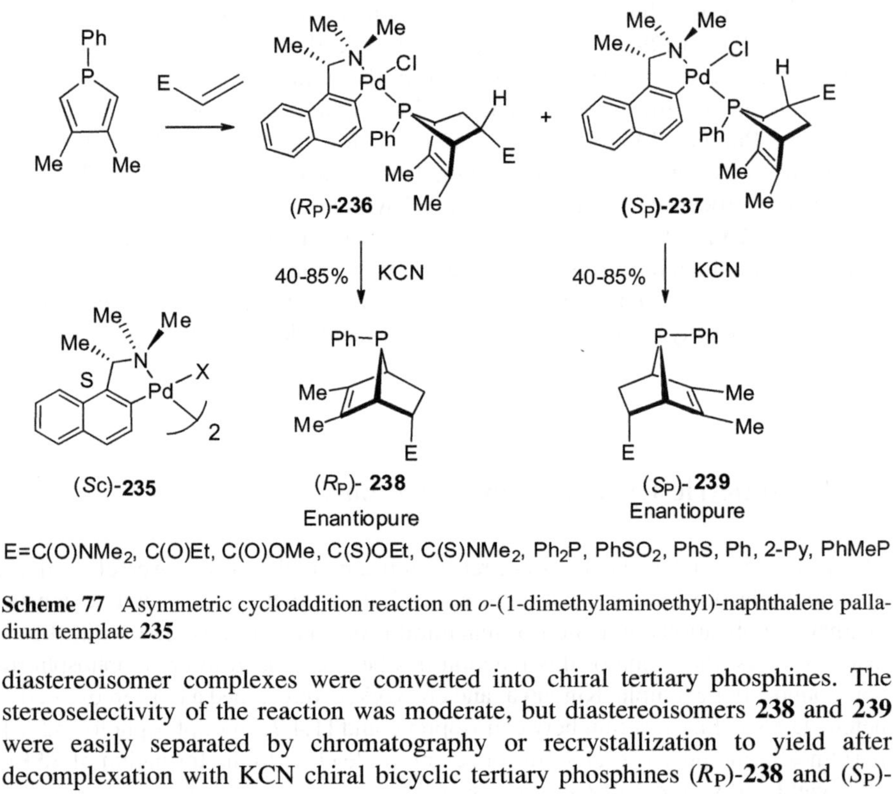

Scheme 77 Asymmetric cycloaddition reaction on *o*-(1-dimethylaminoethyl)-naphthalene palladium template **235**

diastereoisomer complexes were converted into chiral tertiary phosphines. The stereoselectivity of the reaction was moderate, but diastereoisomers **238** and **239** were easily separated by chromatography or recrystallization to yield after decomplexation with KCN chiral bicyclic tertiary phosphines (R_P)-**238** and (S_P)-**239** (Scheme 77).

The stereochemical course of the cycloaddition depends on the presence of silver perchlorate or tetrafluoroborate in the reaction medium [155]. Therefore, it is possible to select either the *exo*- or the *endo*-cycloaddition reaction pathways by controlling the number of coordination sites on the *ortho*-palladated naphthylamine template. In the *endo*-cycloaddition pathway, the kinetically stable chloro ligand is coordinated to the neutral template, but in the *exo*-cycloaddition pathway, the kinetically labile perchlorato ligand forms a cationic intermediate which

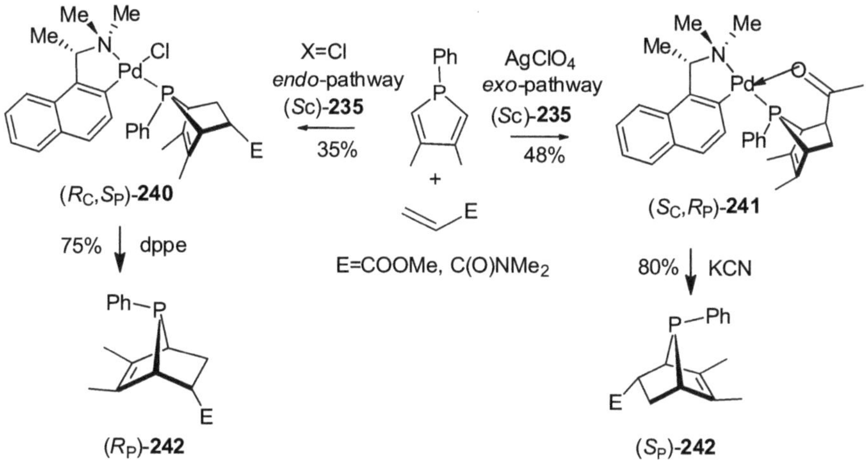

Scheme 78 Effect of $AgClO_4$ on the course of cycloaddition reaction

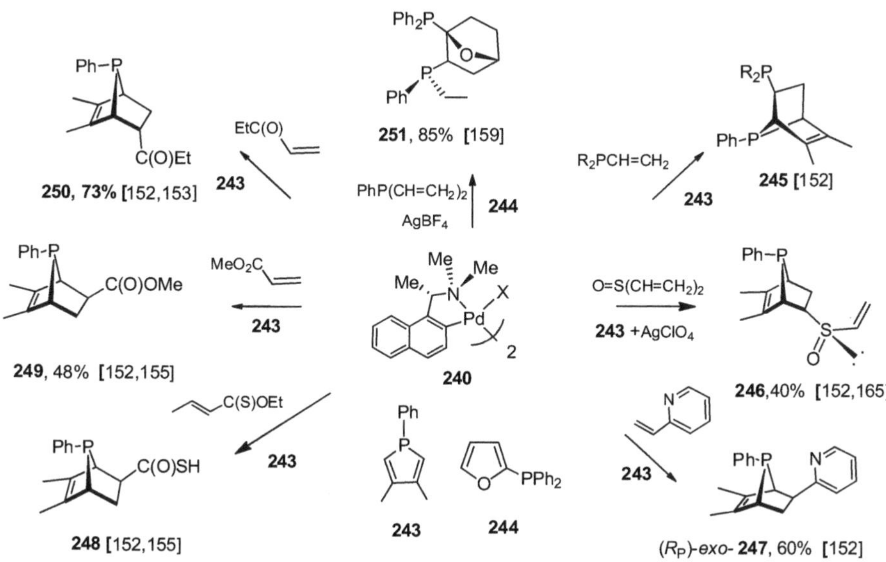

Scheme 79 The [4+2]-cycloaddition reactions of phospholes with various dienophiles

coordinates simultaneously onto the chiral template during the course of cycloaddition reaction. Therefore, the reaction of (S_C)-**235** with 3,4-dimethyl-1-phenylphosphole in the presence of $AgClO_4$ led to the formation of cycloaddition products (S_C, R_P)-**241** and (S_P)-**242** (Scheme 78) [154].

The [4+2]-cycloaddition reaction of phospholes **243** and furan derivatives **244** with various dienophiles catalyzed by palladium complexes **240** led to the formation of a number of chiral bicyclic phosphines **245–251**, which are interesting as chiral *P*-ligands (Scheme 79) [146, 153].

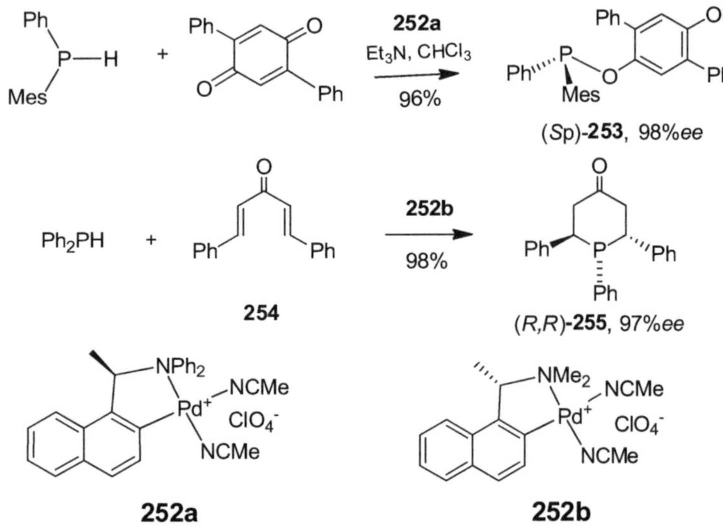

Scheme 80 Examples of asymmetric cycloaddition reaction

The reaction of phenyl(2,4,6-trimethylphenyl)phosphine with a substituted benzoquinone in the presence of a chiral phosphapalladacycle complex **252a** as a catalyst and triethylamine in chloroform at −45°C proceeded in a new type of addition manner to give a high yield of a 4-hydroxyphenyl phenyl(2,4,6-trimethylphenyl)phosphinite (S_P)-**253** with 98% enantioselectivity, which is a versatile intermediate readily convertible into various phosphines and their derivatives with high enantiomeric purity [168]. The asymmetric stepwise double hydrophosphination reaction of bis(enones) **254** with phenylphosphine and **252b** allowed intermolecular construction of chiral tertiary bulky *P*-heterocycles **255** in one pot in high yields (Scheme 80) [169].

Hey-Hawkins reported that the asymmetric Diels–Alder reactions of 2*H*-phospholes with the dienophile (5*R*)-(l-menthyloxy)-(5*H*)-furanone allowed the generation of multiple stereogenic centers in 1-phosphanorbornadienes **256**. The cycloaddition products were converted to their air-stable sulfur derivatives, which were isolated and the *endo*- and *exo*-isomers were separated by column chromatography (Scheme 81). In this case, the principle of face differentiation for a P=C bond is a synthetic tool for highly selective and efficient synthesis of *P*-chiral phosphanes from readily available starting materials. The observed selectivity was explained by the transition states of the two main isomers *exo*-A and *endo*-B (Scheme 82). The two molecules approach mainly in an *endo* fashion, which in consequence leads to the main diastereomer. Because of the favorable secondary orbital interactions between the C=O functionality and the diene system, the *endo* product is kinetically favored, as is known for such systems with normal electron demand [170].

Enantioenriched benzopyrano- and naphthopyrano-fused helical phosphafluorenes **258** were synthesized by the rhodium catalyzed enantioselective

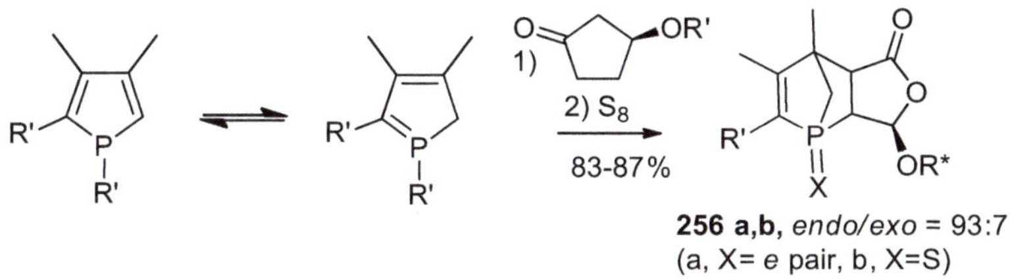

Scheme 81 Asymmetric Diels–Alder reactions of 2*H*-phospholes with the (5*R*)-(l-menthyloxy)-(5H)-furanone

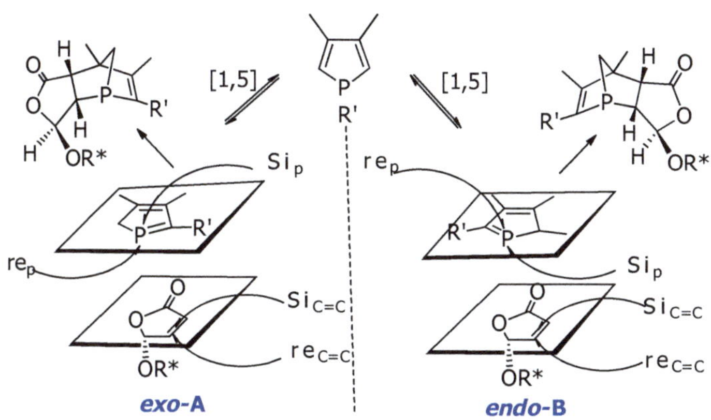

Scheme 82 Asymmetric Diels–Alder reactions of 2*H*-phospholes with the (5*R*)-(l-menthyloxy)-(5*H*)-furanone

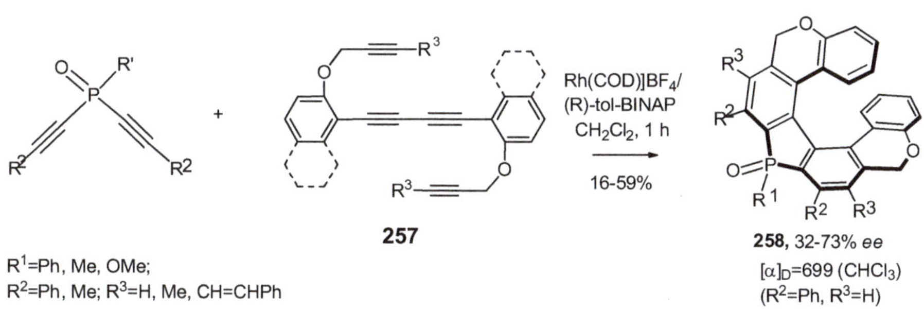

Scheme 83 Synthesis of phosphonohelicenes **258**

double [2+2+2]-cycloaddition of dialkynyl phosphorus compounds with phenol- or naphthol-linked tetraynes **257**. The desired reaction proceeded at room temperature by using the cationic rhodium(I)/tol-BINAP [2,2′-bis(di-tolylphosphino)-1, 1′-binaphthyl] catalyst to yield **258** in good yield. The photophysical properties of phosphorus-containing helicenes **258** were also studied. The phosphafluorenes showed large red shifts of absorption and emission maxima (Scheme 83) [171].

8 Biocatalytic Synthesis of *P*-Chiral Phosphorus Compounds

Efficient and general methods for the synthesis of chiral phosphine oxides and related compounds are a permanent subject for research of organic chemists. Therefore, the application of biocatalytic methods for the preparation of optically active *P*-chiral compounds has attracted great attention and a number of successful syntheses were described [172–178].

For example, Mikołajczyk and Kiełbasiński [172–175] studied the acylation of phosphine boranes **259** using CAL (Chirazyme®) lipase from *Candida Antarctica* and Lipase AK from *Pseudomonas fluorescencs* (Scheme 84). The best enantioselectivity was attained in the lipase AK-catalyzed acylation of **259** in cyclohexane solution with vinyl butyrate as an acyl donor (99% *ee*) for unreacted hydroxyphosphinate **259** and 43% *ee* for the acylated product **260**. The E-values were on the level of 15. The enzymatic resolution of alkoxy (hydroxymethyl)phenyl-phosphine boranes (*R*/*S*)-**261** was achieved by *trans*-esterification with vinyl acetate in the presence of CALB, Amano AK, Amano PS, Amano AH, and LPL in various solvents. The best enantioselectivity of unreacted alcohol **261** and acylated product **262** was attained in cyclohexane (37% *ee*, conversion ~50%). Kiełbasiński [176] recently reported some additional data, including theoretical calculations and more accurate chemical correlation, which proved that the borane reduction of acyclic phosphine oxides proceeded with inversion of configuration at the phosphorus center. On this basis, the stereochemistry of the enzymatic reaction was ultimately determined (Scheme 85).

Lipase-catalyzed acylation of ethyl (1-hydroxyalkyl)phenylphosphinates afforded a single diastereomer in high enantiomeric excess. The substituent effect of the alkyl group toward the acylation using CAL (Chirazyme®) was larger than that of an immobilized lipase AK from *Pseudomonas fluorescens*. The kinetic separation of the major phosphinates (S_P,*S*)-**263** and (R_P,*R*)-**264** was carried out by CAL and lipase AK catalyzed acylation using vinyl acetate as an acyl donor. The influence of the alkyl substituent R in the biocatalyst on the acylation process was more pronounced for CAL compared to AK (Scheme 86) [179, 180].

A synthesis of chiral α-hydroxy-*H*-phosphinates, bearing two asymmetric centers, was achieved via a lipase-catalyzed hydrolysis of acetate precursors. From a

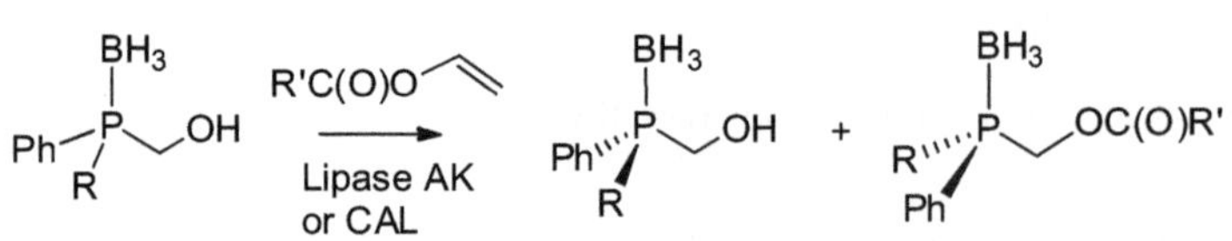

(R/S)-**259** (*R*)-**259**, 56-99% ee 39-51% yields (*S*)-**260**, 50-86% *ee* 48-49% yields

R=Et, Bu, c-C_6H_{11}, t-Bu; R'=Me, Et, Pr. i-Pr

Scheme 84 Enzymatic resolution of racemic tertiary phosphines **259**

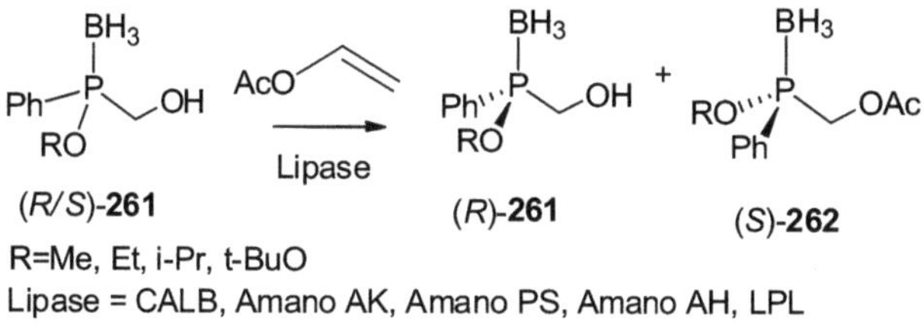

Scheme 85 Enzymatic resolution of alkoxy (hydroxymethyl)phenyl-phosphine boranes **261**

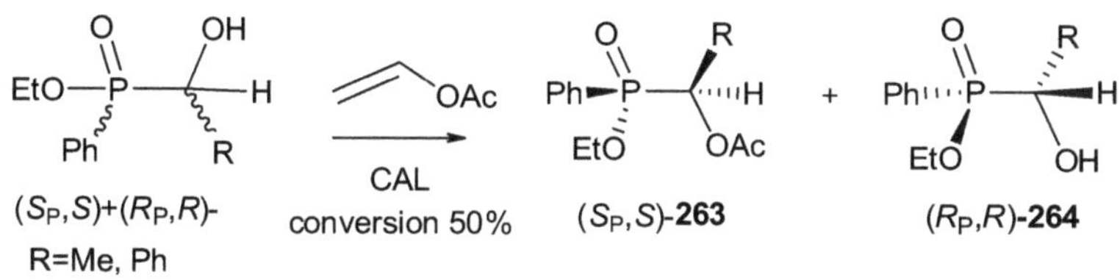

R	Lipase	*ee* (%) of **263**	*ee* (%) of **264**
Me	CAL	98	98
Et	CAL	98	28
Me	AK	98	98
Me	AK	98	98
Et	AK	98	98
Pr	AK	98	2

Scheme 86 Lipase-catalyzed acylation of ethyl (1-hydroxyalkyl)phenylphosphinates

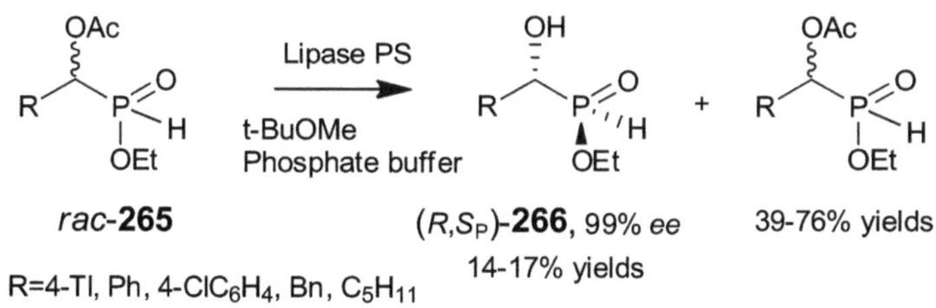

Scheme 87 Enzymatic synthesis of chiral α-hydroxy-*H*-phosphinates

mixture of four kinds of stereoisomers (two enantiomers and two diastereomers) of α-acetoxy-*H*-phosphinates **265**, one isomer of α-hydroxy-*H*-phosphinates (R,S_P)-**266** was obtained with moderate yields and stereoselectivity up to 99% *ee* (Scheme 87) [181].

The kinetic resolution of 1,1-diethoxyethyl(hydroxymethyl)phosphinate *rac*-**267** possessing chirality at the phosphorus atom was achieved via a lipase-catalyzed acylation. The product **269** was transformed into the corresponding amine **270**,

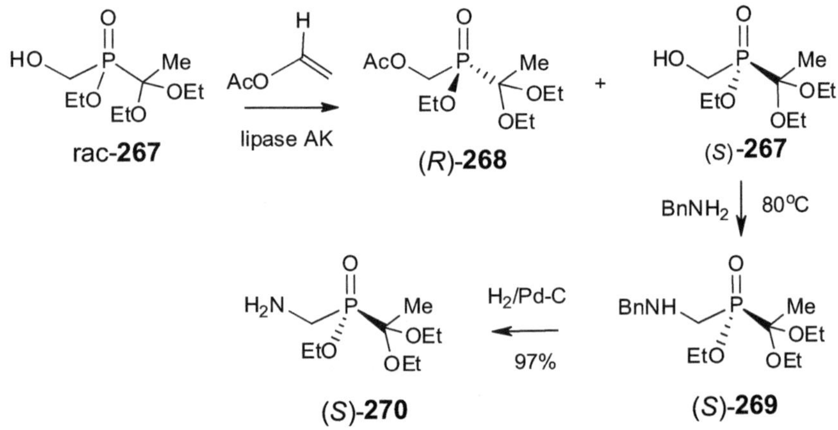

Scheme 88 Kinetic resolution of 1,1-diethoxyethyl(hydroxymethyl)phosphinate *rac*-**267**

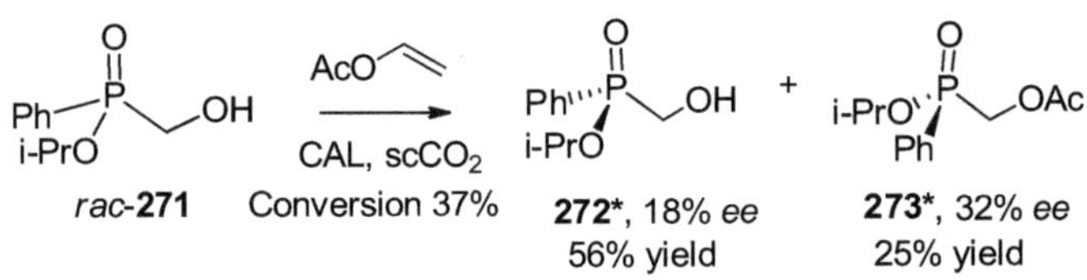

Scheme 89 Biocatalytic kinetic resolution of racemic hydroxymethylphosphinates **271**

which is a useful precursor for the preparation of phosphinyl dipeptide isosteres. Acylation of **267** with vinyl acetate as an acyl donor in the presence of lipase Amano AK led to the formation of (*R*)-**268** (yield 59%, 88% *ee*) and (*S*)-**269** (yield 35%, 92% *ee*), which were separated by column chromatography. The enantiomeric purity of (R_P)-**268** was increased to 99% *ee* by an enzymatic double resolution under the same conditions (Scheme 88) [182].

Biocatalytic kinetic resolution of racemic hydroxymethylphosphinates **271** via their lipase-promoted acetylation in supercritical carbon dioxide as the reaction medium was investigated. The reaction was fastest when pressure was closer to the critical pressure: at 11 MPa the reaction rate reached its maximum when the pressure was increased to 15 MPa. The optimal conditions were obtained at 13 MPa (yields ~50%, ~30% *ee*). The stereoselectivity of the reaction depended on solvent, substituents at phosphorus, and solubility of substrates in $scCO_2$. The best results were obtained with the *Candida antarctica* lipase (Novozym 435) (Scheme 89) [183, 184].

The biocatalytic desymmetrization of various C_2-symmetric tertiary phosphine oxides was used for the preparation of *P*-chiral phosphines. Mikolajczyk et al. [183–185] studied the desymmetrization of bis-functional phosphinates and phosphine oxides. The hydrolysis of prochiral bis(methoxycarbonylmethyl) phenylphosphine oxide **274** was carried out in phosphate buffer in the presence of

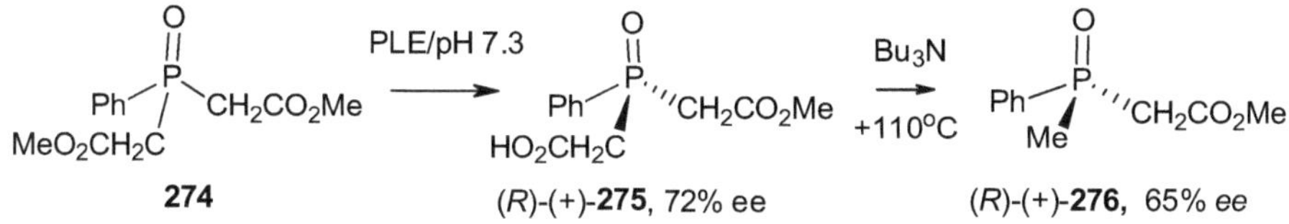

Scheme 90 The biocatalytic desymmetrization of bis-functional phosphinates

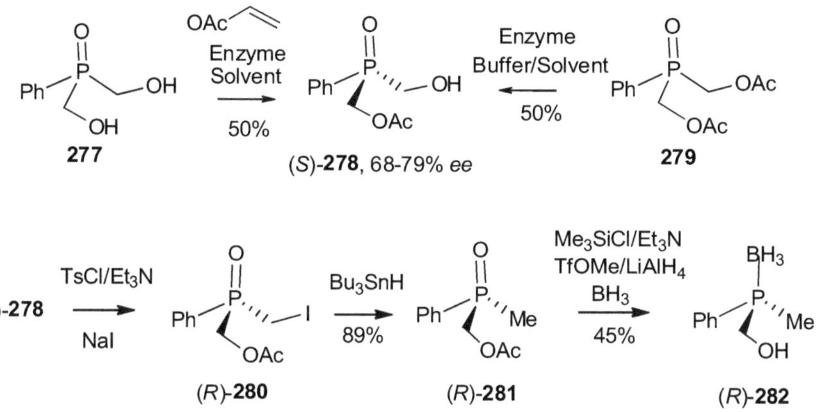

Scheme 91 Enzymatic preparation of tertiary (*R*)- or (*S*) phosphine oxides

porcine liver esterase (PLE) to give a chiral monoacetate (*R*)-**275** in 92% yield and 72% *ee*. The chiral monoacetate (*R*)-**275** by decarboxylation was converted into chiral phosphine oxide (*R*)-**276** (Scheme 90) [185].

The biocatalytic acetylation of prochiral bis(hydroxymethyl)phenylphosphine oxide **277** and the biocatalytic hydrolysis of prochiral bis(methoxycarbonylmethyl) phenylphosphine oxide **279** was subjected to hydrolysis in a phosphate buffer in the presence of several hydrolases (PLE, PFL, AHS, Amano-AK, and Amano-PS), of which only porcine liver esterase (PLE) proved to be efficient. The best results were attained with *Pseudomonas fluorescens* lipase (PFL) in chloroform which allowed the compound **278** to be obtained in yields up to 76% and with *ee* up to 79%. Absolute configuration of the (*S*)-**278** was determined by means of chemical correlation to the earlier described compound (*R*)-**282**, as shown in Scheme 91 [185].

Desymmetrization of the prochiral diol **283** was attained, using vinyl acetate as an acetylating agent and several lipases (CAL, AK, AH, PS, LPL, PFL), of which only *Pseudomonas fluorescens* lipase proved efficient. It was found that the use of various solvents led to opposite enantiomers of the product **284** and substantially influenced the stereoselectivity of the process. For example, the replacement of chloroform by isopropyl ether led to the formation of the optical antipode of **284** [175, 176]. Wiktelius [177] reported that the *Candida antarctica* lipase B (Novozym 435) afforded better results in the desymmetrization of prochiral

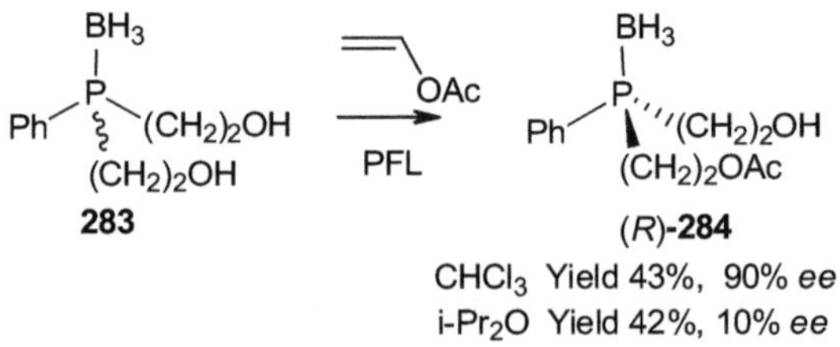

Scheme 92 Enzymatic desymmetrization of the prochiral diol **283**

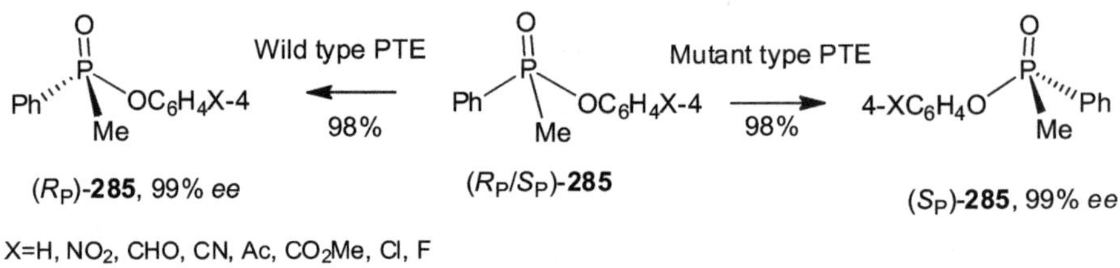

Scheme 93 Hydrolysis of phenoxy (S_P)-phosphinates with wild- or mutant-type PTE

phosphine boranes than Amano ANL, Amano PS, PFL (Amano AK), and PPL (Fluka) lipases (Scheme 92).

Raushel et al. [186–191] have resolved racemic phosphinates **285**, bearing a phenol leaving group at phosphorus by means of fermentative hydrolysis catalyzed by phosphotriesterase (PTE) from *Pseudomonas diminuta*. The hydrolysis led to the formation of chiral phosphinates with optical purity >99.8%, quantified by chiral electrophoresis. The catalyzed PTE hydrolysis of phosphinates (R_P/S_P)-**285** led to the formation of (S_P)-methylphenylphosphinate and unreacted phosphinate (R_P)-**285** [186]. It was found that wild-type phosphotriesterase (PTE) preferably hydrolyzed (S_P)-phosphinates with formation of (R_P)-phosphinates **285**, and mutant-type PTE (TAGW) preferably hydrolyzed (R_P)-phosphinate with formation of (S_P)-phosphinates with enantiomeric factor E = 17. The stereoselectivity of wild-type PTE was affected by the pK_a value of the phenol leaving group. For the wild-type enzyme, the stereoselectivity has been enhanced in excess of three orders of magnitude PTE catalyzed hydrolysis of the most acidic phenolic substituents from an organophosphate triester. (R_P)-Stereoisomers **285** were purified by chromatography on silica gel and were obtained in 98% yield and with 99% *ee* (Scheme 93) [187–189].

The hydrolysis of phosphonate **286** (R=Me) by wild-type PTE yielded mainly thioacids (S)-**287**, whereas the hydrolysis by mutant-type PTE led to the formation of (R)-thioacids **287**. In contrast, the hydrolysis of thioacid triesters **286** by either wild- or mutant-type PTE in many cases yielded thioacids (S)-**287** with 99% *ee*. The chiral thiophosphates synthesized by these enzymatic methods, were proposed as precursors for the synthesis of organic and organophosphorus compounds (Scheme 94) [190].

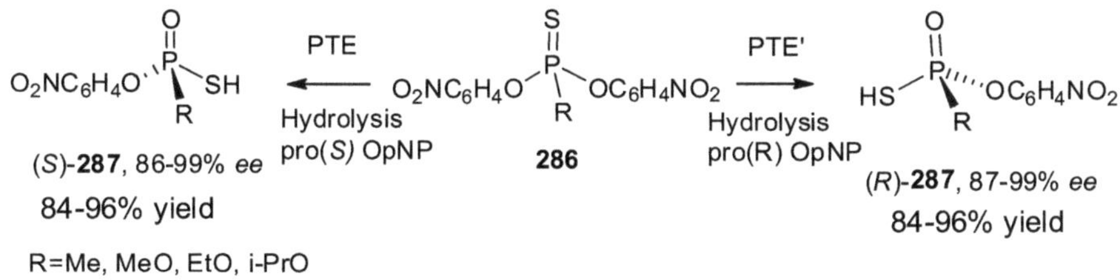

Scheme 94 The hydrolysis of thioacid esters **286** by wild- or mutant-type PTE

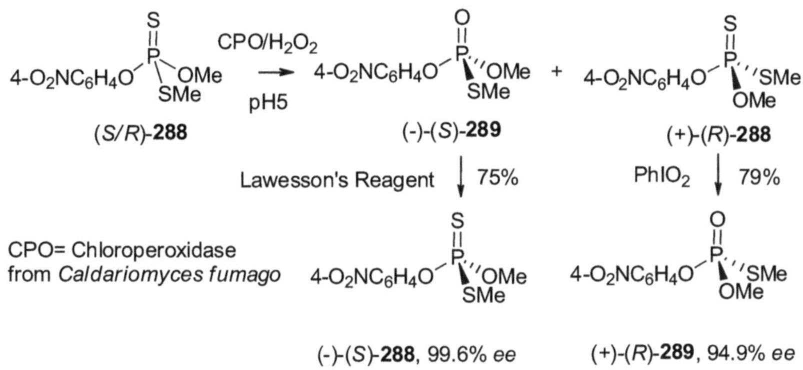

Scheme 95 The biocatalytic oxidation of racemic *O,S*-dimethyl *O-P*-nitrophenyl phosphorodithioate **288**

The biocatalytic oxidation of racemic *O,S*-dimethyl *O-p*-nitrophenyl phosphorodithioate (*S*/*R*)-**288** catalyzed by chloroperoxidase from *Caldariomyces fumago* led to the formation of the corresponding (−)-(*S*)-thiophosphate **289** and unoxidized substrate (+)-(*R*)-**288**. The thionoester (*S*/*R*)-**288** was subjected to oxidation with hydrogen peroxide in the presence of chloroperoxidase (CPO) in a mixture of citrate buffer, pH 5, and ethanol. Both compounds were prepared with 99.6% and 97% *ee*, respectively. The thionation of the (−)-(*S*)-phosphate **289** with Lawesson's reagent gave (−)-(*S*)-phosphorodithioate **288** with full stereospecificity, while the oxidation of unreacted substrate (+)-(*R*)-**288** with iodoxybenzene resulted in the formation of (+)-(*R*)-**289** with 94.9% *ee* (Scheme 95) [192].

Prochiral bis(cyanomethyl)phenylphosphine oxide **290** has been successfully transformed into the corresponding optically active monoamide **291** and monoacid **292** with enantiomeric excesses ranging from low (15%) to very high (up to 99%) using a broad spectrum of nitrile-hydrolyzing enzymes [193]. Enzymatic hydrolysis of prochiral bis(cyanomethyl) phenylphosphine oxide **290** was achieved using nitrile-converting enzymes under mild conditions (buffer solution of pH 7.2, 30°C) with formation of cyanomethylphenyl-phosphinylacetamide **291** and cyanomethylphenyl-phosphinylacetic acid **292** in different proportions and various enantioselectivities ranging from 15% to 99% *ee*. For example, the hydrolysis with Nitrilase 106 led to the formation of products (*S*)-**291** and (*S*)-**292** in yields of 10.8% and 51.0% and with 99% and 70% *ee* (Scheme 96)

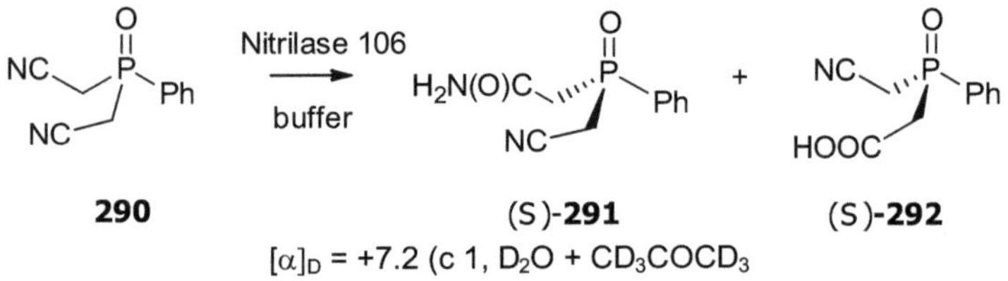

Scheme 96 Enzymatic hydrolysis of bis(cyanomethyl) phenylphosphine oxide **290**

9 Miscellaneous Reactions

Gouverneur et al. showed that prochiral phosphinates and phosphine oxides could be desymmetrized by asymmetric ring-closing metathesis to create a stereogenic phosphorus atom. Enantiomerically enriched *P*-stereogenic phosphinates and phosphine oxides have been prepared by this methodology with 23–98% *ee* and good yields. This methodology allowed access to five-, six-, and seven-membered *P*-stereogenic heterocycles in good to excellent enantioselectivity (Scheme 98). This study revealed that chiral molybdenum-based metathesis catalysts **293** containing identical chiral diol ligands of the same (*S*)-absolute configuration and differing only in their achiral imido ligand led to opposite enantiomers of the product upon ARCM (Scheme 97) [194].

The *P*-chirogenic organocatalysts **294** and **295** were found to promote the enantioselective aza-Morita–Baylis–Hillman reaction of ketimines derived from acyclic α-keto esters. In the *P*-chirogenic organocatalyzed aza-Morita–Baylis–Hillman reactions, α,α-disubstituted α-amino acid derivatives **296** were obtained in high yields and with high enantioselectivities (up to 97% *ee*) (Scheme 98)˙[195].

The addition of methyl magnesium chloride to *P*-*tert*-butyl-*P*-phenyl-*N*-phosphinoyl benzaldimine (*R*)-**296** led to the formation of chiral phosphinoylamines in good yields and moderate stereoselectivity. The amidophosphates were prepared earlier by oxidation of aminophosphines and are identical to compounds **48** and **49** (Scheme 99) [196, 197].

The diastereoselective synthesis of a *P*-chirogenic β-aminophosphine ligand **300** by carbon–carbon bond formation of the ethano bridge in a 3:1 ratio via reaction of an α-metallated *P*-chiral phosphine borane (*S*)-**297** with a benzaldimine was described. The major diastereoisomeric β-aminophosphine borane (S_P)-**298** was separated and decomplexed into the corresponding β-aminophosphine (S_P)-**300** under neutral conditions and without epimerization by heating at reflux in EtOH (Scheme 100) [198].

Phosphine-containing allylic alcohols **301** undergo facile [2,3]-sigmatropic rearrangements with chlorophosphines, furnishing highly enantioenriched (dr 25:1) or enantiopure, crystalline diphosphine monoxides boranes **302**, which were deprotected by treatment with DABCO with formation of **303**. The configuration at the newly-formed stereocenter is opposite to that expected based on prior studies, and an ab initio computational evaluation of the possible transition states was performed to explain the stereochemical course of the reaction (Scheme 101) [199].

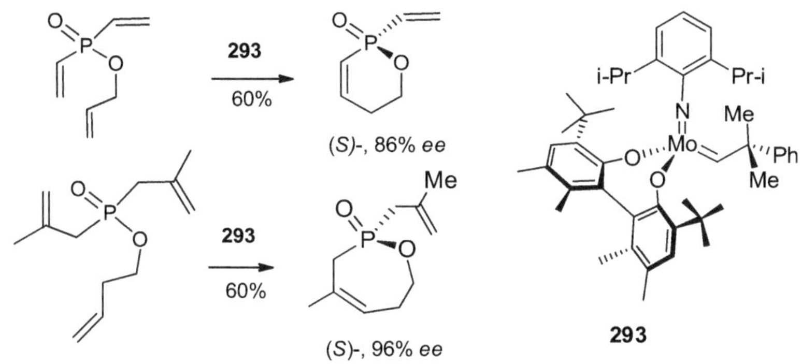

Scheme 97 Enantioselective synthesis of *P*-stereogenic phosphinates

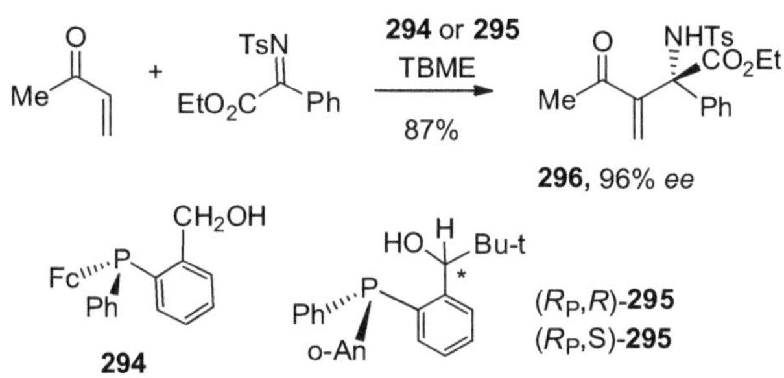

Scheme 98 The organocatalytic enantioselective aza-Morita–Baylis–Hillman reaction

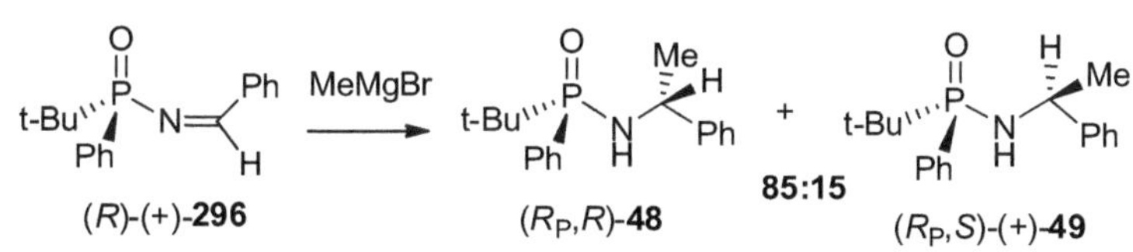

Scheme 99 The addition of Grignard reagent to *N*-phosphinoyl benzaldimine (*R*)-**296**

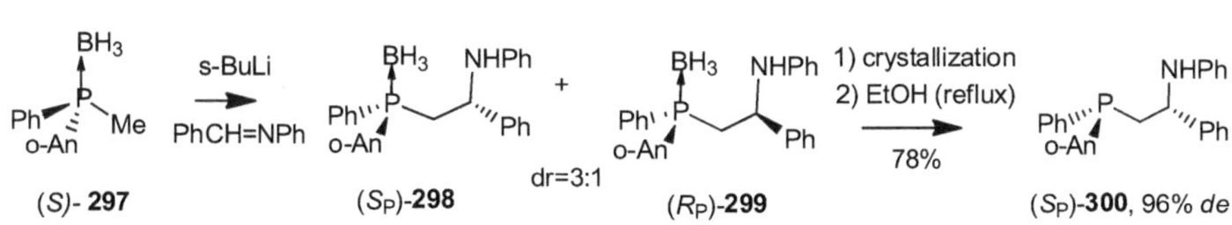

Scheme 100 The diastereoselective synthesis of a *P*-chirogenic β-aminophosphine ligand (S_P)-**300**

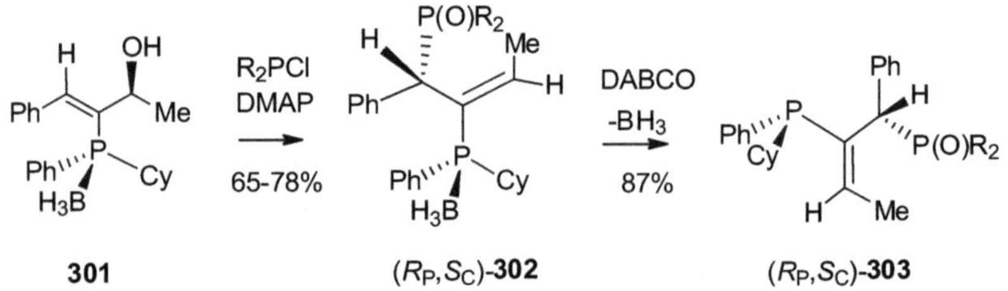

Scheme 101 Synthesis of diphosphine monoxides **303**

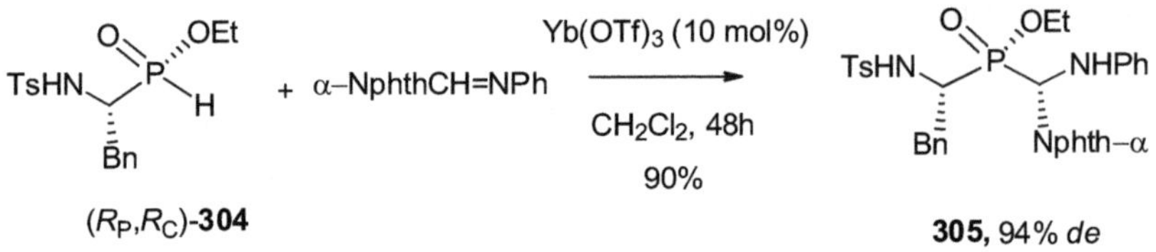

Scheme 102 Preparation of α,α′-diaminophosphinic derivatives **305**

Kaboudin and Yokomatsu reported the diastereoselective addition of α-substituted α-amino-*H*-phosphinates **304** to imines catalyzed by Lewis acids. Among Lewis acids, $Yb(OTf)_3$ was found to be the best catalyst. With this catalyst the α,α′-diaminophosphinic derivatives **305** were obtained with *de*s ranging from 10% to 95%. The structure of product **305** was determined by X-ray analysis as shown in Scheme 102. The reaction proceeded with retention of configuration at the phosphorus atom [200, 201].

Novel *P*-stereogenic PCP pincer Pd complexes **308** were designed and prepared in short steps from optically pure *tert*-butylmethylphosphine borane **306**. The optically active Pd-complexes **308** were successfully applied in asymmetric addition of diarylphosphines to nitroalkenes with high yields and good enantioselectivity (Scheme 103) [202].

Novel synthesis of *P*-chiral 1,3-oxaphospholane **309** starting from optically pure propylene oxide was described. The structure was defined by X-ray crystallographic analysis (Scheme 104) [203].

The cascade reactions starting from the tertiary phosphine borane **310**, providing the biphenyl-fused phosphorinane *P*-boranes **311** and the radical-initiated dearomatizing spirocyclization with formation of benzophospholane *P*-boranes **314**, represent a new methodology for the synthesis of *P*-stereogenic cyclic phosphines. The regiodivergent asymmetric route to *P*-stereogenic five- or six- membered benzophosphacycles, depended on reaction conditions: radical (oxidative addition) or anionic (SNAr) benzannulation The decomplexation of (S_P)-**311** with Et_2NH furnished the corresponding homochiral dibenzophosphorinanes (S_P)-**312** in 93–99% yields. The treatment of (S_P)-310 with *s*-BuLi and $CuCl_2$ gave the chiral spiro-structure (S_P)-314. This spiro benzophospholane borane 313 arose from

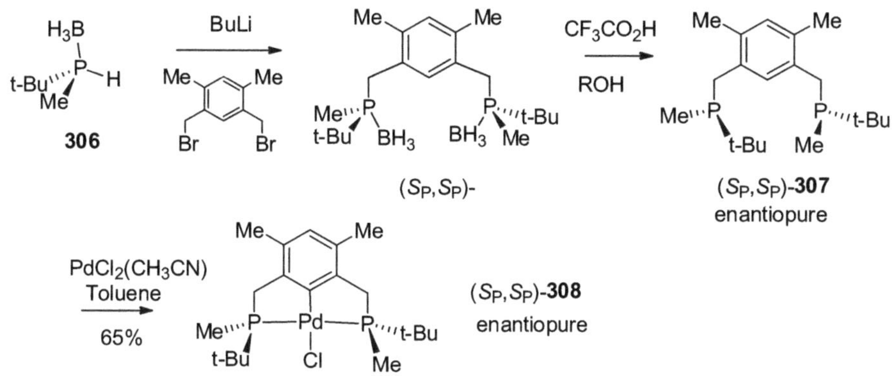

Scheme 103 *P*-Stereogenic PCP pincer Pd complexes **308**

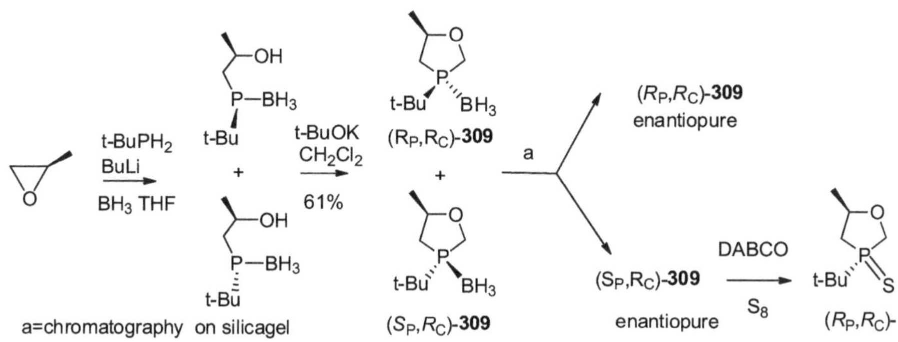

Scheme 104 Synthesis of *P*-chiral 1,3-oxaphospholane **309**

P-α-radical (S_P)-**313** trapping by the neighboring 2,6-dimethoxyphenyl ring, leading to its dearomatization (Scheme 105) [204].

Pereira et al. [205] have reviewed a large scale synthesis of enantiopure and thermally stable (R)- and (S)-BINOL phosphines and phosphinites, which were employed as ligands in palladium-catalyzed hydrosilylation of styrene, affording the corresponding alcohols in high yield and enantiomeric excess.

n-Propoxy-3-methyl-3-phospholene 1-oxide **314** was prepared in optically active form by extending resolution methods applying TADDOL and spiro-TADDOL, as well as the acidic and neutral Ca^{2+} salts of (−)-*O*,*O*′-dibenzoyl- and (−)-*O*,*O*′-di-*P*-toluoyl-(2*R*,3*R*)-tartaric acid. The absolute *P*-configuration of the enantiomers of the phospholene oxide was determined by CD spectroscopy and by single crystal X-ray analysis [206]. Racemic 3-methoxyphenyl(1-naphthyl) phenylphosphine was also effectively resolved via an oxidative resolution procedure utilizing L-menthyl bromoacetate as the resolving agent to give enantiopure 3-methoxyphenyl(1-naphthyl)phenylphosphine oxide (*R*)-**315** in 41% yield (Scheme 106) [207, 208].

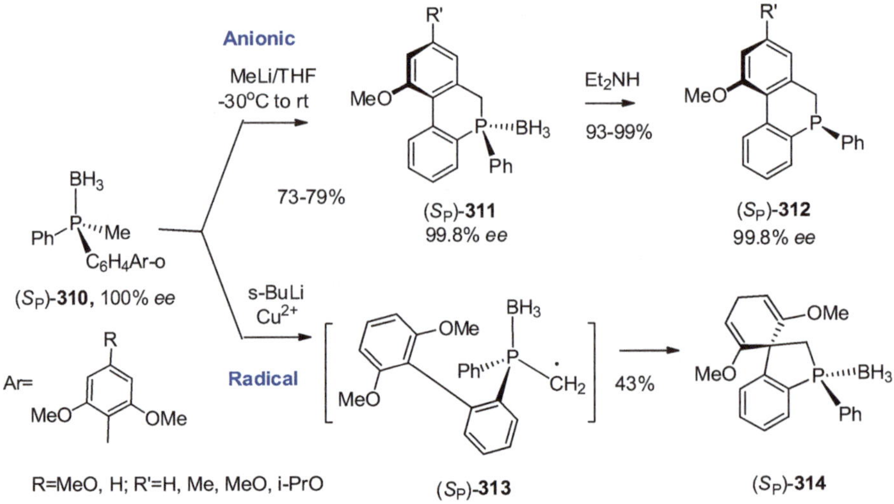

Scheme 105 Synthesis of *P*-stereogenic five- or six-membered benzophosphacycles **312** and **313**

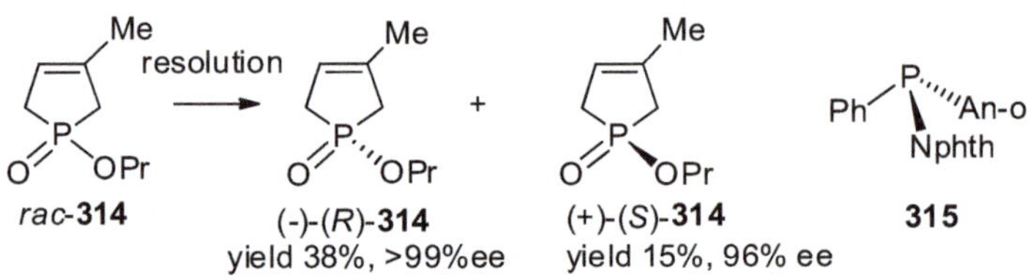

Scheme 106 Resolution of 3-phospholene 1-oxide **314** and tertiary phosphine **315**

Conclusion and Future Directions

It is hoped that this review devoted to *P*-chiral organophosphorus compounds and their application in asymmetric catalysis will be useful to chemists interested in various aspects of organic chemistry and stereochemistry. It should be noted that, despite the impressive progress achieved in the synthesis and studies of properties of *P*-chiral compounds, not all problems have been solved. The problem of the development of enantioselective methods giving easy access to both optical antipodes of chiral tertiary phosphines still remains. The creation of highly effective catalysts for the asymmetric synthesis of *P*-chirogenic compounds, or for the creation of chiral organophosphorus synthons, is an important problem, which is currently awaiting a solution.

It is easy to predict that the basic efforts in studying the chemistry of chiral tertiary phosphines and phosphine oxides will be concentrated in this

(continued)

direction and new enantioselective reactions are promising for solving these problems. The prospects of chemical modification of chiral tertiary phosphines with the introduction of new, more and more complex groups, including those with definite configuration to *P*-centers, are far from exhausted. The actual problem is the resolution of enantiomers and purification of chiral tertiary phosphines and phosphine oxides. The exact structure and of absolute configuration can only be successfully solved in a limited number of cases.

Looking to the future, it seems that tertiary phosphines will be the subject of intensive studies, especially in their range of application as ligands for metal complex catalysts. The most interesting opportunities lie in the development of the application of reagents and catalysts, allowing us to obtain tertiary phosphines by the most accessible methods. It is believed that the area of the application of tertiary phosphines will be expanded greatly.

References

1. Darcel C, Uziel J, Jugé S (2008) In: Börner A (ed) Phosphorus ligands in asymmetric catalysis, vol 3. Wiley-VCH, Weinheim, pp 1211–1233
2. Erre G, Enthaler S, Junge K, Gladiali S, Beller M (2008) Synthesis and application of chiral monodentate phosphines in asymmetric hydrogenation. Coord Chem Rev 252:471–491
3. Harvey JS, Gouverneur V (2010) Catalytic enantioselective synthesis of *P*-stereogenic compounds. J Chem Soc Chem Commun 46:7477–7485
4. Grabulosa (ed) (2011) *P*-Stereogenic ligands in enantioselective catalysis. Royal Society of Chemistry, Cambridge, 501 pp
5. Bayardon J, Juge S (2012) P-Chiral ligands. In: Kamer PCJ, van Leeuwen PWNM (eds) Phosphorus(III) ligands in homogeneous catalysis: design and synthesis. Wiley, Hoboken, pp 355–389
6. Kolodiazhnyi OI (2012) Recent developments in the asymmetric synthesis of *P*-chiral phosphorus compounds. Tetrahedron Asymmetry 23:1–46
7. Grabulosa A, Granell J, Muller G (2007) Preparation of optically pure *P*-stereogenic trivalent phosphorus compounds. Coord Chem Rev 251:25–90
8. Kolodiazhnyi OI (2005) Asymmetric synthesis of hydroxyphosphonates. Tetrahedron Asymmetry 16:3295–3341
9. Wolf C (2008) Dynamic stereochemistry of chiral compounds: principles and applications. RSC, Cambridge, 512 pp
10. Guerrero Rios I, Rosas-Hernandez A, Martin E (2011) Recent advances in the application of chiral phosphine ligands in Pd-catalysed asymmetric allylic alkylation. Molecules 16:970–1010
11. Nemoto T (2008) Transition metal-catalyzed asymmetric reactions using *P*-chirogenic diaminophosphine oxides: DIAPHOXs. Chem Pharm Bull 56:1213–1228
12. Gatineau D, Giordano L, Buono G (2011) Bulky, optically active *P*-stereogenic phosphineboranes from pure H-menthylphosphinates. J Am Chem Soc 133:10728–10731
13. Moraleda D, Gatineau D, Martin D, Giordano L, Buono G (2008) A simple route to chiral phosphinous acid–boranes. J Chem Soc Chem Commun 3031–3033
14. Hoge G (2004) Stereoselective cyclization and pyramidal inversion strategies for *P*-chirogenic phospholane synthesis. J Am Chem Soc 126:9920–9921

15. Kolodiazhnyi OI (1998) Asymmetric synthesis of organophosphorus compounds. Tetrahedron Asymmetry 9:1279–1332
16. Humbel S, Bertrand C, Darcel C, Bauduin C, Jugé S (2003) Configurational stability of chlorophosphines. Inorg Chem 42:420–427
17. Hayakawa Y, Hyodo M, Kimura K, Kataoka M (2003) The first asymmetric synthesis of trialkyl phosphates on the basis of dynamic kinetic resolution in the phosphite method using a chiral source in a catalytic manner. J Chem Soc Chem Commun pp 1704–1705
18. Reichl KD, Ess DH, Radosevich AT (2013) Catalyzing pyramidal inversion: configurational lability of P-stereogenic phosphines via single electron oxidation. J Am Chem Soc 135:9354–9357
19. Imamoto T, Tamura K, Ogura T, Ikematsu Y, Mayama D, Sugiya M (2010) Improved synthetic routes to methylene-bridged *P*-chiral diphosphine ligands via secondary phosphine–boranes. Tetrahedron Asymmetry 21:1522–1528
20. Montchamp J-L (2013) Organophosphorus synthesis without phosphorus trichloride: the case for the hypophosphorous pathway. Phosphorus Sulfur Silicon 188:66–75
21. Cain MF, Glueck DS, Golen JA, Rheingold AL (2012) Asymmetric synthesis and metal complexes of a C_3-symmetric *P*-stereogenic triphosphine, (*R*)-MeSi(CH_2PMe(t-Bu))$_3$ (MT-Siliphos). Organometallics 31:775–778
22. Korpiun O, Lewis RA, Chickos J, Mislow K (1968) Synthesis and absolute configuration of optically active phosphine oxides and phosphinates. J Am Chem Soc 90:4842–4846
23. Crépy KVL, Imamoto TP (2003) Chirogenic phosphine ligands In: Majoral J-P (ed) Topics in Current Chemistry, vol 229. Springer-Verlag, Berlin, Heidelberg, pp 1–41
24. Xu Q, Zhao C-Q, Han L-B (2008) Stereospecific nucleophilic substitution of optically pure *H*-phosphinates: a general way for the preparation of chiral *P*-stereogenic phosphine oxides. J Am Chem Soc 130:12648
25. Leyris A, Bigeault J, Nuel D, Giordano L, Buono G (2007) Enantioselective synthesis of secondary phosphine oxides from (R_P)-(−)-menthyl hydrogenophenylphosphinate. Tetrahedron Lett 48:5247–5250
26. Kolodiazhnyi OI (2005) (1*R*,2*S*,5*R*)-Menthyl phosphinate and its properties. J Russ Gen Chem 75:656–657
27. Kolodiazhna AO (2009) Asymmetric synthesis of hydroxyphosphonates and their derivatives with potential bioilogical activity. PhD Thesis IBONCH, Kiev, 150 pp http://disser.com.ua/content/351636.html
28. Fisher HC, Prost L, Montchamp J-L (2013) Organophosphorus chemistry without PCl_3: a bridge from hypophosphorous acid to *H*-phosphonate diesters. Eur J Org Chem 7973–7978
29. Bravo-Altamirano K, Coudray L, Deal EL, Montchamp J-L (2010) Strategies for the asymmetric synthesis of *H*-phosphinate esters. Org Biomol Chem 8:5541–5551
30. Berger O, Montchamp J-L (2013) A general strategy for the synthesis of *P*-stereogenic compounds. Angew Chem Int Ed 52:11377–11380
31. Coudray L, Montchamp J-L (2008) Green, palladium-catalyzed synthesis of benzylic *H*-phosphinates from hypophosphorous acid and benzylic alcohols. Eur J Org Chem 4101–4103
32. Cardellicchio C, Naso F, Annunziata M, Capozzi M (2004) A convenient route to the phosphorus and sulfur stereoisomers of ethyl menthyl (methylsulfinyl)methylphosphonate. Tetrahedron Asymmetry 15:1471–1476
33. Kolodiazhnyi OI, Sheiko S, Grishkun EV (2000) C_3-Symmetric trialkyl phosphites as starting compounds of asymmetric synthesis. Heteroatom Chem 11:138–143
34. Oliana M, King F, Horton PN, Hursthouse MB, Hii KK (2006) Practical synthesis of chiral vinylphosphine oxides by direct nucleophilic substitution. Stereodivergent synthesis of aminophosphine ligands. J Org Chem 71:2472–2479
35. Cristau H-J, Monbrun J, Schleiss J, Virieux D, Pirat J-L (2005) First synthesis of *P*-arylphosphinosugars, organophosphorus analogues of C-arylglycosides. Tetrahedron Lett 46:3741–3744

36. Ferry A, Malik G, Retailleau P, Guinchard X, Crich D (2013) Alternative synthesis of P-chiral phosphonite-borane complexes: application to the synthesis of phostone – phostone dimers. J Org Chem 78:6858–6867
37. Kolodiazhnyi OI, Gryshkun EV, Andrushko NV, Freytag M, Jones PG, Schmutzler R (2003) Asymmetric synthesis of chiral N-(1-methylbenzyl)aminophosphines. Tetrahedron Asymmetry 14:181–183
38. Gryshkun EV, Andrushko NV, Kolodiazhnyi OI (2004) Stereoselective reactions of chiral amines with racemic chlorophosphines. Phosphorus Sulfur Silicon 179:1027–1046
39. Kolodiazhnyi OI, Andrushko NV, Gryshkun EB (2004) Stereoselective reactions of optically active derivatives of α-methylbenzylaminophosphine. J Russ Gen Chem 74:515–522
40. Gryshkun EV, Kolodiazhna AO, Kolodiazhnyi OI (2003) Synthesis of chiral tert-butylphenylphosphine oxide. J Russ Gen Chem 73:1823–1824
41. Reves M, Ferrer C, Leon T, Doran S, Etayo P, Vidal-Ferran A, Riera A, Verdaguer X (2010) Primary and secondary aminophosphines as novel *P*-stereogenic building blocks for ligand synthesis. Angew Chem Int Ed 49:9452–9455
42. Len T, Parera M, Roglans A, Riera A, Verdaguer X (2012) Chiral N-phosphino sulfinamide ligands in rhodium(I)-catalyzed [2+2+2]-cycloaddition reactions. Angew Chem Int Ed 51:6951–6955
43. Grabulosa A, Doran S, Brandariz G, Muller G, Benet-Buchholz J, Riera A, Verdaguer X (2014) Nickel(II) and palladium(II) complexes of the small-bite-angle P-stereogenic diphosphine ligand MaxPHOS and its monosulfide. Organometallics 33:692–701
44. Brun S, Parera M, Pla-Quintana A, Roglans A, León T, Achard T, Solà J, Verdaguer X, Riera A (2010) *P*-Stereogenic secondary iminophosphorane ligands and their rhodium (I) complexes: taking advantage of NH/PH tautomerism. Tetrahedron 66:9032–9040
45. Kimura T, Murai T (2005) Enantiomerically pure *P*-chiral phosphinoselenoic chlorides: inversion of configuration at the *P*-chirogenic center in the synthesis and reaction of these substances. J Chem Soc Chem Commun 4077–4079
46. Kimura T, Murai T (2004) *P*-Chiral phosphinoselenoic chlorides and optically active *P*-chiral phosphinoselenoic amides: synthesis and stereospecific interconversion with extrusion and addition reactions of the selenium atom. Chem Lett 33:878–879
47. Casimiro M, Roces L, Garcıa-Granda S, Iglesias MJ, Lopez Ortiz F (2013) Directed ortho-lithiation of aminophosphazenes: an efficient route to the stereoselective synthesis of P-chiral compounds. Org Lett 15:2378–2381
48. Adams H, Collins RC, Jones S, Warner CJA (2011) Enantioselective preparation of *P*-chiral phosphine oxides. Org Lett 13:6576–6579
49. Han ZS, Goyal N, Herbage MA, Sieber JD, Qu B, Xu Y, Li Z, Reeves JT, Desrosiers J-N, Ma S, Grinberg N, Lee H, Mangunuru HPR, Zhang Y, Krishnamurthy D, Lu BZ, Song JJ, Wang G, Senanayake CH (2013) Efficient asymmetric synthesis of P-chiral phosphine oxides via properly designed and activated benzoxazaphosphinine-2-oxide agents. J Am Chem Soc 135:2474–2477
50. Nemoto T, Hamada Y (2007) Pd-catalyzed asymmetric allylic substitution reactions using *P*-chirogenic diaminophosphine oxides: DIAPHOXs. Chem Record 7:150–158
51. Maienza F, Spindler F, Thommen M, Pugin B, Malan C, Mezzetti A (2002) Exploring stereogenic phosphorus: synthetic strategies for diphosphines containing bulky, highly symmetric substituents. J Org Chem 67:5239–5249
52. Jugé S (2008) Enantioselective synthesis of *P*-chirogenic phosphorus compounds via the ephedrine-borane complex methodology. Phosphorus Sulfur Silicon 183:233–248
53. Rippert J, Linden A, Hansen HJ (2000) Formation of diastereoisomerically pure oxazaphospholes and their reaction to chiral phosphane-borane adducts. Helv Chim Acta 83:311–321
54. Johansson JM, Kann N, Larsson K (2004) (2R,4S,5R)-3,4-Dimethyl-5-phenyl-2-[4-(trifluoromethyl)phenyl]-1,3,2-oxazaphospholidine(*P*-B)borane. Acta Cryst E60:o287–o288

55. Leon T, Riera A, Verdaguer X (2011) Stereoselective synthesis of *P*-stereogenic aminophosphines: ring opening of bulky oxazaphospholidines. J Am Chem Soc 133:5740–5743
56. Zijlstra H, León T, de Cózar A, Fonseca Guerra C, Byrom D, Riera A, Verdaguer X, Bickelhaupt FM (2013) Stereodivergent S_N2@P reactions of borane oxazaphospholidines: experimental and theoretical studies. J Am Chem Soc 135:4483–4491
57. Khiri N, Bertrand E, Ondel-Eymin M-J, Rousselin Y, Bayardon J, Harvey PD, Juge S (2010) Enantioselective hydrogenation catalysis aided by a σ-bonded calix[4]arene to a *P*-chirogenic aminophosphane phosphinite rhodium complex. Organometallics 29:3622–3631
58. Imamoto T, Saitoh Y, Koide A, Ogura T, Yoshida K (2007) Synthesis and enantioselectivity of *P*-chiral phosphine ligands with alkynyl groups. Angew Chem Intern Ed 46:8636–8639
59. Salomon C, Dal Molin S, Fortin D, Mugnier Y, Boere RT, Juge S, Harvey PD (2010) The first unpaired electron placed inside a C3-symmetry *P*-chirogenic cluster. J Chem Soc Dalton Trans 39:10068–10075
60. Bauduin C, Moulin D, Kaloun EB, Darcel C, Jugé S (2003) Highly enantiomerically enriched chlorophosphine boranes: synthesis and applications as *P*-chirogenic electrophilic blocks. J Org Chem 68:4293–4301
61. Darcel C, Moulin D, Henry JC, Lagrelette M, Richard P, Harvey PD, Jugé S (2007) Modular *P*-chirogenic aminophosphane-phosphinite ligands for Rh-catalyzed asymmetric hydrogenation: a new model for prediction of enantioselectivity. Eur J Org Chem 13:2078–2090
62. Colby EA, Jamison TF (2003) *P*-Chiral, monodentate ferrocenyl phosphines, novel ligands for asymmetric catalysis. J Org Chem 68:156–166
63. Imamoto T, Tamura K, Zhang Z, Horiuchi Y, Sugiya M, Yoshida K, Yanagisawa A, Gridnev ID (2012) Rigid *P*-chiral phosphine ligands with tert-butylmethylphosphino groups for rhodium-catalyzed asymmetric hydrogenation of functionalized alkenes. J Am Chem Soc 134:1754–1769
64. Vinci D, Mateus N, Wu X, Hancock F, Steiner A, Xiao J (2006) Oxazaphospholidine-oxide as an efficient ortho-directing group for the diastereoselective deprotonation of ferrocene. Org Lett 8:215–218
65. Grabulosa A, Muller G, Ordinas JI, Mezzetti A, Maestro MA, Font-Bardia M, Solans X (2005) Allylpalladium complexes with *P*-stereogenic monodentate phosphines. application in the asymmetric hydrovinylation of styrene. Organometallics 24:4961–4973
66. Lam H, Horton PN, Hursthouse MB, Aldous DJ, Hii KK (2005) Synthesis of *P*-chirogenic diarylphosphinoacetic acids and their proline derivatives for palladium-catalysed allylic alkylation reactions. Tetrahedron Lett 46:8145–8148
67. Leyris A, Nuel D, Giordano L, Achard M, Buono G (2005) Enantioselective synthesis of both enantiomers of tert-butylphenylphosphine oxide from (*S*)-prolinol. Tetrahedron Lett 46:8677–8680
68. Ngono CJ, Constantieux T, Buono G (2006) Diastereoselective synthesis of new *P*-stereogenic (ortho-hydroxyaryl)-diazaphospholidine–borane complexes by a totally stereoselective P–O to P–C migration rearrangement. Eur J Org Chem 1499–1507
69. Toselli N, Fortrie R, Martin D, Buono G (2010) New *P*-stereogenic triaminophosphines and their derivatives: synthesis, structure, conformational study, and application as chiral ligands. Tetrahedron Asymmetry 21:1238–1245
70. Delapierre G, Achard M, Buono G (2002) New *P*-stereogenic triaminophosphines and their derivatives: synthesis, structure, conformational study, and application as chiral ligands. Tetrahedron Lett 43:4025–4028
71. Rajendran KV, Gilheany DG (2012) Identification of a key intermediate in the asymmetric Appel process: one pot stereoselective synthesis of *P*-stereogenic phosphines and phosphine boranes from racemic phosphine oxides. J Chem Soc Chem Commun 48:10040–10042
72. Bergin E, O'Connor CT, Robinson SB, McGarrigle EM, O'Mahony CP, Gilheany DG (2007) Synthesis of *P*-stereogenic phosphorus compounds. Asymmetric oxidation of phosphines under Appel conditions. J Am Chem Soc 129:9566–9567

73. Rajendran KV, Kennedy L, Gilheany DG (2010) P-Stereogenic phosphorus compounds: effect of aryl substituents on the oxidation of arylmethylphenylphosphanes under asymmetric Appel conditions. Eur J Org Chem 5642–5649
74. Rajendran KV, Kudavalli JS, Dunne KS, Gilheany DG (2012) A U-turn in the asymmetric Appel reaction: stereospecific reduction of diastereomerically enriched alkoxyphosphonium salts allows the asymmetric synthesis of *P*-stereogenic phosphanes and phosphane boranes. Eur J Org Chem 2720–2723
75. Rajendran KV, Kennedy L, O'Connor CT, Bergin E, Gilheany DG (2013) Systematic survey of positive chlorine sources in the asymmetric Appel reaction: oxalyl chloride as a new phosphine activator. Tetrahedron Lett 54:7009–7012
76. Carr DJ, Kudavalli JS, Dunne KS, Müller-Bunz H, Gilheany DG (2013) Synthesis of 2,3-dihydro-1-phenylbenzo[b]phosphole (1-phenylphosphindane) and its use as a mechanistic test in the asymmetric Appel reaction: decisive evidence against involvement of pseudorotation in the stereoselecting step. J Org Chem 78:10500–10505
77. Nikitin K, Rajendran KV, Müller-Bunz H, Gilheany DG (2014) Turning regioselectivity into stereoselectivity: efficient dual resolution of *P*-stereogenic phosphine oxides through bifurcation of the reaction pathway of a common intermediate. Angew Chem Int Ed 53:1906–1909
78. Muci AR, Campos KR, Evans DA (1995) Enantioselective deprotonation as a vehicle for the asymmetric synthesis of C2-symmetric *P*-chiral diphosphines. J Am Chem Soc 117:9075–9076
79. Strohmann C, Strohfeldt K, Schildbach D, McGrath MJ, O'Brien P (2004) Crystal structures of (+)-sparteine surrogate adducts of methyllithium and phenyllithium. Organometallics 23:5389–5391
80. Strohmann C, Seibel T, Strohfeldt K (2003) Monomeric butyllithium compound [t-BuLi · (−)-sparteine]: molecular structure of the first monomeric butyllithium compound. Angew Chem Int Ed 42:4531–4533
81. Vestergren M, Eriksson J, Hilmersson G, Håkansson MJ (2003) Giving phenyllithium a right-handed double-helical twist. Syntheses and crystal structures of enantiopure alkyl-, aryl-, and amidolithium aggregates. J Organomet Chem 682:172–179
82. Genet C, Canipa SJ, O'Brien P, Taylor S (2006) Catalytic asymmetric synthesis of ferrocenes and *P*-stereogenic bisphosphines. J Am Chem Soc 128:9336–9337
83. Granander J, Secci F, Canipa SJ, O'Brien P, Kelly B (2011) One-ligand catalytic asymmetric deprotonation of a phosphine borane: synthesis of *P*-stereogenic bisphosphine ligands. J Org Chem 76:4794–4799
84. Imamoto T, Nishimura M, Koide A, Yoshida K (2007) t-Bu-QuinoxP* ligand: applications in asymmetric Pd-catalyzed allylic substitution and Ru-catalyzed hydrogenation. J Org Chem 72:7413–7416
85. Wu X, O'Brien P, Ellwood S, Secci F, Kelly B (2013) Synthesis of P-stereogenic phospholene boranes via asymmetric deprotonation and ring-closing metathesis. Org Lett 15:192–195
86. Gammon JJ, Canipa SJ, O'Brien P, Kelly B, Taylor S (2008) Catalytic asymmetric deprotonation of phosphine boranes and sulfides as a route to *P*-stereogenic compounds. J Chem Soc Chem Commun 3750–3752
87. Johansson MJ, Berglund S, Hu Y, Andersson KHO, Kann N (2012) Parallel and modular synthesis of *P*-chirogenic P,O-ligands. ACS Comb Sci 14:304–308
88. Dolhem F, Johansson MJ, Antonsson T, Kann N (2006) *P*-Chirogenic α-carboxyphosphine boranes as effective pre-ligands in palladium-catalyzed asymmetric reactions. Synlett 3389–3394
89. Johansson MJ, Schwartz L, Amedjkouh M, Kann N (2004) New chiral amine ligands in the desymmetrization of prochiral phosphine boranes. Tetrahedron Asymmetry 15:3531–3538
90. Morisaki Y, Imoto H, Tsurui K, Chujo Y (2009) Practical synthesis of *P*-stereogenic diphosphacrowns. Org Lett 11:2241–2244

91. Morisaki Y, Imoto H, Hirano K, Hayashi T, Chujo Y (2011) Synthesis of enantiomerically pure *P*-stereogenic diphosphacrowns and their palladium complexes. J Org Chem 76:1795–1803
92. Morisaki Y, Kato R, Chujo Y (2013) Synthesis of enantiopure *P*-stereogenic diphosphacrowns using *P*-stereogenic secondary phosphines. J Org Chem 78:2769–2774
93. Canipa SJ, O'Brien P, Taylor S (2009) Catalytic asymmetric deprotonation of a phosphine borane: comparison of two-ligand and one-ligand catalysis. Tetrahedron Asymmetry 20:2407–2412
94. Johansson MJ, Schwartz LO, Amedjkouh M, Kann NC (2004) Desymmetrization of prochiral phosphanes using derivatives of (−)-cytisine. Eur J Org Chem 1894–1896
95. Johansson MJ, Andersson KHO, Kann N (2008) Modular asymmetric synthesis of *P*-chirogenic-amino phosphine boranes. J Org Chem 73:4458–4463
96. Morisaki Y, Imoto H, Ouchi Y, Nagata Y, Chujo Y (2008) Stereospecific construction of a *trans*-1,4-diphosphacyclohexane skeleton. Org Lett 10:1489–1492
97. Imoto H, Morisaki Y, Chujo Y (2010) Synthesis and coordination behaviors of *P*-stereogenic polymers. J Chem Soc Chem Commun 46:7542–7544
98. Ouchi Y, Morisaki Y, Chujo Y (2006) Synthesis of photoresponsive polymers having *P*-chiral phosphine in the main chain. Polym Preprints 47:708–709
99. Ouchi Y, Morisaki Y, Ogoshi T, Chujo Y (2007) Synthesis of a stimuli-responsive *P*-chiral polymer with chiral phosphorus atoms and azobenzene moieties in the main chain. Chem Asian J 2:397–402
100. Oohara N, Katagiri K, Imamoto T (2003) A novel P-chirogenic phosphine ligand, (S, S)-1,2-bis-[(ferrocenyl)methylphosphino]ethane: synthesis and use in rhodium-catalyzed asymmetric hydrogenation and palladium-catalyzed asymmetric allylic alkylation. Tetrahedron Asymmetry 14:2171–2175
101. Hoge G (2003) Synthesis of both enantiomers of a *P*-chirogenic 1,2-bisphospholanoethane ligand via convergent routes and application to rhodium-catalyzed asymmetric hydrogenation of CI-1008 (Pregabalin). J Am Chem Soc 125:10219–10227
102. Tang W, Wang W, Zhang X (2003) Phospholane–oxazoline ligands for Ir-catalyzed asymmetric hydrogenation. Angew Chem Int Ed 42:943–946
103. Liu D, Zhang X (2005) Practical *P*-chiral phosphane ligand for Rh-catalyzed asymmetric hydrogenation. Eur J Org Chem 646–649
104. Tang W, Liu D, Zhang X (2003) Asymmetric hydrogenation of itaconic acid and enol acetate derivatives with the Rh-TangPhos catalyst. Org Lett 5:205–207
105. Gammon JJ, Gessner VH, Barker GR, Granander J, Whitwood AC, Strohmann C, O'Brien P, Kelly B (2010) Synthesis of *P*-stereogenic compounds via kinetic deprotonation and dynamic thermodynamic resolution of phosphine sulfides: opposite sense of induction using (−)-sparteine. J Am Chem Soc 132:13922–13927
106. Miyazaki T, Sugawara M, Danjo H, Imamoto T (2004) Dihydroboronium derivatives of (S, S)-1,2-bis(t-butylmethylphosphino)ethane as convenient chiral ligand precursors. Tetrahedron Lett 45:9341–9344
107. Imamoto T, Crepy KVL, Katagiri K (2004) Optically active 1,10-di-tert-butyl-2,20-dibenzophosphetenyl: a highly strained *P*-stereogenic diphosphine ligand. Tetrahedron Asymmetry 75:2213–2218
108. Gammon JJ, O'Brien P, Kelly B (2009) Regioselective lithiation of silyl phosphine sulfides: asymmetric synthesis of *P*-stereogenic compounds. Org Lett 11:5022–5025
109. Glueck DS (2008) Catalytic asymmetric synthesis of chiral phosphanes. Chem Eur J 14:7108–7117
110. Glueck DS (2008) Applications of ^{31}P NMR spectroscopy in development of M(Duphos)-catalyzed asymmetric synthesis of *P*-stereogenic phosphines (M=Pt or Pd). Coord Chem Rev 252:2171–2179
111. Glueck DS (2007) Metal-catalyzed asymmetric synthesis of *P*-stereogenic phosphines. Synlett 2627–2634

112. Moncarz JR, Brunker TJ, Jewett JC, Orchowski M, Glueck DS, Sommer RD, Lam K-C, Incarvito CD, Concolino TE, Ceccarelli C, Zakharov LN, Rheingold AL (2003) Palladium-catalyzed asymmetric phosphination. enantioselective synthesis of PAM*P*-BH3, ligand effects on catalysis, and direct observation of the stereochemistry of transmetalation and reductive elimination. Organometallics 22:3205–3221
113. Glueck DS (2010) Recent advances in metal-catalyzed C–P bond formation. Topics Organomet Chem 31:65–100
114. Moncarz JR, Brunker TJ, Glueck DS, Sommer RD, Rheingold AL (2003) Stereochemistry of palladium-mediated synthesis of PAMP$-BH_3$: retention of configuration at P in formation of Pd$-$P and P$-$C bonds. J Am Chem Soc 125:1180–1181
115. Brunker TJ, Anderson BJ, Blank NF, Glueck DS, Rheingold AL (2007) Enantioselective synthesis of *P*-stereogenic benzophospholanes via palladium-catalyzed intramolecular cyclization. Org Lett 9:1109–1112
116. Guino-o MA, Zureick AH, Blank NF, Anderson BJ, Chapp TW, Kim Y, Glueck DS, Rheingold AL (2012) Synthesis and structure of platinum bis(phospholane) complexes Pt (diphos*)(R)(X), catalyst precursors for asymmetric phosphine alkylation. Organometallics 31:6900–6910
117. Chapp TW, Glueck DS, Golen JA, Moore CE, Rheingold AL (2010) Platinum-catalyzed asymmetric alkylation of bis(isitylphosphino)ethane: stereoselectivity reversal in successive formation of two P–C bonds. Organometallics 29:378–388
118. Scriban C, Glueck DS, DiPasquale AG, Rheingold AL (2006) Chiral platinum duphos terminal phosphido complexes: synthesis, structure, phosphido transfer, and ligand behavior. Organometallics 25:5435–5448
119. Chapp TW, Schoenfeld AJ, Glueck DS (2010) Effects of linker length on the rate and selectivity of platinum-catalyzed asymmetric alkylation of the bis(isitylphosphino)alkanes $IsHP(CH_2)_nPHIs$ ($Is = 2,4,6\text{-}(i\text{-}Pr)_3C_6H_2$, $n = 1\text{–}5$). Organometallics 29:2465–2473
120. Scriban C, Glueck DS, Golen JA, Rheingold AL (2007) Platinum-catalyzed asymmetric alkylation of a secondary phosphine: mechanism and origin of enantioselectivity. Organometallics 26:1788–1800
121. Scriban C, Glueck DS (2006) Platinum-catalyzed asymmetric alkylation of secondary phosphines: enantioselective synthesis of P-stereogenic phosphines. J Am Chem Soc 128:2788–2789
122. Scriban C, Glueck DS, Golen JA, Rheingold AL (2007) Platinum-catalyzed asymmetric alkylation of a secondary phosphine: mechanism and origin of enantioselectivity. Organometallics 26:5124
123. Anderson BJ, Guino-o MA, Glueck DS, Golen JA, DiPasquale AG, Liable-Sands LM, Rheingold AL (2008) Platinum-catalyzed enantioselective tandem alkylation/arylation of primary phosphines. Asymmetric synthesis of *P*-stereogenic 1-phosphaacenaphthenes. Org Lett 10:4425–4428
124. Scriban C (2009) Catalytic asymmetric phosphorus-carbon bond formation with platinum-phosphido complexes. Ph.D. thesis, pp 1–464. http://search.proquest.com/docview/304879651
125. Scriban C, Kovacik I, Glueck DS (2005) A protic additive suppresses formation of byproducts in platinum-catalyzed hydrophosphination of activated olefins. Evidence for P-C and C-C bond formation by Michael addition. Organometallics 24:4871–4874
126. Anderson BJ, Glueck DS, DiPasquale AG, Rheingold AL (2008) Substrate and catalyst screening in platinum-catalyzed asymmetric alkylation of bis(secondary) phosphines. Synthesis of an enantiomerically pure C2-symmetric diphosphine. Organometallics 27:4992–5001
127. Scriban C, Glueck DS (2006) Asymmetric catalytic synthesis of *P*-stereogenic phosphines via a nucleophilic ruthenium phosphido complex. J Am Chem Soc 128:2786–2787

128. Chan VS, Chiu M, Bergman RG, Toste FD (2009) Development of ruthenium catalysts for the enantioselective synthesis of *P*-stereogenic phosphines via nucleophilic phosphido intermediates. J Am Chem Soc 131:6021–6032
129. Gschwend B (2009) *P*-Chiral phosphorus ligands: synthesis and application in asymmetric hydrogenation inaugural dissertation, Basel, http://edoc.unibas.ch/1066/1/Dissertation_Bj%C3%B6rn_Gschwend.pdf
130. Blank NF, McBroom KC, Glueck DS, Kassel WS, Rheingold AL (2006) Chirality breeding via asymmetric phosphination. Palladium-catalyzed diastereoselective synthesis of a *P*-stereogenic phosphine. Organometallics 25:1742–1748
131. Moncarz JR, Laritcheva NF, Glueck DS (2002) Palladium-catalyzed asymmetric phosphination: enantioselective synthesis of a *P*-chirogenic phosphine. J Am Chem Soc 124:13356–13357
132. Blank NF, Moncarz JR, Brunker TJ, Scriban C, Anderson BJ, Amir O, Glueck DS, Zakharov LN, Golen JA, Incarvito CD, Rheingold AL (2007) Palladium-catalyzed asymmetric phosphination. Scope, mechanism, and origin of enantioselectivity. J Am Chem Soc 129:6847–6858
133. Chan VS, Bergman RG, Toste FD (2007) Pd-catalyzed dynamic kinetic enantioselective arylation of silylphosphines. J Am Chem Soc 129:15122–15123
134. Pican S, Gaumont A-C (2005) Palladium catalysed enantioselective phosphination reactions using secondary phosphine-boranes and aryl iodide. J Chem Soc Chem Commun 2393–2395
135. Scriban C, Glueck DS (2012) Method for enantioselective synthesis of phosphorus-stereogenic phosphines. US 8,193,392 B2 http://www.lens.org/lens/patent/US_8193392_B2
136. Zhou SJ, Huang Z (2013) Chiral phosphines for palladium-catalyzed asymmetric alpha-arylation of ester enolates to produce tertiary stereocenters in high enantioselectivity. WO/2013/028132 http://www.google.com/patents/WO2013028132A1?cl=en
137. Korff C, Helmchen G (2004) Preparation of chiral triarylphosphines by Pd-catalysed asymmetric P–C cross-coupling. J Chem Soc Chem Commun 530–531
138. Join B, Mimeau D, Delacroix O, Gaumont A-C (2006) Pallado-catalysed hydrophosphination of alkynes: access to enantioenriched *P*-stereogenic vinyl phosphine–boranes. J Chem Soc Chem Commun 3249–3251
139. Lu J, Ye J, Duan W-L (2013) Palladium-catalyzed asymmetric addition of diarylphosphines to α, β-unsaturated sulfonic esters for the synthesis of chiral phosphine sulfonate compounds. Org Lett 15:5016–5019
140. Rosenberg L (2013) Mechanisms of metal-catalyzed hydrophosphination of alkenes and alkynes. ACS Catal 3:2845–2855
141. Lebel H, Morin S, Paquet V (2003) Alkylation of phosphine boranes by phase-transfer catalysis. Org Lett 5:2347–2349
142. Reynolds SC, Hughes RP, Glueck DS, Rheingold AL (2012) Synthesis, reactivity, and resolution of a C2 symmetric, *P*-stereogenic benzodiphosphetane, a building block for chiral bis(phosphines). Org Lett 14:4238–4241
143. Dai L-X, Hou X-L (eds) (2010) Chiral ferrocenes in asymmetric catalysis: synthesis and applications. Wiley-VCH, Hoboken, p 431
144. Colacot TJ (2003) A concise update on the applications of chiral ferrocenyl phosphines in homogeneous catalysis leading to organic synthesis. Chem Rev 103:3101–3118
145. Chen WP, Mbafor W, Roberts SM, Whittall J (2006) A very simple, highly stereoselective and modular synthesis of ferrocene-based *P*-chiral phosphine ligands. J Am Chem Soc 128:3922–3923
146. Boaz NW, Mackenzie EB, Debenham SD, Large SE, Ponasik JA (2005) Synthesis and application of phosphinoferrocenylaminophosphine ligands for asymmetric catalysis. J Org Chem 70:1872–1880
147. Gschwend B, Pugin B, Bertogg A, Pfaltz A (2009) *P*-Chiral ferrocenephospholanes: synthesis, reactivity, metal complex chemistry and application in the asymmetric hydrogenation of olefins. Chem Eur J 15:12993–13007

148. Chen WP, McCormack PJ, Mohammed K, Mbafor W, Roberts SM, Whittall J (2007) Stereoselective synthesis of ferrocene-based c2-symmetric diphosphine ligands: application to the highly enantioselective hydrogenation of α-substituted cinnamic acids. Angew Chem Int Ed 46:4141–4144
149. Chen WP, Roberts SM, Whittall J, Steiner A (2006) An efficient and highly stereoselective synthesis of new *P*-chiral 1,5-diphosphanylferrocene ligands and their use in enantioselective hydrogenation. J Chem Soc Chem Commun 2916–2918
150. Chen W, Spindler F, Pugin B, Nettekoven U (2013) ChenPhos: highly modular *P*-stereogenic C1-symmetric diphosphine ligands for the efficient asymmetric hydrogenation of α-substituted cinnamic acids. Angew Chem Intern Ed 52:8652–8656
151. Bansal RK, Karaghiosoff K, Gupta N, Gandhi N, Kumawata SK (2005) Diastereo- and regioselectivity in Diels–Alder reaction of [1,4,2] diazaphospholo[4,5-a]pyridines. Tetrahedron 61:10521–10528
152. Leung PH (2004) Asymmetric synthesis and organometallic chemistry of functionalized phosphines containing stereogenic phosphorus centers. Acc Chem Res 37:169–177
153. Ng JK-P, Chen S, Li Y, Tan G-K, Koh L-L, Leung P-H (2007) Cyclopalladation of the prochiral (di-tert-butyl)(diphenylmethyl)phosphine: kinetic lability of the corresponding (+)-phosphapalladacyclic Pd–C bond and the reluctance of the phosphine to bind in a monodentate fashion. Inorg Chem 46:5100–5109
154. Chen S, Pullarkat SA, Li Y, Leung P-H (2011) Synthesis of homo- and hetero-bimetallic arsenic complexes by means of regioselective monoinsertion of alkynylarsane into the Pd–C bond of a palladacycle. Eur J Inorg Chem 3111–3121
155. Qin Y, Lang H, Vittal JJ, Tan G-K, Selvaratnam S, White AJP, Williams DJ, Leung P-H (2003) Asymmetric synthesis and coordination chemistry of bidentate *P*-stereogenic phosphines containing ester and thionoester functionalities. Organometallics 22:3944–3950
156. Bungabong ML, Tan KW, Li Y, Selvaratnam SV, Dongol KG, Leung P-H (2007) A novel asymmetric hydroarsination reaction promoted by a chiral organopalladium complex. Inorg Chem 46:4733–4736
157. Ma M, Pullarkat SA, Yuan M, Zhang N, Li Y, Leung P-H (2009) Metal effects on the asymmetric cycloaddition reaction between 3,4-dimethyl-1-phenylarsole and diphenylvinylphosphine oxide. Organometallics 28:4886–4889
158. Li Y, Selvaratnam S, Vittal JJ, Leung PH (2003) A rational approach to the design and synthesis of chiral organopalladium-amine complexes. Inorg Chem 42:3229–3236
159. Yeo W-C, Chen S, Tan G-K, Leung P-H (2007) Synthesis of *P*-chiral phosphines via chiral metal template promoted asymmetric furan Diels–Alder reaction. J Organomet Chem 692:2539–2547
160. Yeo W-C, Tang L, Yan B, Tee S-Y, Koh LL, Tan G-K, Leung P-H (2005) Chiral metal template induced asymmetric synthesis of a mixed phosphine-phosphine oxide ligand. Organometallics 24:5581–5585
161. Yuan M, Pullarkat SA, Li Y, Lee Z-Y, Leung P-H (2010) Novel synthesis of chiral 1,3-diphosphines via palladium template promoted hydrophosphination and functional group transformation reactions. Organometallics 29:3582–3588
162. Yuan M, Zhang NA, Pullarkat S, Li Y, Liu F, Pham P-T, Leung P-H (2010) Asymmetric synthesis of functionalized 1,3-diphosphines via chiral palladium complex promoted hydrophosphination of activated olefins. Inorg Chem 49:989–996
163. Yeo WC, Vittal JJ, Koh LL, Tan GK, Leung PH (2004) Chiral metal template promoted asymmetric pyrrole Diels–Alder reaction between N-(diphenylphosphino)pyrrole and diphenylvinylphosphine. Organometallics 23:3474–3482
164. Chen S, Pullarkat SA, Li Y, Leung P-H (2011) Asymmetric synthesis of *P*-stereogenic homo- and heterobimetallic complexes via selective monoinsertion of dialkynylphosphine into the Pd–C bond of a palladacycle. Organometallics 30:1530–1550

165. Leung PH, Siah SY, White AJP, Williams DJ (1998) Asymmetric syntheses, structures and reactions of palladium(II) complexes containing thiolato- and sulfinyl-substituted P chiral phosphines. J Chem Soc Dalton Trans 893–899
166. Chen K, Pullarkat SA, Li Y, Leung PH (2012) Chiral cyclopalladated complex promoted asymmetric synthesis of diester-substituted P, N-ligands via stepwise hydrophosphination and hydroamination reactions. J Chem Soc Dalton Trans 41:5391–5400
167. Chen S, Ng JK-P, Pullarkat SA, Liu F, Li Y, Leung P-H (2010) Asymmetric synthesis of new diphosphines and pyridylphosphines via a kinetic resolution process promoted and controlled by a chiral palladacycle. Organometallics 29:3374–3386
168. Huang Y, Li Y, Leung P-H, Hayashi T (2014) Asymmetric synthesis of *P*-stereogenic diarylphosphinites by palladium-catalyzed enantioselective addition of diarylphosphines to benzoquinones. J Am Chem Soc 136:4865–4868
169. Huang Y, Pullarkat SA, Teong S, Chew RJ, Li Y, Leung PH (2012) Palladacycle-catalyzed asymmetric intermolecular construction of chiral tertiary *P*-heterocycles by stepwise addition of H–P–H bonds to bis(enones). Organometallics 31:4871–4875
170. Möller T, Sárosi MB, Hey-Hawkins E (2012) Asymmetric phospha-Diels–Alder reaction: a stereoselective approach towards *P*-chiral phosphanes through diastereotopic face differentiation. Chem Eur J 18:16604–16607
171. Fukawa N, Osaka T, Noguchi K, Tanaka K (2010) Asymmetric synthesis and photophysical properties of benzopyrano- or naphthopyrano-fused helical phosphafluorenes. Org Lett 12:1324
172. Krasinski G, Cypryk M, Kwiatkowska M, Mikołajczyk M, Kiełbasinski P (2012) Molecular modeling of the lipase-catalyzed hydrolysis of acetoxymethyl(i-propoxy)phenylphosphine oxide and its *P*-borane analogue. J Mol Graph Model 38:290–297
173. Kiełbasinski P, Mikołajczyk M (2007) Chiral heteroatom-containing compounds. In: Matsuda T (ed) Future directions in biocatalysis. Elsevier, Amsterdam, pp 159–203
174. Shioji K, Kurauchi Y, Okuma K (2003) Novel synthesis of *P*-chiral hydroxymethylphosphine–boranes through lipase-catalyzed optical resolution. Bull Chem Soc Jpn 76:833–834
175. Kiełbasinski P, Albrycht M, Zurawinski R, Mikołajczyk M (2006) Lipase-mediated kinetic resolution of racemic and desymmetrization of prochiral organophosphorus *P*-boranes. J Mol Catal B 39:45–49
176. Kwiatkowska M, Krasiński G, Cypryk M, Cierpiał T, Kiełbasiński P (2011) Lipase-mediated stereoselective transformations of chiral organophosphorus *P*-boranes revisited: revision of the absolute configuration of alkoxy(hydroxymethyl)phenylphosphine *P*-boranes. Tetrahedron Asymmetry 22:1581–1590
177. Wiktelius D, Johansson MJ, Luthman K, Kann N (2005) A biocatalytic route to *P*-chirogenic compounds by lipase-catalyzed desymmetrization of a prochiral phosphine-borane. Org Lett 7:4991–4994
178. Kolodiazhnyi OI (2011) Enzymatic synthesis of organophosphorus compounds. Russ Chem Rev 80:883–910
179. Shioji K, Tashiro A, Shibata S, Okuma K (2003) Synthesis of bifunctional *P*-chiral hydroxy phosphinates; lipase-catalyzed stereoselective acylation of ethyl (1-hydroxyalkyl) phenylphosphinates. Tetrahedron Lett 44:1103–1105
180. Majewska P, Kafarski P, Lejczak B (2006) Simple and effective method for the deracemization of ethyl 1-hydroxyphosphinate using biocatalysts with lipolytic activity. Tetrahedron Asymmetry 17:2870–2875
181. Yamagishi T, Miyamae T, Yokomatsu T, Shibuya S (2004) Lipase-catalyzed kinetic resolution of α-hydroxy-*H*-phosphinates. Tetrahedron Lett 45:6713–6716
182. Yamagishi T, Mori J-I, Haruki T, Yokomatsu T (2011) A chemo-enzymatic synthesis of optically active 1,1-diethoxyethyl(aminomethyl)phosphinates: useful chiral building blocks for phosphinyl dipeptide isosteres. Tetrahedron Asymmetry 22:1358–1363

183. Albrycht M, Kiełbasinski P, Drabowicz J, Mikołajczyk M, Matsuda T, Harada T, Nakamura K (2005) Supercritical carbon dioxide as a reaction medium for enzymatic kinetic resolution of *P*-chiral hydroxymethanephosphinates. Tetrahedron Asymmetry 16:2015–2018
184. Shioji K, Ueyama T, Ueda N, Mutoh E, Kurisaki T, Wakita H, Okuma K (2008) Evaluation of enantioselectivity in lipase-catalyzed acylation of hydroxyalkylphosphine oxides. J Mol Cat B Enzym 55:146–151
185. Kiełbasinski P, Zurawinski R, Albrycht M, Mikołajczyk M (2003) The first enzymatic desymmetrizations of prochiral phosphine oxides. Tetrahedron Asymmetry 14:3379–3384
186. Li Y, Aubert SD, Maes EG, Raushel FM (2004) Enzymatic resolution of chiral phosphinate esters. J Am Chem Soc 126:8888–8889
187. Tsai P-C, Bigley A, Li Y, Ghanem E, Cadieux CL, Kasten SA, Reeves TE, Cerasoli DM, Raushel FM (2010) Stereoselective hydrolysis of organophosphate nerve agents by the bacterial phosphotriesterase. Biochemistry 49:7978–7987
188. Kim J, Tsai P-C, Chen S-L, Himo F, Almo SC, Raushel FM (2008) Structure of diethyl phosphate bound to the binuclear metal center of phosphotriesterase. Biochemistry 47:9497–9504
189. Li Y, Aubert SD, Raushel FM (2003) Operational control of stereoselectivity during the enzymatic hydrolysis of racemic organophosphorus compounds. J Am Chem Soc 125:7526–7527
190. Li WS, Li Y, Hill CM, Lum KT, Raushel FM (2002) Enzymatic synthesis of chiral organophosphothioates from prochiral precursors. J Am Chem Soc 124:3498–3499
191. Nowlan C, Li Y, Hermann JC, Evans T, Carpenter J, Ghanem E, Shoichet BK, Raushel FM (2006) Resolution of chiral phosphate phosphonate and phosphinate esters by an enantioselective enzyme library. J Am Chem Soc 128:15892–15902
192. Mikołajczyk M, Łuczak J, Kiełbasinski P, Colonna S (2009) Biocatalytic oxidation of thiophosphoryl compounds: a new chemo-enzymatic approach to enantiomeric insecticidal thionophosphates and their oxons. Tetrahedron Asymmetry 20:1948–1951
193. Kiełbasiński P, Rachwalski M, Kwiatkowska M, Mikołajczyk M, Wieczorek WM, Szyrej M, Sieroń L, Rutjes FPJT (2007) Enzyme-promoted desymmetrisation of prochiral bis (cyanomethyl)phenylphosphine oxide. Tetrahedron Asymmetry 18:2108–2112
194. Harvey JS, Malcolmson SJ, Dunne KS, Meek SJ, Thompson AL, Schrock RR, Hoveyda AH, Gouverneur V (2009) Enantioselective synthesis of *P*-stereogenic phosphinates and phosphine oxides by molybdenum-catalyzed asymmetric ring-closing metathesis. Angew Chem Int Ed Engl 48:762–766
195. Takizawa S, Rémond E, Arteaga FA, Yoshida Y, Sridharan V, Bayardon J, Jugé S, Sasai H (2013) *P*-Chirogenic organocatalysts: application to the aza-Morita–Baylis–Hillman (aza-MBH) reaction of ketimines. J Chem Soc Chem Commun 49:8392–8394
196. Francesco IN, Wagner A, Colobert F (2010) Stereoselective addition of Grignard reagents to new *P*-chirogenic N-phosphinoylimines. J Chem Soc Chem Commun 46:2139–2141
197. Francesco IN, Egloff C, Wagner A, Colobert F (2011) Stereoselective addition of Grignard reagents to new *P*-chirogenic N-phosphinoylbenzaldimines: effect of the phosphorus substituents on the stereoselectivity. Eur J Org Chem 4037–4045
198. Camus J-M, Andrieu J, Richard P, Poli R, Darcel C, Jugé S (2004) A *P*-chirogenic β-aminophosphine synthesis by diastereoselective reaction of the α-metallated PAMP–borane complex with benzaldimine. Tetrahedron Asymmetry 15:2061–2065
199. Busacca CA, Qu B, Farber E, Haddad N, Gret N, Saha AK, Eriksson MC, Wu J-P, Fandrick KR, Han S, Grinberg N, Ma S, Lee H, Li Z, Spinelli M, Gold A, Wang Z, Wang G, Wipf P, Senanayake CH (2013) [2,3]-Sigmatropic rearrangements of 2-phosphineborane 2-propen-1-ols: rapid access to enantioenriched diphosphine monoxide derivatives. Org Lett 15:1136–1139
200. Kaboudin B, Alaie S, Yokomatsu T (2011) Resolution of enantiomers of [α-hydroxy-(o-chlorophenyl)methyl]phosphinic acid via diastereomeric salt formation with enantiopure 1-phenylethylamines. Tetrahedron Asymmetry 22:1813–1816

201. Kaboudin B, Haruki T, Yamagishi T, Yokomatsu T (2007) Diastereoselective addition of α-substituted α-amino-H-phosphinates to imines using $Yb(OTf)_3$ as an efficient Lewis acid catalyst. Tetrahedron 63:8199–8205
202. Ding B, Zhang Z, Xu Y, Liu Y, Sugiya M, Imamoto T, Zhang W (2013) *P*-Stereogenic PCP pincer–Pd complexes: synthesis and application in asymmetric addition of diarylphosphines to nitroalkenes. Org Lett 15:5476–5479
203. Huang K, Emge TJ, Zhang X (2014) Synthesis of a novel *P*-chiral 1,3-oxaphospholane from optically pure propylene oxide. Heteroatom Chem 25:131–134
204. Mohar B, Čusak A, Modec B, Stephan M (2013) *P*-Stereogenic phospholanes or phosphorinanes from o-biarylylphosphines: two bridges not too far. J Org Chem 78:4665–4673
205. Pereira MM, Calvete MJF, Carrilho RMB, Abreua AR (2013) Synthesis of binaphthyl based phosphine and phosphite ligands. Chem Soc Rev 42:6990–7027
206. Bagi P, Kállay M, Hessz D, Kubinyic M, Holczbauer T, Czugler M, Fogassy E, Keglevich G (2014) Resolution of 1-n-propoxy-3-methyl-3-phospholene 1-oxide by diastereomeric complex formation using TADDOL derivatives and calcium salts of O, O′-dibenzoyl-(2R,3R)- or O, O′-di-P-toluoyl-(2R,3R)-tartaric acid. Tetrahedron Asymmetry 25:318–326
207. Dziuba K, Flis A, Szmigielska A, Pietrusiewicz KM (2010) Efficient oxidative resolution of a *P*-stereogenic triarylphosphine and asymmetric synthesis of a *P*-stereogenic atropoisomeric biphenyl diphosphine dioxide. Tetrahedron Asymmetry 21:1401–1405
208. Holt J, Majc AM, Schudde EP, Pietrusiewicz KM, Sieroń L, Wieczorek W, Jerphagnon T, Arends IWCE, Hanefeld U, Minnaard AJ (2009) On the resolution of secondary phosphine oxides via diastereomeric complex formation: the case of tert-butylphenylphosphine oxide. Synthesis 2061–2065

Index

MIX
Papier aus verantwortungsvollen Quellen
Paper from responsible sources
FSC® C105338

If you have any concerns about our products,
you can contact us on
ProductSafety@springernature.com

In case Publisher is established outside the EU,
the EU authorized representative is:
Springer Nature Customer Service Center GmbH
Europaplatz 3, 69115 Heidelberg, Germany

Printed by Libri Plureos GmbH
in Hamburg, Germany